DISTRIBUTED NETWORKS

Intelligence, Security, and Applications

Edited by **Qurban A. Memon**

DISTRIBUTED NETWORKS

Intelligence, Security, and Applications

CRC Press
Taylor & Francis Group
Boca Raton   London   New York

CRC Press is an imprint of the
Taylor & Francis Group, an **informa** business

MATLAB® is a trademark of The MathWorks, Inc. and is used with permission. The MathWorks does not warrant the accuracy of the text or exercises in this book. This book's use or discussion of MATLAB® software or related products does not constitute endorsement or sponsorship by The MathWorks of a particular pedagogical approach or particular use of the MATLAB® software.

CRC Press
Taylor & Francis Group
6000 Broken Sound Parkway NW, Suite 300
Boca Raton, FL 33487-2742

First issued in paperback 2017

© 2014 by Taylor & Francis Group, LLC
CRC Press is an imprint of Taylor & Francis Group, an Informa business

No claim to original U.S. Government works

Version Date: 20130613

ISBN 13: 978-1-138-07701-0 (pbk)
ISBN 13: 978-1-4665-5957-8 (hbk)

This book contains information obtained from authentic and highly regarded sources. Reasonable efforts have been made to publish reliable data and information, but the author and publisher cannot assume responsibility for the validity of all materials or the consequences of their use. The authors and publishers have attempted to trace the copyright holders of all material reproduced in this publication and apologize to copyright holders if permission to publish in this form has not been obtained. If any copyright material has not been acknowledged please write and let us know so we may rectify in any future reprint.

Except as permitted under U.S. Copyright Law, no part of this book may be reprinted, reproduced, transmitted, or utilized in any form by any electronic, mechanical, or other means, now known or hereafter invented, including photocopying, microfilming, and recording, or in any information storage or retrieval system, without written permission from the publishers.

For permission to photocopy or use material electronically from this work, please access www.copyright.com (http://www.copyright.com/) or contact the Copyright Clearance Center, Inc. (CCC), 222 Rosewood Drive, Danvers, MA 01923, 978-750-8400. CCC is a not-for-profit organization that provides licenses and registration for a variety of users. For organizations that have been granted a photocopy license by the CCC, a separate system of payment has been arranged.

Trademark Notice: Product or corporate names may be trademarks or registered trademarks, and are used only for identification and explanation without intent to infringe.

Library of Congress Cataloging-in-Publication Data

Distributed networks : intelligence, security, and applications / editor, Qurban Ali Memon.
       pages cm
   Includes bibliographical references and index.
   ISBN 978-1-4665-5957-8 (hardback)
    1. Electronic data processing--Distributed processing. 2. Internetworking (Telecommunication) 3. Distributed databases. 4. Computer networks--Security measures. I. Memon, Qurban Ali.

QA76.9.D5D55626 2013
005.8--dc23   2013020080

Visit the Taylor & Francis Web site at
http://www.taylorandfrancis.com

and the CRC Press Web site at
http://www.crcpress.com

Dedicated to my parents, to whom I owe everything.

Contents

Preface ix
Editor xiii
Contributors xv

Part I   Distributed Network Intelligence and Systems

1. Cooperative Regression-Based Forecasting in Distributed Traffic Networks 3
   Jelena Fiosina and Maksims Fiosins

2. A Sensor Data Aggregation System Using Mobile Agents 39
   Tomoki Yoshihisa, Yuto Hamaguchi, Yoshimasa Ishi, Yuuichi Teranishi, Takahiro Hara and Shojiro Nishio

3. Underlay-Aware Distributed Service Discovery Architecture with Intelligent Message Routing 67
   Haja M. Saleem, Seyed M. Buhari, Mohammad Fadzil Hassan and Vijanth S. Asirvadam

4. System-Level Performance Simulation of Distributed Embedded Systems via ABSOLUT 91
   Subayal Khan, Jukka Saastamoinen, Jyrki Huusko, Juha Korpi, Juha-Pekka Soininen and Jari Nurmi

5. Self-Organizing Maps: The Hybrid SOM–NG Algorithm 119
   Mario J. Crespo, Iván Machón, Hilario López and Jose Luis Calvo

6. A Semi-Supervised and Active Learning Method for Alternatives Ranking Functions 151
   Faïza Dammak, Hager Kammoun and Abdelmajid Ben Hamadou

Part II   Distributed Network Security

7. Tackling Intruders in Wireless Mesh Networks 167
   Al-Sakib Khan Pathan, Shapla Khanam, Habibullah Yusuf Saleem and Wafaa Mustafa Abduallah

8. Semi-Supervised Learning BitTorrent Traffic Detection 191
   Raymond Siulai Wong, Teng-Sheng Moh and Melody Moh

9. Developing a Content Distribution System over a Secure Peer-to-Peer Middleware 211
   Ana Reyna, Maria-Victoria Belmonte and Manuel Díaz

Part III Applications and Trends in Distributed Enterprises

10. **User Activity Recognition through Software Sensors** 243
    Stephan Reiff-Marganiec, Kamran Taj Pathan and Yi Hong

11. **Multi-Agent Framework for Distributed Leasing-Based Injection Mould Remanufacturing** 267
    Bo Xing, Wen-Jing Gao and Tshilidzi Marwala

12. **The Smart Operating Room: smartOR** 291
    Marcus Köny, Julia Benzko, Michael Czaplik, Björn Marschollek, Marian Walter, Rolf Rossaint, Klaus Radermacher and Steffen Leonhardt

13. **Distributed Online Safety Monitor Based on Multi-Agent System and AADL Safety Assessment Model** 317
    Amer Dheedan

14. **State of the Art of Service-Level Agreements in Cloud Computing** 347
    Mohammed Alhamad

15. **Used Products Return Service Based on Ambient Recommender Systems to Promote Sustainable Choices** 359
    Wen-Jing Gao, Bo Xing and Tshilidzi Marwala

Index 379

Preface

In the current decade, computing usage has evolved from a static activity to one that is mobile and has in fact turned out to be a de facto usage paradigm in computing. Additionally, the current and ongoing distribution of storage and processing is likely to move the computer as a stand-alone system into the background, leaving it as a mere computing device. This scenario opens the door for the consumer electronics industry with a challenge full of opportunities by replacing the disappearing computer with a new user experience through the addition of networked and ambient intelligence. Currently, in our homes and commercial places, we have started seeing a distributed network of devices that provide information and entertainment. In the near future, we will see intelligence within these devices that would enable them with communication capabilities.

Intelligent systems have numerous civilian, homeland security and military applications; however, all these applications carry certain requirements like communication bandwidth, sensing range, power constraints and stealth requirements prior to managing a centralised command and control. The alternative to centralised command and control is distributed and networked coordination, which is proving to be more promising in terms of scalability, flexibility and robustness.

Distributed and networked intelligence introduces problems, requires models and raises issues such as collective periodic coordination, collective tracking with a dynamic leader and containment control with multiple leaders, as well as exploring ideas for their solution.

The Objective of This Book

The objective of this book is to provide an internationally respected recent collection of scientific research in distributed network intelligence, security and novel applications. The book aims at providing the readers with a better understanding of how distributed intelligent systems and services can be successfully implemented to incorporate recent trends and advances in theory and applications of network intelligence. The key benefits of the collected work in this book can be summarised as follows:

- Enhance the understanding of distributed network intelligence concepts.
- Inspire new R&D directions within the distributed computing and networking society.
- Facilitate professional development in advanced technologies.
- Put together new concepts and theories that affect security on distributed platform.

The Structure

The book is a collection of 15 exclusive chapters written by leading scholars and experts in the field and is organised into three parts. Each part consists of a set of chapters focusing on the respective subject outline to provide readers with an in-depth understanding of concept and technology related to that area. Moreover, special emphasis is laid on a tutorial style of chapters concerning the development process of complex distributed networked systems and services, thus revisiting the difficult issue of knowledge engineering of such systems.

The book is structured as follows:

- Part I: Distributed Network Intelligence and Systems
- Part II: Distributed Network Security
- Part III: Applications and New Trends in Distributed Enterprises

Part I comprises six chapters on distributed network intelligence and systems.

Chapter 1 presents forecasting in distributed traffic networks. The authors use distributed linear and kernel-based regression models for travelling time forecasting on the basis of available values of affecting factor values.

Chapter 2 discusses a data aggregation system using mobile agents for integrated sensor networks. The approach presented by authors helps in describing the queries that include operations over every spatial and temporal unit. Thus, the users do not need to submit multiple queries to each sensor network site that has different sensor network architecture.

Chapter 3 presents distributed service (or resource) discovery (DSD) in service-oriented computing. The research reported in this chapter proposes to move the query routing process from the overlay layer to the Internet Protocol layer, as opposed to peer-to-peer applications.

In Chapter 4, the authors present the system-level performance evaluation of distributed embedded systems. The approach enables the integration of new communication protocol models with lightweight modelling effort and spans multiple domains of distributed systems.

Chapter 5 discusses the Hybrid SOM-NG algorithm and its applications. A hybrid algorithm is presented and a step-by-step explanation is given of how a self-organising map algorithm can be created. A number of examples are illustrated as applications.

In Chapter 6, the authors extend the supervised RankBoost algorithm to combine labelled and unlabelled data. The goal of the research presented in this chapter is to understand how combining labelled and unlabelled data may change the ranking behaviour, and how RankBoost can, with its character, improve ranking performance.

Part II comprises three chapters on distributed network security.

Chapter 7 takes a different approach to tackle an intruder rather than purging it out of the network unless it is marked as a direct threat to the network's operation. The chapter presents an intrusion tackling model for wireless mesh networks with the idea of utilising the resources of an intruder before taking a final decision of removing it from the network. Thus, the model works on intrusion tackling rather than direct exclusion by detection or prevention.

In Chapter 8, the authors first survey major peer-to-peer traffic detection methods, including port-based, DPI (Deep Packet Inspection)-based, DFI (Deep Flow Inspection)-based, the combination of DPI and DFI, and others. Then a new detection method is presented, which is based on an intelligent combination of DPI and DFI with semi-supervised learning.

Chapter 9 presents the development of a content distribution system based on coalitions. The system uses an incentive mechanism to fight against malicious peers or selfish behaviours that could hurt a system's performance while promoting cooperation among users. The authors choose SMEPP (A Secure Middleware for Embedded Peer-to-Peer systems) to develop the proposed content distribution system. They explain how to develop a complex application taking advantage of SMEPP middleware, proving the suitability of using P2P middleware in the development of a real-life complex distributed application.

Part III comprises six chapters on applications and trends in distributed enterprises.

In Chapter 10, Reiff-Marganiec et al. have introduced software sensors as an inexpensive alternative to hardware sensors. They propose a complimentary sensor technology through the use of software sensors. The software sensors are essentially based on monitoring SOAP messages and inserting data for further reasoning and querying into a semantic context model.

The design and manufacturing of moulds represent a significant link in the entire production chain, especially in the automotive industry, consumer electronics and consumer goods industries, because nearly all produced discrete parts are formed using production processes that employ moulds. In Chapter 11, the authors look at the mould industry to illustrate the challenges of the emerging business model (i.e., leasing-based model) that can be found in many other industries as well.

The development of the smartOR standard is driven by a medical motivation for a standardised human–machine interaction and the need for a manufacturer independent network in the operating room. Chapter 12 addresses the development with two main focuses: the OSCB protocol, which aims to build a modular and flexible protocol for manufacturer independent networking, and new innovative concepts for human–machine interaction in the operating room.

Critical systems, such as nuclear power plants, industrial chemical processes and means of transportation, may fail catastrophically, resulting in death or a tremendous loss of assets. Chapter 13 presents a model produced *via* application of the offline safety assessment tool, Architecture Analysis and Design Language (AADL), and is brought forward to the online stage and exploited to inform the reasoning of a number of collaborative agents and develop a state-of-the-art distributed online safety monitor that integrates the delivery of the safety tasks. The presented model shows that with appropriate knowledge about dynamic behaviour, the monitor can also determine the functional effects of low-level failures and provide a simplified and easier-to-comprehend functional view of the failure.

Cloud computing technology has created many challenges for service providers and customers, especially for those users who already own complicated legacy systems. Chapter 14 reviews the challenges related to the concepts of trust, service-level agreement (SLA) management and cloud computing. The existing frameworks of SLAs in different domains such as web services and grid computing are surveyed. The advantages and limitations of current performance measurement models for SOA, distributed systems, grid computing and cloud services are also discussed.

Product recovery requires that end-of-life (EoL) products are acquired from the end users so that the value-added operations can be processed and the products (or parts from them) can be resold in the market. Chapter 15 essentially aims at improving the bridge between end users and EoL product disposal flows to form a seamless, synchronous network functioning, so as to support the product recovery process from the end user's point of view. Further, the proposed model behaves as an added personalised recommender to the end users in their return choices. The model is based on multi-agent paradigm for designing and implementing ambient intelligence (AmI) recommender systems.

What to Expect from This Book

This book will help in creating interest among researchers and practitioners towards realising safe and distributed interactive multimedia systems. This book is useful to researchers, professors, research students and practitioners as it reports novel research work on challenging topics in the area of distributed network intelligence. It enhances the understanding of concepts and technologies in distributed intelligence and discusses security solutions and applications for distributed environment.

What Not to Expect from This Book

This book is written with the objective of collecting latest research works in distributed network computing and intelligence. The book does not provide detailed introductory information on the basic topics in this field, as it is not written in a textbook style. However, it can be used as a good reference to help graduate students and researchers by providing the latest concepts, tools and advancements in distributed network intelligence and the computing field. The readers require at least some basic knowledge about the field.

Special Thanks

First of all, we would like to thank God for giving us the courage and the necessary strength to complete this book. We would also like to thank all authors who contributed to this book by authoring related chapters in specific areas of this field. Their cooperation, prompt responses, adherence to guidelines and timely submission of chapters has helped us to meet the book preparation deadlines. A total of 51 authors from 11 countries contributed to this book. We hope that the preparation of this book, based on the latest developments in the field, will prove beneficial to professionals and researchers in this field and, at the same time, be instrumental in generating new applications and case studies in related fields.

Dr. Qurban A. Memon
UAE University, Al-Ain, United Arab Emirates

MATLAB® is a registered trademark of The MathWorks, Inc. For product information, please contact:

The MathWorks, Inc.
3 Apple Hill Drive
Natick, MA 01760-2098 USA
Tel: 508 647 7000
Fax: 508-647-7001
E-mail: info@mathworks.com
Web: www.mathworks.com

Editor

Qurban A. Memon received his MS from the University of Gainesville, Florida, USA in 1993, and his PhD from the Department of Electrical and Computer Engineering, University of Central Florida, Orlando, USA in 1996. He is currently associate professor at the Electrical Engineering Department of College of Engineering, UAE University, Al-Ain, United Arab Emirates.

In his professional career, he has contributed at levels of teaching, research and service in the area of electrical and computer engineering. He has authored and coauthored over 80 research papers during an academic career of about 15 years. His publications include two book chapters: 'Intelligent Network System for Process Control: Applications, Challenges, Approaches' published in a book titled *Robotics, Automation and Control*, ISBN 978-953-7619-39-8, 2009; and 'Smarter Health-Care Collaborative Network' published in a book titled *Building Next-Generation Converged Networks: Theory and Practice*, ISBN 1466507616, 2013. His research interests include intelligent systems and networks.

Dr. Memon takes interest in motivating students to participate and present research in student conferences. His research project involving students won the Chancellor's Undergraduate Research Award (CURA) in 2009. During his academic years, he has served as a teacher, team member and researcher. For the last seven consecutive years, his research contributions have been rated as excellent. He has executed research grants and development projects in the area of microcontroller-based systems, intelligent systems and networks. He has served as a reviewer of many international journals and conferences and session chair at various conferences.

His profile is multidisciplinary in engineering and sciences in terms of curriculum and systems development; teamwork at the institutional level; and participation at various levels in international conferences and forums. Dr. Memon has an industrial experience of five years (1987–1991) working as an engineer to instal, test and commission telecom and embedded systems equipment.

Contributors

Wafaa Mustafa Abduallah
Department of Computer Science
International Islamic University
Kuala Lumpur, Malaysia

Mohammed Alhamad
Curtin University
Western Australia, Australia

Vijanth S. Asirvadam
Universiti Teknology PETRONAS
Perak, Malaysia

Maria-Victoria Belmonte
University of Málaga
Málaga, Spain

Julia Benzko
RWTH Aachen University
Aachen, Germany

Seyed M. Buhari
Universiti Brunei Darussalam
Bandar Seri Begawan, Brunei Darussalam

Jose Luis Calvo
Universidad de A Coruña
A courña, Spain

Mario J. Crespo
Universidad de Oviedo
Asturias, Spain

Michael Czaplik
Department of Anesthesiology
University Hospital RWTH Aachen
Aachen, Germany

Faïza Dammak
Laboratory MIRACL–ISIMS SFAX
Sfax, Tunisia

Amer Dheedan
Department of Computer Science
Delmon University
Manama, Kingdom of Bahrain

Manuel Díaz
University of Málaga
Málaga, Spain

Jelena Fiosina
Clausthal University of Technology
Clausthal-Zellerfeld, Germany

Maksims Fiosins
Clausthal University of Technology
Clausthal-Zellerfeld, Germany

Wen-Jing Gao
Faculty of Engineering and the Built Environment (FEBE)
University of Johannesburg
Johannesburg, South Africa

Abdelmajid Ben Hamadou
Laboratory MIRACL–ISIMS SFAX
Sfax, Tunisia

Yuto Hamaguchi
Mitsubishi Electric Corp.
Tokyo, Japan

Takahiro Hara
Osaka University
Osaka, Japan

Mohammad Fadzil Hassan
Universiti Teknology PETRONAS
Perak, Malaysia

Yi Hong
Department of Computer Science
University of Leicester
Leicester, United Kingdom

Jyrki Huusko
VTT Technical Research Centre of Finland
Espoo, Finland

Yoshimasa Ishi
Osaka University
Osaka, Japan

Hager Kammoun
Laboratory MIRACL–ISIMS SFAX
Sfax, Tunisia

Subayal Khan
VTT Technical Research Centre of Finland
Espoo, Finland

Shapla Khanam
Department of Computer Science
International Islamic University
Kuala Lumpur, Malaysia

Marcus Köny
RWTH Aachen University
Aachen, Germany

Juha Korpi
VTT Technical Research Centre of Finland
Espoo, Finland

Steffen Leonhardt
RWTH Aachen University
Aachen, Germany

Hilario López
Universidad de Oviedo
Asturias, Spain

Iván Machón
Universidad de Oviedo
Asturias, Spain

Björn Marschollek
RWTH Aachen University
Aachen, Germany

Tshilidzi Marwala
Faculty of Engineering and the Built
    Environment (FEBE)
University of Johannesburg
Johannesburg, South Africa

Melody Moh
Department of Computer Science
San Jose State University
San Jose, California

Teng-Sheng Moh
Department of Computer Science
San Jose State University
San Jose, California

Shojiro Nishio
Osaka University
Osaka, Japan

Jari Nurmi
Department of Computer Systems
Tampere University of Technology
Tampere, Finland

Al-Sakib Khan Pathan
Department of Computer Science
International Islamic University
Kuala Lumpur, Malaysia

Kamran Taj Pathan
Department of Computer Science
University of Leicester
Leicester, United Kingdom

Klaus Radermacher
RWTH Aachen University
Aachen, Germany

Stephan Reiff-Marganiec
Department of Computer Science
University of Leicester
Leicester, United Kingdom

Ana Reyna
University of Málaga
Málaga, Spain

Rolf Rossaint
Department of Anesthesiology
University Hospital, RWTH Aachen
Aachen, Germany

Jukka Saastamoinen
VTT Technical Research Centre of Finland
Espoo, Finland

Habibullah Yusuf Saleem
Department of Computer Science
International Islamic University
Kuala Lumpur, Malaysia

Haja M. Saleem
Universiti Tunku Abdul Rahman
Perak, Malaysia

Juha-Pekka Soininen
VTT Technical Research Centre of Finland
Espoo, Finland

Yuuichi Teranishi
National Institute of Information and Communications Technology
Tokyo, Japan

Marian Walter
RWTH Aachen University
Aachen, Germany

Raymond Siulai Wong
Department of Computer Science
San Jose State University
San Jose, California

Bo Xing
Faculty of Engineering and the Built Environment (FEBE)
University of Johannesburg
Johannesburg, South Africa

Tomoki Yoshihisa
Osaka University
Osaka, Japan

Part I

Distributed Network Intelligence and Systems

1

Cooperative Regression-Based Forecasting in Distributed Traffic Networks

Jelena Fiosina and Maksims Fiosins

CONTENTS

1.1 Introduction 3
1.2 Data Processing and Mining in Distributed Traffic Networks 5
    1.2.1 MAS Architectures for Distributed Networks 5
    1.2.2 Distributed Data Processing and Mining 8
    1.2.3 Travelling Time Forecasting Models 9
    1.2.4 Problem Formulation 10
1.3 Distributed Cooperative Recursive Forecasting Based on Linear Regression 11
    1.3.1 Forecasting with Multivariate Linear Regression 11
    1.3.2 Local Recursive Parameter Estimation 13
    1.3.3 Cooperative Linear Regression-Based Learning Algorithm 14
    1.3.4 Numerical Example 15
1.4 Distributed Cooperative Forecasting Based on Kernel Regression 17
    1.4.1 Kernel Density Estimation 17
    1.4.2 Kernel-Based Local Regression Model 19
    1.4.3 Cooperative Kernel-Based Learning Algorithm 20
    1.4.4 Numerical Example 21
1.5 Case Studies 24
    1.5.1 Problem Formulation 24
    1.5.2 Centralised Architecture 26
    1.5.3 Decentralised Uncoordinated Architecture 27
    1.5.4 Decentralised Coordinated Architecture 29
1.6 Discussion 32
1.7 Conclusions 34
Acknowledgements 35
References 35

1.1 Introduction

The problems in distributed networks, which often appear in complex applications, such as sensor, traffic or logistics networks, can be solved using multi-agent system (MAS) architectures. Currently, research on MAS and distributed networks is focused mainly on decision making, whereas little attention is paid to the problems of data processing and mining. However, adequate decision making requires appropriate analysis and processing of input data. In this context, the development of methods for intelligent data analysis and mining of distributed sources is very timely.

In MAS, the individual and collective behaviours of the agents depend on the observed data from distributed sources. In a typical distributed environment, analysing distributed data is a nontrivial problem because of several constraints such as limited bandwidth (in wireless networks), privacy-sensitive data, distributed computing nodes and so on.

Traditional centralised data processing and mining typically requires central aggregation of distributed data, which may not always be feasible because of the limited network bandwidth, security concerns, scalability problems and other practical issues. Distributed data processing and mining (DDPM) carries out communication and computation by analysing data in a distributed fashion [1]. The DDPM technology offers more efficient solutions in such applications. The DDPM field deals with these challenges by analysing distributed data and offers many algorithmic solutions for data analysis, processing and mining using different tools in a fundamentally distributed manner that pays careful attention to the resource constraints [2].

In this study, we focus on the urban traffic domain, where many traffic characteristics such as travelling time, travelling speed, congestion probability and so on can be evaluated by autonomous vehicle-agents in a distributed manner. Numerous data processing and mining techniques were suggested for forecasting travelling time in a centralised and distributed manner. Statistical methods, such as regression and time series, and artificial intelligence methods, such as neural networks, are successfully implemented for similar problems. However, travelling time is affected by a range of different factors. Thus, its accurate forecasting is difficult and needs considerable amount of traffic data. Understanding the factors affecting travelling time is essential for improving forecasting accuracy [3].

We focus on regression analysis, which is a powerful statistical tool for solving forecasting problems. We compare the well-known multivariate linear regression technique with modern nonparametric computationally intensive kernel-based regression.

We propose two distributed algorithms based on multivariate linear and kernel-based regression for forecasting the travelling time using real-world data from southern Hanover. We assume that each agent autonomously estimates the parameters of its local multivariate linear and kernel-based regression models, paying special attention to the implementation of these models for real-time streaming data. If an agent cannot estimate the travelling time accurately due to a lack of data, that is, if there are no data near the point of interest (because kernel-based estimation uses approximations), it cooperates with other agents. We suggest two algorithms for multi-agent cooperative learning based on the transmission of the observations (kernel regression) or parameters (linear regression) in response to requests from agents experiencing difficulties. After obtaining the necessary data from other agents, the agent can forecast the travelling time autonomously. We also propose the use of a combination of the estimates to achieve more accurate forecasting. The travelling time estimated during the DDPM stage can be the input for the next stage of distributed decision making by intelligent agents [4,5].

The study contributes to the following: (1) the development of the structure of a DDPM module for an autonomous agent, which allows travelling time forecasting and cooperation with other agents; (2) the development of a distributed regression-based (kernel and linear) multi-agent cooperative learning algorithm for real-time streaming data; (3) the development of the structure of a regression model for travelling time forecasting on the basis of available real-world observations; (4) the experimental comparison of different MAS architectures and forecasting methods, including a combination of linear and kernel-based estimates on the basis of real-world data.

The remainder of this chapter is organised as follows. Section 1.2 describes previous work related to MAS applications for solving problems in distributed networks, the DDPM field in MASs, travelling time forecasting models, and formulates the research question. In Section 1.3, we present a distributed cooperative recursive forecasting algorithm based on multivariate linear regression. This section concludes with a simple tutorial example to illustrate the proposed technique. Section 1.4 presents the multivariate kernel-based regression model used for streaming data and proposes a cooperative learning algorithm for optimal forecasting in a distributed MAS architecture. This section also concludes with a simple tutorial study. Section 1.5 presents case studies using data from Hanover (Germany) and also provides a comparison of different MAS architectures and forecasting methods using real-world data. Section 1.6 presents the discussion of the results and the final section concludes on perspectives of future work.

1.2 Data Processing and Mining in Distributed Traffic Networks

In this section, we provide some theoretical background and discuss work related to the problems of DDPM when making forecasts about traffic flows.

1.2.1 MAS Architectures for Distributed Networks

Complex modern information systems are usually characterised by a large number of relatively autonomous components with a large number of interactions [6]. System designers can usually identify three aspects related to these components: behaviour, environment and interactions. In most cases, the components act and interact in flexible ways in specific environments to achieve their objectives. The representation of a complex system in this manner means demands on the adoption of a multi-agent approach during its engineering [7].

MASs are one of the most popular paradigms for modelling these types of complex systems. They contain a set of autonomous intelligent components (agents) with the following important properties [8].

1. *Autonomy*. An agent makes its own decision, without external influences. During decision making, it considers its goals and its knowledge about the current status in the environment.

2. *Local views*. An agent stores information about the current state of the environment, although its knowledge may be partial because it lacks global knowledge in the overall system. Agents may cooperate to exchange available information to produce a more complete picture so that they have sufficient information on decision making.

3. *Decentralisation*. A 'central' agent does not control the overall system. There can be a central authority, but it does not control the overall system and it does not store information in a centralised manner. Thus, agents make their own decisions, and cooperate and coordinate their actions when necessary.

When considering multi-agent representations of complex systems it is important to distinguish between distributed and decentralised systems. In a distributed system, a central authority provides services or performs systems management, which involves autonomous components. This central authority can be virtual or distributed among several components. However, system-wide data sets or decisions are modelled by the distributed systems. By contrast, a decentralised system does not assume any type of centralised regulation (even virtual). The fully autonomous components of decentralised systems act to achieve their goals and cooperate when necessary.

Distributed systems are usually more expensive because they require additional costs for a central authority. However, decentralised systems experience more problems with work organisation and system goal achievement. In our study, we compared distributed approach with a central authority (called 'centralised') and decentralised coordinated/uncoordinated approach to travelling time forecasting and we reached conclusions about the effect of a central authority and coordination in different situations.

The agents in MAS usually have two important modules: a DDPM module and a distributed decision support (DDS) module [4,5,9]. The DDPM module is responsible for receiving observations of the environment and discovering patterns or dependencies that may facilitate the decision-making process. The DDS module is responsible for making decisions on the basis of the information received from the DDPM module and executing appropriate actions in the environment. The general structure of an agent is shown in Figure 1.1.

The agent acts in the environment and makes observations, which contain data and rewards (measures of the agent efficiency from the environmental perspective). The DDPM module pre-processes incoming data using filtering/aggregating data streams and, if necessary, stores the data for the agent. The necessary information is used to estimate the parameters of data models, which are essential, but not obvious, characteristics of the environment. At this stage, information can be exchanged with other agents (or central authority) to improve the quality of estimation.

Finally, these parameters are used to form the state of the agent. Decision making is based on the current state, which contains information about different system features (usually a vector). The state, rewards and other essential information are transferred to the DDS module.

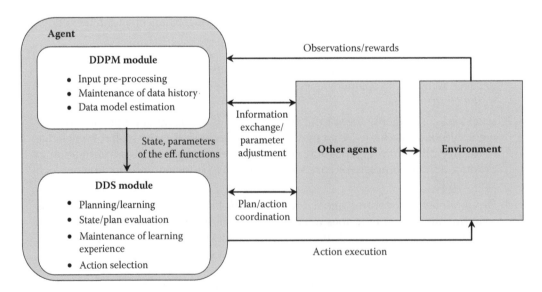

FIGURE 1.1
Structure of DDPM and DDS modules in an autonomous agent.

The DDS module formulates a plan for goal achievement from the current state. Each plan is evaluated with respect to the goals and capabilities of the agent to check whether it allows an agent to achieve its goal in a realistic manner. This is usually achieved based on efficiency functions, which can be fixed or improved ('learnt') during the planning process, and the agent's operations. The plan or a set of alternative plans can be coordinated with other agents (or central authority). The learning results are also stored in the experience storage. Based on the actual plan, an action is selected and it is executed in the environment. This affects the environment (and other agents) and its results can be observed by the DDPM module.

A road traffic system is a typical complex system, which involves self-interested autonomous participants. Information transfer and action/plan coordination between traffic participants can significantly increase the safety and quality of a traffic system. This is why many researchers have recently applied MAS architectures for modelling different services in traffic networks. In these architectures, the active components (e.g. vehicles, pedestrians and traffic lights) are modelled as interacting autonomous agents [10–12].

The following example illustrates an application of multi-agent technology to the design of an intelligent routing support system.

Example: Intelligent Routing Support

We describe the general structure of an advanced driver support system. This device supports the vehicle's driver during the selection of an optimal route in a street network and has the ability to collect actual information and communicate with other vehicles or a central authority if available.

The traffic system is modelled as an MAS. We model the vehicles (more precisely: intelligent devices in the vehicles) as agents, which corresponds to the architecture shown in Figure 1.1. The architecture of an intelligent routing device is shown in Figure 1.2.

The inputs used in this device (agent) are sensor data that indicate the current situation on a road (e.g. the position, speed and distances to other vehicles). The user input includes the destination and driving preferences. A digital map contains an underlying road map in the form of a graph.

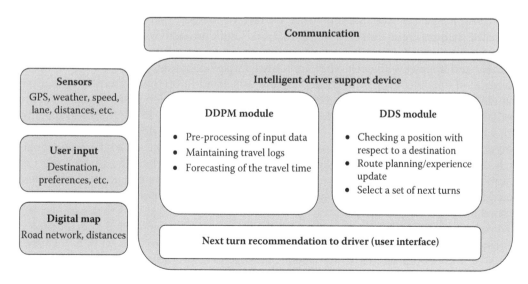

FIGURE 1.2
Structure of an intelligent driver support device that provides route recommendations.

These data are pre-processed to determine average values for important characteristics, which are stored in the travel log and transferred to the data model for future analysis. The most important task of the DDPM module is travelling time forecasting. An appropriate data model (e.g. regression, neural network, time series and probabilistic forecasting) is used for this purpose.

The state of the agent includes its current position on the map. The agent produces a set of alternative plans (paths) to travel from the current position to the destination. The travelling times estimated by the DDPM module are used as parameters by the efficiency functions and, together with the preferences and learning experience, are used to evaluate each plan. The plans can be coordinated with other agents during cooperative driving or when driving in convoys. The best possible plan (route) is recommended to the driver.

1.2.2 Distributed Data Processing and Mining

Recent progress in automated data collection and transmission has created a need for better interpretation and exploitation of vast data sources. Thus, the IT society requires the development of new methods, models and tools to extract useful information from data. It should also be taken into account that data sources and data processing are distributed, which is problematic for DDPM.

DDPM provides algorithmic solutions for data analysis in a distributed manner to detect hidden patterns in data and extract the knowledge necessary for decentralised decision making [13,14]. A promising research area is currently investigating the possibility of coupling MAS with DDPM, by exploiting DDPM methods to improve agent intelligence and MAS performance [2]. This will facilitate highly scalable data architectures, online data processing, hidden data pattern interpretation and analysis and information extraction from distributed environments. Furthermore, the coupling of MAS with DDPM may be described in terms of ubiquitous intelligence [15], with the aim of fully embedding information processing into everyday life. This concept is very similar to the architecture of data clouds, where data and services are virtualised and provided on demand.

A strong motivation for implementing DDPM for MAS is given by da Silva et al. [2], where the authors argue that DDPM in MAS deals with pro-active and autonomous agents that perceive their environment, dynamically reason out actions on the basis of the environment and interact with each other. In many complex domains, the knowledge of agents is a result of the outcome of empirical data analysis, in addition to the preexisting domain knowledge. The DDPM of agents often involves detecting hidden patterns, constructing predictive and clustering models, identifying outliers and so on. In MAS, this knowledge is usually collective. This collective 'intelligence' of MAS must be developed by the distributed domain knowledge and analysis of the distributed data observed by different agents. Such distributed data analysis may be a nontrivial problem when the underlying task is not completely decomposable and computational resources are constrained by several factors such as limited power supply, poor bandwidth connection, privacy-sensitive multi-party data and so on.

Klusch et al. [16] conclude that autonomous data mining agents, as a special type of information agents, may perform various kinds of mining operations on behalf of their user(s) or in collaboration with other agents. Systems of cooperative information agents for data mining in distributed, heterogeneous or homogeneous, and massive data environments appear to be quite a natural progression for the current systems to be realised in the near future.

A common characteristic of all approaches is that they aim at integrating the knowledge that is extracted from data at different geographically distributed network nodes with minimum network communication and maximum local computations [2].

Local computation is carried out on each node, and either a central node communicates with each distributed node to compute the global models or a peer-to-peer architecture is used. In the case of the peer-to-peer architecture, individual nodes might communicate with a resource-rich centralised node, but they perform most tasks by communicating with neighbouring nodes through messages passing over an asynchronous network [2].

According to da Silva et al. [2], a distributed system should have the following features for the efficient implementation of DDPM:

1. The system consists of multiple independent data sources, which communicate only through message passing.
2. Communication between peers is expensive.
3. Peers have resource constraints (e.g. battery power).
4. Peers have privacy concerns.

Typically, communication involves bottlenecks. Since communication is assumed to be carried out exclusively by message passing, the primary goal of several DDPM methods, as mentioned in the literature, is to minimise the number of messages sent. Building a monolithic database in order to perform non-DDPM may be infeasible or simply impossible in many applications. The costs of transferring large blocks of data may be very expensive and result in very inefficient implementations [1].

Moreover, sensors must process continuous (possibly fast) streams of data. The resource-constrained distributed environments of sensor networks and the need for a collaborative approach to solve many problems in this domain make MAS architecture an ideal candidate for application development.

In our study, we deal with homogeneous data. However, a promising approach to agent-based parameter estimation for partially heterogeneous data in sensor networks is suggested in Ref. [17]. Another decentralised approach for homogeneous data was suggested in Ref. [18] to estimate the parameters of a wireless network by using a parametric linear model and stochastic approximations.

1.2.3 Travelling Time Forecasting Models

Continuous traffic jams indicate that the maximum capacity of a road network is met or even exceeded. In such a situation, the modelling and forecasting of traffic flow is one of the important techniques that needs to be developed [19]. Nowadays, knowledge about travelling time plays an important role in transportation and logistics, and it can be applied in various fields and purposes. From the travellers' point of view, knowledge about the predicted travelling time helps to reduce the actual travelling time and improves reliability through better selection of travelling routes. In logistics, accurate travelling time estimation could help to reduce transport delivery costs and to increase the service quality of commercial delivery by delivering goods within the required time window by avoiding congested sections. For traffic managers, travelling time is an important index for traffic system operation efficiency [3].

There are several studies in which a centralised approach is used to predict the travelling time. The approach was used in various intelligent transport systems, such as in-vehicle route guidance and advanced traffic management systems. A good overview is given in Ref. [3]. To make the approach effective, agents should cooperate with each other to achieve their common goal via the so-called gossiping scenarios. The estimation of the actual travelling time using vehicle-to-vehicle communication without MAS architecture was described in Ref. [20].

On the other hand, a multi-agent architecture is better suited for distributed traffic networks, which are complex stochastic systems.

Further, by using centralised approaches the system cannot adapt quickly to situations in real time, and it is very difficult to transmit a large amount of information over the network. In centralised approaches, it is difficult or simply physically impossible to store and process large data sets in one location. In addition, it is known from practice that most drivers rely mostly on their own experience; they use their historical data to forecast the travelling time [4,5].

Thus, decentralised MASs are fundamentally important for the representation of these networks [19]. We model our system with autonomous agents to allow vehicles to make decisions autonomously using not only the centrally processed available information, but also their historical data.

Traffic information generally goes through the following three stages: data collection and cleansing, data fusion and integration and data distribution. The system presented in Ref. [21] consists of three components, namely a Smart Traffic Agent, the Real-time Traffic Information Exchange Protocol and a centralised Traffic Information Centre (TIC) that acts as the backend. A similar architecture is used in this study, but the forecasting module, incorporated into the Start Traffic Agent (vehicle agent), is different.

In our study, we do not focus on the transmission protocol describing only the information, which should be sent from one node to another, without the descriptions of protocol packets. The centralised TIC in our model is used only for storing system information. A combination of centralised and decentralised agent-based approaches to the traffic control is presented in Ref. [22]. In this approach, the agents maintain and share the 'local weights' for each link and turn, exchanging this information with a centralised TIC. The decentralised MAS approach for urban traffic network is also considered in Ref. [23], where the authors forecast the traversal time for each link of the network separately. Two types of agents were used for vehicles and links, and a neural network was used as the forecasting model.

Travelling time forecasting for the similar decentralised urban traffic network based on the local linear regression model is presented in Ref. [24] and for that based on the kernel regression model is presented in Ref. [25]. The comparison of parametric and nonparametric approaches for traffic-flow forecasting is made in Ref. [26], which shows the efficiency of the nonparametric kernel-based regression approach.

1.2.4 Problem Formulation

We consider a traffic network with several vehicles, represented as autonomous agents, which predict their travelling time on the basis of their current observations and history. Each agent estimates locally the parameters of the same traffic network. To make a forecast, each agent constructs a local regression model, which explains the manner in which different explanatory variables (factors) influence the travelling time (local initial forecasting). A detailed overview of such factors is provided in Ref. [3]. The following information is important for predicting the travelling time [27]: average speed before the current segment, number of stops, number of left turns, number of traffic light, average travelling time

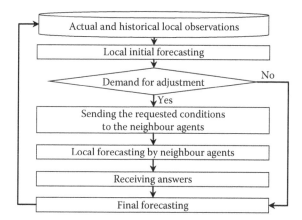

FIGURE 1.3
Algorithm for local travelling time forecasting by an individual agent.

estimated by TIC. We should also take into account the possibility of an accident, network overload (rush hour) and the weather conditions.

In the case when the vehicle has no or little experience of driving in specific conditions, its local forecasting will be inaccurate. For more accurate travelling time estimation, the vehicle needs adjustment of its models (demand for adjustment). It contacts other traffic participants, which share their experience in the requested conditions. In this step, the focal agent that experiences difficulties with an initial forecast sends the description of the requested conditions to other traffic participants in the transmission radius. Each of the neighbour agents tries to make a forecast by itself. In the case of a successful forecast, it shares its experience by sending its parameter estimates (parametric linear regression) or observations that are nearest to the requested point (nonparametric kernel-based regression). After receiving the data from the other agents, the focal agent aggregates the obtained results, adjusting the model and increasing its experience, and makes a forecast autonomously (final forecasting). The forecasting procedure of one such vehicle is shown in Figure 1.3.

In short, decentralised travelling time forecasting consists of four steps: (1) local initial forecasting; (2) in the case of unsuccessful initial forecasting: sending the requested conditions to the neighbour agents; (3) local forecasting by neighbour agents and sending answers; (4) aggregation of the answers and final forecasting.

1.3 Distributed Cooperative Recursive Forecasting Based on Linear Regression

1.3.1 Forecasting with Multivariate Linear Regression

Regression analysis constitutes several methods widely employed for modelling the dependency between the dependent variable (usually denoted by **Y**) and independent variables (factors, usually denoted by **X**). It describes the dependence of conditional expectation $E[Y|X]$ on the values of several independent variables (factors) **X**.

To use a regression model, the following classical assumptions should be satisfied [28]:

1. The error is a random variable with zero expectation.
2. The independent variables are linearly independent (no variable can be expressed as a linear combinations of others).
3. The errors are uncorrelated.
4. The variance of the errors is constant.

A linear regression model supposes that the dependent variable **Y** is a linear combination of the factors **X**. The linear regression model in the matrix form can be expressed as

$$\mathbf{Y} = \mathbf{X}\boldsymbol{\beta} + \boldsymbol{\varepsilon}, \tag{1.1}$$

where **Y** is an $n \times 1$ vector of dependent variables; **X** is an $n \times d$ matrix of explanatory (dependent) variables; $\boldsymbol{\beta}$ is a $d \times 1$ vector of unknown coefficients, which are parameters of the system to be estimated; $\boldsymbol{\varepsilon}$ is an $n \times 1$ vector of random errors. The rows of the matrix **X** correspond to observations and the columns correspond to factors.

The regression assumptions can be expressed in this case as follows. We suppose that the components $\{\varepsilon_i\}$ of the error vector $\boldsymbol{\varepsilon}$ are mutually independent, have zero expectation, $E[\boldsymbol{\varepsilon}] = 0$, zero covariances $cov[\varepsilon_i, \varepsilon_j] = 0$, $i \neq j$ and equal variances, $Var[\boldsymbol{\varepsilon}] = \sigma^2 \mathbf{I}$, where **I** is an $n \times n$ identity matrix.

The well-known least-squares estimator (LSE) **b** of $\boldsymbol{\beta}$ is

$$\mathbf{b} = (\mathbf{X}^T \mathbf{X})^{-1} \mathbf{X}^T \mathbf{Y}. \tag{1.2}$$

After the estimation of the parameters $\boldsymbol{\beta}$, we make a forecast for certain t-th future values of the factors $\mathbf{x}_t$:

$$E[Y_t] = \mathbf{x}_t \mathbf{b}. \tag{1.3}$$

For quality estimation of a linear regression model, we separate the total sum of squares (SST) of the dependent variable **Y** into two parts: the sum of squares explained by the regression (SSR) and the sum of squares of errors (SSE):

$$SSR = (\mathbf{Xb} - \bar{Y})^T (\mathbf{Xb} - \bar{Y}),$$

$$SST = (\mathbf{Y} - \bar{Y})^T (\mathbf{Y} - \bar{Y}),$$

$$SSE = SST - SSR.$$

An important measure of model quality is the coefficient of determination R^2, which is defined as a relative part of the sum of squares explained by the regression:

$$R^2 = \frac{SSR}{SST} = \frac{(\mathbf{Xb} - \bar{Y})^T(\mathbf{Xb} - \bar{Y})}{(\mathbf{Y} - \bar{Y})^T(\mathbf{Y} - \bar{Y})} = \frac{\sum_{i=1}^{n}(\mathbf{x}_i \mathbf{b} - \bar{Y})^2}{\sum_{i=1}^{n}(Y_i - \bar{Y})^2}, \tag{1.4}$$

where $\bar{Y} = \sum_{i=1}^{n} Y_i / n$ when there is an intercept term and $\bar{Y} = 0$ when there is no intercept term in the model.

Precision of the forecast of Y_t can be measured by a confidence interval. The confidence interval with a confidence level $1 - \alpha$ for the forecast of Y_t is given by

$$[\mathbf{x}_t\mathbf{b} - \Delta(\mathbf{x}_t), \mathbf{x}_t\mathbf{b} + \Delta(\mathbf{x}_t)], \qquad (1.5)$$

where $\Delta(\mathbf{x}_t)$ is the width of the confidence interval with a confidence level $1 - \alpha$ for a forecast at the point $\mathbf{x}_t$. It can be calculated as

$$\Delta(\mathbf{x}_t) = t_{1-(\alpha/2), n-d-1} \sqrt{\frac{SSE}{n-d-1}(1 + \mathbf{x}_t(\mathbf{X}^T\mathbf{X})^{-1}\mathbf{x}_t^T)}, \qquad (1.6)$$

where $t_{1-(\alpha/2), n-d-1}$ is a $1 - (\alpha/2)$-level quantile of t-distribution with $n - d - 1$ degrees of freedom.

We also use other measures of efficiency of the estimates denoted by the average forecasting error

$$AFE = \frac{1}{n}\sum_{i=1}^{n}|Y_i - \mathbf{x}_i\mathbf{b}| \qquad (1.7)$$

and the relative forecasting error for the point $\mathbf{x}_t$

$$RFE_t = \frac{|Y_t - \mathbf{x}_t\mathbf{b}|}{Y_t}. \qquad (1.8)$$

1.3.2 Local Recursive Parameter Estimation

The described estimation procedure, (1.2), requires information about all observations, that is, the complete matrix $\mathbf{X}$. In practice, for real-time streaming data, the estimation is performed iteratively, being updated after each new observation. The recurrent iterative method for LSE was suggested in Refs. [29,30]. This method assumes the recalculation of system parameters for each new observation.

We briefly describe the key aspects of this algorithm. Let $\mathbf{b}_t$ be the estimate after t first observations. After receiving the $(t + 1)$-th observation, we recalculate the estimates of β (Y_{t+1}-value of a dependent variable; and $\mathbf{x}_{t+1}$-values of explanatory variables):

$$\mathbf{b}_{t+1} = \mathbf{b}_t + \mathbf{K}_{t+1}(Y_{t+1} - \mathbf{x}_{t+1}\mathbf{b}_t), \qquad t = 0,1,\ldots, \qquad (1.9)$$

where $\mathbf{K}_{t+1}$ is a $d \times 1$ vector of proportionality, smoothness or compensation. From Equation 1.9, one can observe that $\mathbf{b}_{t+1}$ is represented as a sum of the previous estimate $\mathbf{b}_t$ and the correction term $\mathbf{K}_{t+1}(Y_{t+1} - \mathbf{x}_{t+1}\mathbf{b}_t)$. Formula (1.9) is based on exponential smoothness, an adaptive forecasting method [30].

To calculate $\mathbf{K}_{t+1}$, we need values of matrices $\mathbf{A}_t$ and $\mathbf{B}_t$, obtained after the last t-th iteration. $\mathbf{A}_t$ and $\mathbf{B}_t$ are square $d \times d$ matrices. $\mathbf{B}_t$ is equal to $(\mathbf{X}^T\mathbf{X})^{-1}$ if this matrix exists, else it is equal to a pseudo-inverse matrix. Matrix $\mathbf{A}_t$ is a projection matrix; therefore, if $\mathbf{x}_{t+1}$ is the linear combination of the rows of matrix $\mathbf{X}_t$, its projection is equal to zero: $\mathbf{x}_{t+1}\mathbf{A}_t = 0$. Starting the algorithm, we set the following initial values $\mathbf{A}_0 = \mathbf{I}$, $\mathbf{B}_0 = 0$, $\mathbf{b}_0 = 0$.

If the condition $x_{t+1}A_t = 0$ is satisfied, then

$$B_{t+1} = B_t - (1 + x_{t+1}B_t x_{t+1}^T)^{-1} B_t x_{t+1}^T (B_t x_{t+1}^T)^T,$$

$$A_{t+1} = A_t, \quad K_{t+1} = (1 + x_{t+1}B_t x_{t+1}^T)^{-1} B_t x_{t+1}^T,$$

Otherwise,

$$B_{t+1} = B_t - (x_{t+1}A_t x_{t+1}^T)^{-1}((B_t x_{t+1}^T)(A_t x_{t+1}^T)^T + (A_t x_{t+1}^T)(B_t x_{t+1}^T)^T)$$

$$+ (x_{t+1}A_t x_{t+1}^T)^{-2}(1 + x_{t+1}B_t x_{t+1}^T)(A_t x_{t+1}^T)(A_t x_{t+1}^T)^T,$$

$$A_{t+1} = A_t - (x_{t+1}A_t x_{t+1}^T)^{-1} A_t x_{t+1}^T (A_t x_{t+1}^T)^T,$$

$$K_{t+1} = (x_{t+1}A_t x_{t+1}^T)^{-1} A_t x_{t+1}^T.$$

1.3.3 Cooperative Linear Regression-Based Learning Algorithm

Suppose we have an MAS consisting of s autonomous agents $Ag = \{Ag^{(1)}, Ag^{(2)}, \ldots, Ag^{(s)}\}$, and each of them contains a local regression model, which is estimated on the basis of its experience. We now introduce a parameter adjustment algorithm, which allows the agents to exchange their model parameters in order to improve the forecasts.

We introduce the following notations. We use superscript i for the variables in formula (1.1) to refer to the model of the agent $Ag^{(i)}$:

$$Y^{(i)} = X^{(i)}\beta + \varepsilon^{(i)}, \quad i = 1, \ldots, s. \tag{1.10}$$

$Ag^{(i)}$ calculates the estimates $b_t^{(i)}$ of β on the basis of its experience using Equation 1.9. It predicts $E[Y_{t+1}^{(i)}]$ using Equation 1.3 at a future time moment $t + 1$ for the values of factors $x_{t+1}^{(i)}$.

After forecasting, $Ag^{(i)}$ checks if it needs to adjust its locally estimated parameters $b_t^{(i)}$ with other agents. This is done when the agent considers a forecast $E[Y_{t+1}^{(i)}]$ as not reliable. One possible approach is to compare the width of the confidence interval of the forecast with a forecast value. The forecast $E[Y_{t+1}^{(i)}]$ is considered to be not reliable if its value is sufficiently smaller than its confidence interval width:

$$\frac{\Delta^{(i)}(x_{t+1}^{(i)})}{x_{t+1}^{(i)} b_t^{(i)}} > p,$$

where p is an agent's parameter representing the maximal ratio of the confidence interval width to the forecast, after which a coordination takes place; $\Delta^{(i)}(x)$ is the confidence interval for factors x based on the agent's data.

Now, let us describe the parameter adjustment procedure. First, $Ag^{(i)}$ sends a request to other agents within its transmission radius; this request contains a set of factors $x_{t+1}^{(i)}$ as well as a threshold $Tr^{(i)}(x_{t+1}^{(i)})$, which is set equal to the confidence interval width: $Tr^{(i)}(x_{t+1}^{(i)}) = \Delta^{(i)}(x_{t+1}^{(i)})$.

Each agent $Ag^{(j)}$ in the transmission radius calculates the confidence interval $\Delta^{(j)}(x_{t+1}^{(i)})$ for the requested values of factors $x_{t+1}^{(i)}$ on the basis of its data and compares it with the threshold. If this value is less than the threshold, $\Delta^{(j)}(x_{t+1}^{(i)}) < Tr^{(i)}(x_{t+1}^{(i)})$, $Ag^{(j)}$ replies to $Ag^{(i)}$

by sending its parameters $\mathbf{b}_t^{(j)}$. Let us define $G^{(i)} \subset Ag$ as a group of agents who are able to reply to $Ag^{(i)}$ by sending their parameters, including $Ag^{(i)}$.

Second, $Ag^{(i)}$ receives replies from the group $G^{(i)}$. It assigns weights to each $Ag^{(j)} \in G^{(i)}$ (including itself) $c^{(i)} = \{c_j^{(i)}\}$; these weights vary with time and represent the reliability level of each $Ag^{(j)}$ (including reliability of own experience). In our case, the agents' weights depend on the forecasting experience.

According to the logic of constructing discrete-time consensus, we assume that $c^{(i)}$ is a stochastic vector for all t (the sum of its elements is equal to 1). Then an updated estimate $\tilde{\mathbf{b}}_{t+1}^{(i)}$ is calculated as

$$\tilde{\mathbf{b}}_{t+1}^{(i)} = \sum_{Ag^{(j)} \in G^{(i)}} c_j^{(i)} \mathbf{b}_t^{(j)}. \qquad (1.11)$$

This adjustment procedure aims to increase the reliability of the estimates, especially for insufficient or missing historical data, and to contribute to the overall estimation accuracy [18].

1.3.4 Numerical Example

We illustrate the linear regression algorithm by providing a numerical example. First, let us consider a single agent that maintains a data set of four observations, where variable $\mathbf{Y}$ depends on three factors $\mathbf{X}$. The corresponding matrices $\mathbf{X}$ and $\mathbf{Y}$ are

$$\mathbf{X} = \begin{pmatrix} 5.4 & 3.9 & 2.2 \\ 1.7 & 4.6 & 3.5 \\ 3.2 & 2.3 & 1.2 \\ 4.3 & 2.1 & 3.2 \end{pmatrix}, \quad \mathbf{Y} = \begin{pmatrix} 2.7 \\ 1.5 \\ 2.6 \\ 3.4 \end{pmatrix}.$$

The agent uses formula (1.9):

$$\mathbf{A}_0 = \begin{pmatrix} 1 & 0 & 0 \\ 0 & 1 & 0 \\ 0 & 0 & 1 \end{pmatrix}, \quad \mathbf{B}_0 = \begin{pmatrix} 0 & 0 & 0 \\ 0 & 0 & 0 \\ 0 & 0 & 0 \end{pmatrix}, \quad \mathbf{K}_0 = \begin{pmatrix} 0 \\ 0 \\ 0 \end{pmatrix}, \quad \mathbf{b}_0 = \begin{pmatrix} 0 \\ 0 \\ 0 \end{pmatrix}$$

$$\mathbf{x}_1 = (5.4 \quad 3.9 \quad 2.2), \quad Y_1 = 2.7, \quad \mathbf{x}_1 \mathbf{A}_0 = (5.4 \quad 3.9 \quad 2.2) \neq 0$$

$$\mathbf{A}_1 = \begin{pmatrix} 0.41 & -0.43 & -0.24 \\ -0.43 & 0.69 & -0.17 \\ -0.24 & -0.17 & 0.90 \end{pmatrix}, \quad \mathbf{B}_1 = \begin{pmatrix} 0.01 & 0.01 & 0.01 \\ 0.01 & 0.01 & 0.00 \\ 0.01 & 0.00 & 0.00 \end{pmatrix}, \quad \mathbf{K}_1 = \begin{pmatrix} 0.11 \\ 0.08 \\ 0.05 \end{pmatrix}, \quad \mathbf{b}_1 = \begin{pmatrix} 0.30 \\ 0.21 \\ 0.12 \end{pmatrix}$$

$$\mathbf{x}_2 = (1.7 \quad 4.6 \quad 3.5), \quad Y_2 = 1.5, \quad \mathbf{x}_2 \mathbf{A}_1 = (-2.12 \quad 1.84 \quad 1.94) \neq 0$$

$$\mathbf{A}_2 = \begin{pmatrix} 0.02 & -0.09 & 0.11 \\ -0.09 & 0.40 & -0.48 \\ 0.11 & -0.48 & 0.58 \end{pmatrix}, \quad \mathbf{B}_2 = \begin{pmatrix} 0.09 & -0.04 & -0.05 \\ -0.04 & 0.06 & 0.03 \\ -0.04 & 0.03 & 0.03 \end{pmatrix}, \quad \mathbf{K}_2 = \begin{pmatrix} -0.18 \\ 0.16 \\ 0.17 \end{pmatrix}, \quad \mathbf{b}_2 = \begin{pmatrix} 0.37 \\ 0.15 \\ 0.05 \end{pmatrix}$$

$$\mathbf{x}_3 = (3.2 \quad 2.3 \quad 1.2), \quad Y_3 = 2.6, \quad \mathbf{x}_3\mathbf{A}_2 = (-0.01 \quad 0.04 \quad -0.06) \neq 0$$

$$\mathbf{A}_3 = \begin{pmatrix} 0 & 0 & 0 \\ 0 & 0 & 0 \\ 0 & 0 & 0 \end{pmatrix}, \quad \mathbf{B}_3 = \begin{pmatrix} 6.42 & -25.99 & 30.98 \\ -25.99 & 106.15 & -126.80 \\ 30.98 & -126.80 & 151.59 \end{pmatrix}, \quad \mathbf{K}_3 = \begin{pmatrix} -2.05 \\ 8.81 \\ -10.59 \end{pmatrix}, \quad \mathbf{b}_3 = \begin{pmatrix} -1.70 \\ 9.03 \\ -10.61 \end{pmatrix}$$

$$\mathbf{x}_4 = (4.3 \quad 2.1 \quad 3.2), \quad Y_4 = 3.4, \quad \mathbf{x}_4\mathbf{A}_3 = (0 \quad 0 \quad 0) = 0$$

$$\mathbf{A}_4 = \begin{pmatrix} 0 & 0 & 0 \\ 0 & 0 & 0 \\ 0 & 0 & 0 \end{pmatrix}, \quad \mathbf{B}_4 = \begin{pmatrix} 0.06 & -0.04 & -0.04 \\ -0.04 & 0.20 & -0.19 \\ -0.04 & -0.19 & 0.31 \end{pmatrix}, \quad \mathbf{K}_4 = \begin{pmatrix} 0.09 \\ -0.35 \\ 0.43 \end{pmatrix}, \quad \mathbf{b}_4 = \begin{pmatrix} 0.57 \\ -0.21 \\ 0.43 \end{pmatrix}$$

We can see that the agent obtains the same estimate **b** as they did using formula (1.2):

$$\mathbf{b} = (\mathbf{X}^T\mathbf{X})^{-1}\mathbf{X}^T\mathbf{Y} = \begin{pmatrix} 0.57 \\ -0.21 \\ 0.43 \end{pmatrix}.$$

Now, we consider the parameter adjustment algorithm. Let us denote the above considered agent as $Ag^{(1)}$. Suppose it should make a forecast for a new point $\mathbf{x}_{t+1}^{(1)} = (3.7 \quad 2.8 \quad 1.1)$. $Ag^{(1)}$ makes a forecast $E[Y_{t+1}^{(1)}]$ for the point $\mathbf{x}_{t+1}^{(1)}$:

$$\mathbf{x}_{t+1}^{(1)}\mathbf{b}_t^{(1)} = (3.7 \quad 2.8 \quad 1.1) \begin{pmatrix} 0.57 \\ -0.21 \\ 0.43 \end{pmatrix} = 1.98.$$

The corresponding confidence interval is

$$SSR = (\mathbf{Xb})^T (\mathbf{Xb}) = 27.05$$

$$SST = \mathbf{Y}^T \mathbf{Y} = 27.86$$

$$SSE = SST - SSR = 0.81$$

$$\Delta^{(1)}(\mathbf{x}_{t+1}^{(1)}) = t_{\frac{0.05}{2}, 5-3-1} \sqrt{\frac{SSE}{5-3-1}(1 + \mathbf{x}_{t+1}^{(1)}(\mathbf{X}^T\mathbf{X})^{-1}(\mathbf{x}_{t+1}^{(1)})^T)} = 14.38.$$

Let the maximal ratio $p = 1.5$. The agent checks the ratio

$$\frac{\Delta^{(1)}(\mathbf{x}_{t+1}^{(1)})}{\mathbf{x}_{t+1}^{(1)}\mathbf{b}_t^{(1)}} = \frac{14.38}{1.98} = 7.28 > 1.5$$

and requests parameter adjustment, because it exceeds p. For this purpose, it sends a point $\mathbf{x}_{t+1}^{(1)} = (3.7 \quad 2.8 \quad 1.1)$ as well as the threshold $Tr^{(1)}(\mathbf{x}_{t+1}^{(1)}) = \Delta^{(1)}(\mathbf{x}_{t+1}^{(1)}) = 14.38$.

Consider agents $Ag^{(2)}$ and $Ag^{(3)}$ in the transmission radius with equal experience 4 and the following parameters:

$$\mathbf{b}_t^{(2)} = \begin{pmatrix} 0.58 \\ 0.02 \\ 0.43 \end{pmatrix}, \quad \mathbf{b}_t^{(3)} = \begin{pmatrix} 0.54 \\ -0.004 \\ 0.49 \end{pmatrix}.$$

$Ag^{(2)}$ and $Ag^{(3)}$ calculate their confidence limits $\Delta^{(2)}(\mathbf{x}_{t+1}^{(1)}) = 7.08$, $\Delta^{(3)}(\mathbf{x}_{t+1}^{(1)}) = 0.07$. As both are less than the threshold, they send their parameters $\mathbf{b}_t^{(2)}$ and $\mathbf{b}_t^{(3)}$ as well as their experience 4. The group $G^{(i)} = \{1,2,3\}$.

$Ag^{(1)}$ receives the answers and uses weights proportional to the experience of the agents, $c^{(1)} = <\frac{4}{12}, \frac{4}{12}, \frac{4}{12}> = <\frac{1}{3}, \frac{1}{3}, \frac{1}{3}>$. Now an updated estimate $\tilde{\mathbf{b}}_{t+1}^{(i)}$ can be calculated according to Equation 1.6:

$$\tilde{\mathbf{b}}_{t+1}^{(1)} = \sum_{Ag^j \in G^{(1)}(t+1)} c_j^{(1)} \mathbf{b}_t^{(j)} = 0.33 \begin{pmatrix} 0.56 \\ -0.21 \\ 0.43 \end{pmatrix} + 0.33 \begin{pmatrix} 0.58 \\ 0.02 \\ 0.43 \end{pmatrix} + 0.33 \begin{pmatrix} 0.54 \\ -0.004 \\ 0.49 \end{pmatrix} = \begin{pmatrix} 0.56 \\ -0.06 \\ 0.45 \end{pmatrix}.$$

For a new point $\mathbf{x}_{t+1}^{(1)} = (3.7 \quad 2.8 \quad 1.1)$, an estimate $E[Y_{t+1}^{(1)}]$ after adjustment is equal to

$$\mathbf{x}_{t+1}^{(1)} \tilde{\mathbf{b}}_{t+1}^{(1)} = (3.7 \quad 2.8 \quad 1.1) \begin{pmatrix} 0.56 \\ -0.06 \\ 0.45 \end{pmatrix} = 2.39.$$

Remember that the estimate without adjustment $E[Y_{t+1}^{(1)}] = 1.98$. For this data point, the corresponding true value of $Y_{t+1} = 2.5$. We can see that the adjustment facilitates a better estimate of Y_{t+1}.

1.4 Distributed Cooperative Forecasting Based on Kernel Regression

1.4.1 Kernel Density Estimation

Kernel density estimation is a nonparametric approach for estimating the probability density function of a random variable. Kernel density estimation is a fundamental data-smoothing technique where inferences about the population are made on the basis of a finite data sample. A kernel is a weighting function used in nonparametric estimation techniques.

Let $X_1, X_2, \ldots, X_n$ be an iid sample drawn from some distribution with an unknown density $f(x)$. We attempt to estimate the shape of $f(x)$, whose kernel density estimator is

$$\hat{f}_h(x) = \frac{1}{nh} \sum_{i=1}^{n} K\left(\frac{x - X_i}{h}\right), \tag{1.12}$$

where kernel $K(\bullet)$ is a nonnegative real-valued integrable function satisfying the following two requirements: $\int_{-\infty}^{\infty} K(u)du = 1$ and $K(u) = K(-u)$ for all values of u; $h > 0$ is a smoothing parameter called bandwidth. The first requirement of $K(\bullet)$ ensures that the kernel density estimator is a probability density function. The second requirement of $K(\bullet)$ ensures that the average of the corresponding distribution is equal to that of the sample used [31]. Different kernel functions are available: Uniform, Epanechnikov, Gaussian and so on. They differ in the manner in which they take into account the vicinity observations to estimate the function from the given variables.

An important problem in kernel density estimation is selection of the appropriate bandwidth h. It has an influence on the structure of the neighbourhood in Equation 1.12: the bigger the h that is selected, the more points X_i have a significant influence on the estimator $\hat{f}_h(x)$. In multi-dimensional cases, it also regulates a balance between the factors.

Most of the bandwidth selection methods are based on the minimisation of the integrated squared error $ISE(h)$ (or similar asymptotic or averaged characteristics) with respect to the bandwidth h. The integrated squared error is defined as

$$ISE(h) = \int [\hat{f}_h(x) - f(x)]^2 dx.$$

However, the direct use of this formula is impossible due to the involvement of an unknown true density $f(x)$. So different heuristic approaches are used for this problem.

The methods of bandwidth selection can be divided into two main groups [32]:

1. Plug-in methods
2. Resampling methods

The plug-in methods replace unknown terms in the expression for $ISE(h)$ by their heuristic approximations. One of the well-known approaches is rule-of-thumb, proposed by Silverman [33], which works for Gaussian kernels. In the case of no correlation between explanatory variables, there is a simple and useful formula for bandwidth selection [34]:

$$h_j = n^{-1/(d+4)} \sigma_j, \qquad j = 1, 2, \ldots, d, \qquad (1.13)$$

where σ_j is a variance of the j-th factor, d is a number of factors.

The resampling methods include cross-validation and bootstrap approaches. They use available data in different combinations to obtain approximation of $ISE(h)$. One of the most commonly used approaches is the minimisation of the least-squares cross-validation function $LSCV(h)$ introduced in Ref. [35]

$$LSCV(h) = \int [\hat{f}_h(x)]^2 dx - \frac{2}{n} \sum_{i=1}^{n} \hat{f}_{h,-i}(X_i),$$

where $\hat{f}_{h,-i}(X_i)$ is a kernel density estimation calculated without the i-th observation. $LSCV(h)$ is a d-dimensional function, which can be minimised with respect to h.

A very suitable property of the kernel function is its additive nature. This property makes the kernel function easy to use for streaming and distributed data [2,17,31]. In Ref. [16], the distributed kernel-based clustering algorithm was suggested on the basis of the same property. In this study, kernel density is used for kernel regression to estimate the conditional expectation of a random variable.

1.4.2 Kernel-Based Local Regression Model

The nonparametric approach to estimating a regression curve has four main purposes. First, it provides a versatile method for exploring a general relationship between a dependent variable **Y** and factor **X**. Second, it can predict observations yet to be made without reference to a fixed parametric model. Third, it provides a tool for finding spurious observations by studying the influence of isolated points. Fourth, it constitutes a flexible method of substitution or interpolation between adjacent **X** values for missing observations [31].

Let us consider a nonparametric regression model [36] with a dependent variable **Y** and a vector of d regressors **X**

$$\mathbf{Y} = m(\mathbf{x}) + \varepsilon, \tag{1.14}$$

where ε is a random error such that $E[\varepsilon|\mathbf{X}=\mathbf{x}] = 0$ and $Var[\varepsilon|\mathbf{X}=\mathbf{x}] = \sigma^2(\mathbf{x})$; and $m(\mathbf{x}) = E[\mathbf{Y}|\mathbf{X}=\mathbf{x}]$. Further, let $(\mathbf{X}_i, Y_i)_{i=1}^n$ be the observations sampled from the distribution of $(\mathbf{X}, \mathbf{Y})$. Then the Nadaraya–Watson kernel estimator is

$$\hat{m}_n(\mathbf{x}) = \frac{\sum_{i=1}^n K((\mathbf{x}-\mathbf{X}_i)/h)Y_i}{\sum_{i=1}^n K((\mathbf{x}-\mathbf{X}_i)/h)} = \frac{p_n(\mathbf{x})}{q_n(\mathbf{x})}, \tag{1.15}$$

where h as a vector $h = (h_1, h_2, \ldots, h_d)$, $K(\bullet)$ is the kernel function of R^d and $h = (h_1, h_2, \ldots, h_d)$ is the bandwidth. Kernel functions satisfy the restrictions from Equation 1.12. In our case, we have a multi-dimensional kernel function $K(u) = K(u_1, u_2, \ldots, u_d)$ that can be easily presented with univariate kernel functions as: $K(u) = K(u_1) \cdot K(u_2) \cdots K(u_d)$. We used the Gaussian kernel in our experiments.

From formula (1.15), we can see that each value of the historical forecast Y_i is taken with some weight of the corresponding independent variable value of the same observation $\mathbf{X}_i$. Let us denote these weights in the following form:

$$\omega_i(\mathbf{x}) = K\left(\frac{\mathbf{x}-\mathbf{X}_i}{h}\right), \quad \hat{\omega}_i(\mathbf{x}) = \frac{\omega_i(\mathbf{x})}{\sum_{i=1}^n \omega_i}, \tag{1.16}$$

where $\omega_i(\mathbf{x})$ are the weights of the historical observations in the forecast of **x** and $\hat{\omega}_i(\mathbf{x})$ are the corresponding normalised weights. Then, formula (1.15) can be rewritten as

$$\hat{m}_n(\mathbf{x}) = \frac{\sum_{i=1}^n K((\mathbf{x}-\mathbf{X}_i)/h)Y_i}{\sum_{i=1}^n K((\mathbf{x}-\mathbf{X}_i)/h)} = \frac{\sum_{i=1}^n \omega_i(\mathbf{x})Y_i}{\sum_{i=1}^n \omega_i} = \sum_{i=1}^n \hat{\omega}_i(\mathbf{x})Y_i. \tag{1.17}$$

Standard statistical and data mining methods usually deal with a data set of a fixed size n and corresponding algorithms that are functions of n. However, for streaming data, there is no fixed n: data are continually captured and must be processed as they arrive. It is important to develop algorithms that work with nonstationary data sets, handle the streaming data directly and update their models on the fly [37]. For the kernel density estimator, it is possible to use a simple method that allows recursive estimation of $\hat{m}_n(\mathbf{x})$:

$$\hat{m}_n(\mathbf{x}) = \frac{p_n(\mathbf{x})}{q_n(\mathbf{x})} = \frac{p_{n-1}(\mathbf{x}) + \omega_n(\mathbf{x})Y_n}{q_{n-1}(\mathbf{x}) + \omega_n(\mathbf{x})}. \tag{1.18}$$

The quality of the kernel-based regression model can be verified by calculating R^2 as

$$R^2 = \frac{\left[\sum_{i=1}^{n}(Y_i - \bar{Y})(\hat{m}_n(\mathbf{X}_i) - \bar{Y})\right]^2}{\sum_{i=1}^{n}(Y_i - \bar{Y})^2 \sum_{i=1}^{n}(\hat{m}_n(\mathbf{X}_i) - \bar{Y})^2}, \tag{1.19}$$

where $\bar{Y} = \sum_{i=1}^{n} Y_i / n$. It can be demonstrated that R^2 is identical to the standard measure for the linear regression model (1.4), fitted with least squares, and includes an intercept term.

The corresponding confidence interval for the forecast $\hat{m}_n(\mathbf{x})$ with a confidence level $1 - \alpha$ is represented by

$$\left[\hat{m}_n(\mathbf{x}) - z_{1-\frac{\alpha}{2}}\sqrt{\frac{\hat{\sigma}^2(\mathbf{x})\|K\|_2^2}{nh\hat{f}_{h,-i}(\mathbf{x})}}, \; \hat{m}_n(\mathbf{x}) + z_{1-\frac{\alpha}{2}}\sqrt{\frac{\hat{\sigma}^2(\mathbf{x})\|K\|_2^2}{nh\hat{f}_{h,-i}(\mathbf{x})}}\right],$$

where $\|K\|_2^2 = [\int K^2(u)du]^2$, $\hat{\sigma}^2(\mathbf{x}) = \frac{1}{n}\sum_{i=1}^{n}\hat{\omega}_i(\mathbf{x})[Y_i - \hat{m}_n(\mathbf{x})]^2$ and $z_{1-\alpha/2}$ is a $1 - \alpha/2$-level quantile of a standard normal distribution.

1.4.3 Cooperative Kernel-Based Learning Algorithm

In this section, we describe the cooperation for sharing the forecasting experience among the agents in a network. While working with streaming data, one should take into account two main aspects:

1. The nodes should coordinate their forecasting experience over some previous sampling period.
2. The nodes should adapt quickly to the changes in the streaming data, without waiting for the next coordination action.

As in the case of Section 3.3, we have an MAS consisting of s autonomous agents $Ag = \{Ag^{(1)}, Ag^{(2)}, \ldots, Ag^{(s)}\}$, and each of them has a local kernel-based regression model, which is estimated on the basis of its experience.

Now, we introduce the following notations. Let $D^{(j)} = \{(X_c^{(j)}, Y_c^{(j)}) \mid c = 1, \ldots, n^{(j)}\}$ denote a local data set of $Ag^{(j)}$, where $\mathbf{X}_c^{(j)}$ is a d-dimensional vector of factors on the c-th observation. To highlight the dependence of the forecasting function (1.15) on the local data set of $Ag^{(j)}$, we denote the forecasting function by $\hat{m}_n^{(j)}(x)$.

Consider a case when some agent $Ag^{(i)}$ wants to forecast for some d-dimensional future data point $\mathbf{x}_{n+1}^{(i)}$. Denote the weights and the corresponding normalised weights of formula (1.16) of the c-th observation from $D^{(i)}$ in the forecasting $\mathbf{x}_{n+1}^{(i)}$ as $\omega_c^{(i)}(\mathbf{x}_{n+1}^{(i)})$ and $\hat{\omega}_c^{(i)}(\mathbf{x}_{n+1}^{(i)})$, respectively.

We consider a forecast for $\mathbf{x}_{n+1}^{(i)}$ as not reliable if only one of the observations is taken with a significant weight in this forecast:

$$\max(\hat{\omega}_c^{(i)}(\mathbf{x}_{n+1}^{(i)})) > bp,$$

where bp is an agent's parameter representing the maximal weight, after which a coordination takes place.

In this case, $Ag^{(i)}$ expects from other agents the points that are closer to the requested point than its own points. For this purpose, it sends a request to other traffic participants within its transmission radius by sending the data point $\mathbf{x}_{n+1}^{(i)}$ as well as the threshold for the observation weight. This threshold is set as the weight of the second best observation:

$$Tr^{(i)}(\mathbf{x}_{n+1}^{(i)}) = \omega_{(n-1)}^{(i)}(\mathbf{x}_{n+1}^{(i)}),$$

where $\omega_{(1)}^{(i)}(\mathbf{x}_{n+1}^{(i)}), \omega_{(2)}^{(i)}(\mathbf{x}_{n+1}^{(i)}), \ldots, \omega_{(n)}^{(i)}(\mathbf{x}_{n+1}^{(i)})$ is an ordered sequence of weights in the forecast for $\mathbf{x}_{n+1}^{(i)}$.

Each $Ag^{(j)}$ that receives the request calculates the weights $\omega_c^{(j)}(\mathbf{x}_{n+1}^{(i)})$ on the basis of its own data. If there are observations with weights $\omega_c^{(j)}(\mathbf{x}_{n+1}^{(i)}) > Tr^{(i)}(\mathbf{x}_{n+1}^{(i)})$, it forms a reply $\hat{D}^{(j,i)}$ from these observations (maximum 2) and sends it to $Ag^{(i)}$.

Let us define $G^{(i)} \subset Ag$ as a group of agents who are able to reply to $Ag^{(i)}$ by sending the requested data. All the data $\hat{D}^{(j,i)}, Ag^{(j)} \in G^{(i)}$ received by $Ag^{(i)}$ are verified and duplicated data are discarded. These new observations are added to the data set of $Ag^{(i)}$: $D^{(j)} \leftarrow \bigcup_{Ag^{(j)} \in G^{(i)}} \hat{D}^{(j,i)} \cup D^{(j)}$. Suppose that $Ag^{(i)}$ received r observations. Then, the new kernel function of $Ag^{(i)}$ is updated by considering the additive nature of this function:

$$\tilde{m}_{n+r}^{(i)}(\mathbf{x}) = \frac{p_n^{(i)}(\mathbf{x}) + \sum_{Ag^{(j)} \in G^{(i)}} p^{(i,j)}(\mathbf{x})}{q_n^{(i)}(\mathbf{x}) + \sum_{Ag^{(j)} \in G^{(i)}} q^{(i,j)}(\mathbf{x})}, \quad (1.20)$$

where $p^{(i,j)}(\mathbf{x})$ and $q^{(i,j)}(\mathbf{x})$ are the numerator and denominator of Equation 1.15, respectively, calculated by $\hat{D}^{(j,i)}$.

Finally, $Ag^{(i)}$ can autonomously make its forecast for $\mathbf{x}_{n+1}^{(i)}$ as $\tilde{m}_{n+r}^{(i)}(\mathbf{x}_{n+1}^{(i)})$.

1.4.4 Numerical Example

To illustrate the implementation of kernel-based regression, we provide an example using the same data we used for the linear regression:

$$\mathbf{X} = \begin{pmatrix} 5.4 & 3.9 & 2.2 \\ 1.7 & 4.6 & 3.5 \\ 3.2 & 2.3 & 1.2 \\ 4.3 & 2.1 & 3.2 \end{pmatrix}, \quad \mathbf{Y} = \begin{pmatrix} 2.7 \\ 1.5 \\ 2.6 \\ 3.4 \end{pmatrix}.$$

Let us consider how an agent can calculate its forecasting function from formula (1.15). We take a Gaussian kernel function and for a one-dimensional case, $K(u)$ is a density of the standard normal distribution $N(0,1)$ with the mean being 0 and the variance equal to 1: $K(u) = (1/\sqrt{2\pi})e^{-(1/2)u^2}$.

In a multi-dimensional case, the kernel function is the product of one-dimensional kernels, as follows:

$$K(<u_1, u_2, \ldots, u_d>) = K(u_1) \cdot K(u_2) \cdot \cdots \cdot K(u_d).$$

We calculate the bandwidth $h = (1.30 \quad 1.00 \quad 0.86)$ according to Equation 1.13 and make a forecast according to formula (1.17) for some new point $\mathbf{x}_5 = (3.7 \quad 2.8 \quad 1.1)$ as

$$\hat{m}_4(\mathbf{x}_5) = \frac{\omega_1 \cdot 2.7 + \omega_2 \cdot 1.5 + \omega_3 \cdot 2.6 + \omega_4 \cdot 3.4}{\omega_1 + \omega_2 + \omega_3 + \omega_4},$$

$$\omega_1 = K\left(\frac{\mathbf{x}_5 - \mathbf{X}_1}{h}\right) = K\left(\frac{\mathbf{x}_5 - (5.4 \quad 3.9 \quad 2.2)}{h}\right) = K\left(\frac{3.7 - 5.4}{1.30}\right) K\left(\frac{2.8 - 3.9}{1.00}\right) K\left(\frac{1.1 - 2.2}{0.86}\right)$$

$$= K(-1.31)K(-1.10)K(-1.28) = 0.17 \cdot 0.22 \cdot 0.17 = 0.006,$$

$$\omega_2 = K\left(\frac{\mathbf{x}_5 - \mathbf{X}_2}{h}\right) = K\left(\frac{\mathbf{x}_5 - (1.7 \quad 4.6 \quad 3.5)}{h}\right) = K\left(\frac{3.7 - 1.7}{1.30}\right) K\left(\frac{2.8 - 4.6}{1.00}\right) K\left(\frac{1.1 - 3.5}{0.86}\right)$$

$$= K(1.54)K(-1.80)K(-2.80) = 0.12 \cdot 0.08 \cdot 0.008 = 7.5 \cdot 10^{-5},$$

$$\omega_3 = K\left(\frac{\mathbf{x}_5 - \mathbf{X}_3}{h}\right) = K\left(\frac{\mathbf{x}_5 - (3.2 \quad 2.3 \quad 1.2)}{h}\right) = K\left(\frac{3.7 - 3.2}{1.30}\right) K\left(\frac{2.8 - 2.3}{1.00}\right) K\left(\frac{1.1 - 1.2}{0.86}\right)$$

$$= K(0.39)K(0.50)K(-0.12) = 0.37 \cdot 0.35 \cdot 0.40 = 0.05,$$

$$\omega_4 = K\left(\frac{\mathbf{x}_5 - \mathbf{X}_4}{h}\right) = K\left(\frac{\mathbf{x}_5 - (4.3 \quad 2.1 \quad 3.2)}{h}\right) = K\left(\frac{3.7 - 4.3}{1.30}\right) K\left(\frac{2.8 - 2.1}{1.00}\right) K\left(\frac{1.1 - 3.2}{0.86}\right)$$

$$= K(-0.46)K(0.70)K(-2.45) = 0.36 \cdot 0.31 \cdot 0.02 = 0.002,$$

$$\hat{m}_4(\mathbf{x}_5) = \frac{\omega_1 \cdot 2.7 + \omega_2 \cdot 1.5 + \omega_3 \cdot 2.6 + \omega_4 \cdot 3.4}{\omega_1 + \omega_2 + \omega_3 + \omega_4} = \frac{0.16}{0.06} = 2.64.$$

Next, we illustrate a cooperative learning algorithm. Let us denote the above considered agent as $Ag^{(1)}$. Suppose that the focal agent needs to make a forecast for the point $\mathbf{x}_5^{(1)} = (3.7 \quad 2.8 \quad 1.1)$.

The weights of the observations and the corresponding normalised weights during forecasting are

$$\begin{pmatrix} \omega_1^{(1)} \\ \omega_2^{(1)} \\ \omega_3^{(1)} \\ \omega_4^{(1)} \end{pmatrix} = \begin{pmatrix} 0.006 \\ 7.5 \cdot 10^{-5} \\ 0.05 \\ 0.002 \end{pmatrix}, \begin{pmatrix} \hat{\omega}_1^{(1)} \\ \hat{\omega}_2^{(1)} \\ \hat{\omega}_3^{(1)} \\ \hat{\omega}_4^{(1)} \end{pmatrix} = \begin{pmatrix} 0.107 \\ 0.001 \\ 0.856 \\ 0.036 \end{pmatrix}.$$

The maximal weight $\hat{\omega}_3^{(1)} = 0.856$ is greater than $bp = 0.8$, so the agent requests cooperation. We test the following threshold $Tr^{(1)} = 0.006$.

We suppose that in the neighbourhood of the focal agent are two agents $Ag^{(2)}$ and $Ag^{(3)}$ with data sets:

$$\mathbf{X}^{(2)} = \begin{pmatrix} 4.1 & 2.5 & 1.3 \\ 0.4 & 3.7 & 3.2 \\ 3.1 & 3.4 & 0.7 \\ 5.4 & 0.7 & 0.3 \end{pmatrix}, \mathbf{Y}^{(2)} = \begin{pmatrix} 2.6 \\ 1.8 \\ 2.3 \\ 3.5 \end{pmatrix}, \mathbf{X}^{(3)} = \begin{pmatrix} 5.0 & 2.7 & 3.5 \\ 3.2 & 2.2 & 1.4 \\ 3.3 & 3.4 & 1.7 \\ 0.8 & 4.3 & 1.2 \end{pmatrix}, \mathbf{Y}^{(3)} = \begin{pmatrix} 4.4 \\ 2.4 \\ 2.6 \\ 1.0 \end{pmatrix}.$$

They calculate the weights $\omega^{(j)}$, $j = 2..3$ on the basis of their own historical data in the same manner as the focal agent has done at the beginning of this example. Let the obtained weights be

$$\begin{pmatrix} \omega_1^{(2)} \\ \omega_2^{(2)} \\ \omega_3^{(2)} \\ \omega_4^{(2)} \end{pmatrix} = \begin{pmatrix} 0.06 \\ 8.2 \cdot 10^{-5} \\ 0.04 \\ 1.9 \cdot 10^{-3} \end{pmatrix}, \begin{pmatrix} \omega_1^{(3)} \\ \omega_2^{(3)} \\ \omega_3^{(3)} \\ \omega_4^{(3)} \end{pmatrix} = \begin{pmatrix} 7.5 \cdot 10^{-4} \\ 0.05 \\ 0.04 \\ 1.6 \cdot 10^{-3} \end{pmatrix}.$$

Both agents $Ag^{(2)}$ and $Ag^{(3)}$ have observations with the weights $\omega_i^{(j)} > Tr^{(1)}$. So they send these observations as follows:

$$\hat{D}^{(2,1)} = \left(\begin{bmatrix} 4.1 & 2.5 & 1.3 \\ 3.1 & 3.4 & 0.7 \end{bmatrix}, \begin{bmatrix} 2.6 \\ 2.3 \end{bmatrix} \right), \hat{D}^{(3,1)} = \left(\begin{bmatrix} 3.2 & 2.2 & 1.4 \\ 3.3 & 3.4 & 1.7 \end{bmatrix}, \begin{bmatrix} 2.4 \\ 2.6 \end{bmatrix} \right).$$

Next, $Ag^{(1)}$ updates its matrices:

$$\mathbf{X}^{(1)} = \begin{pmatrix} 5.4 & 3.9 & 2.2 \\ 1.7 & 4.6 & 3.5 \\ 3.2 & 2.3 & 1.2 \\ 4.3 & 2.1 & 3.2 \\ 4.1 & 2.5 & 1.3 \\ 3.1 & 3.4 & 0.7 \\ 3.2 & 2.2 & 1.4 \\ 3.3 & 3.4 & 1.7 \end{pmatrix}, \mathbf{Y}^{(1)} = \begin{pmatrix} 2.7 \\ 1.5 \\ 2.6 \\ 3.4 \\ 2.6 \\ 2.3 \\ 2.4 \\ 2.6 \end{pmatrix}$$

and calculates the new weights and the corresponding normalised weights during forecasting for the data point $\mathbf{x}_5^{(1)} = (3.7 \quad 2.8 \quad 1.1)$:

$$\begin{pmatrix} \omega_1^{(1)} \\ \omega_2^{(1)} \\ \omega_3^{(1)} \\ \omega_4^{(1)} \\ \omega_5^{(1)} \\ \omega_6^{(1)} \\ \omega_7^{(1)} \\ \omega_8^{(1)} \end{pmatrix} = \begin{pmatrix} 0.006 \\ 7.5 \cdot 10^{-5} \\ 0.05 \\ 0.002 \\ 0.06 \\ 0.04 \\ 0.05 \\ 0.04 \end{pmatrix}, \quad \begin{pmatrix} \hat{\omega}_1^{(1)} \\ \hat{\omega}_2^{(1)} \\ \hat{\omega}_3^{(1)} \\ \hat{\omega}_4^{(1)} \\ \hat{\omega}_5^{(1)} \\ \hat{\omega}_6^{(1)} \\ \hat{\omega}_7^{(1)} \\ \hat{\omega}_8^{(1)} \end{pmatrix} = \begin{pmatrix} 0.026 \\ 3.1 \cdot 10^{-4} \\ 0.211 \\ 0.009 \\ 0.230 \\ 0.174 \\ 0.189 \\ 0.161 \end{pmatrix}.$$

Now, we can make the estimation based on the new data:

$$\hat{m}_8(\mathbf{x}) = \frac{\omega_1^{(1)} \cdot 2.7 + \omega_2^{(1)} \cdot 1.5 + \omega_3^{(1)} \cdot 2.6 + \omega_4^{(1)} \cdot 3.4 + \omega_5^{(1)} \cdot 2.6 + \omega_6^{(1)} \cdot 2.3 + \omega_7^{(1)} \cdot 2.4 + \omega_8^{(1)} \cdot 2.6}{\omega_1^{(1)} + \omega_2^{(1)} + \omega_3^{(1)} + \omega_4^{(1)} + \omega_5^{(1)} + \omega_6^{(1)} + \omega_7^{(1)} + \omega_8^{(1)}}$$

$$= \frac{0.619}{0.245} = 2.52.$$

Remember that for this data point the corresponding true value $Y_{t+1} = 2.5$. We can see that the adjustment facilitates a better estimate $\hat{m}_8(\mathbf{x}) = 2.52$ of Y_{t+1} while before adjustment the estimate was $\hat{m}_4(\mathbf{x}) = 2.61$.

1.5 Case Studies

1.5.1 Problem Formulation

We simulated a traffic network in the southern part of Hanover, Germany. The network contained three parallel and five perpendicular streets, which formed 15 intersections with a flow of approximately 5000 vehicles/h. The network is shown in Figure 1.4.

We assumed that only a small proportion of vehicles were equipped with corresponding devices, which helped to solve the travelling time forecasting problem. These devices could communicate with other vehicles and make the necessary calculations used by the DDPM module. In the current case, the DDPM module was implemented for regression-based forecasting models. Other vehicles played the role of the 'environment', creating corresponding traffic flows.

The vehicles received information from TIC about the centrally estimated system variables (such as average speed, number of stops, congestion level, etc.) in this city district, which they combined with their stored historical information, before making adjustments after exchanging information with other vehicles using cooperative learning algorithms.

We assumed that each traffic participant considered the same factors during travelling time forecasting. These factors are shown in Table 1.1. Only significant factors were used, which were tested specifically for the linear regression model using the t-distribution.

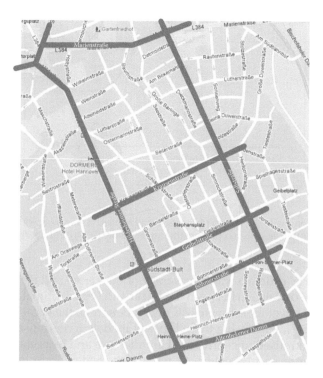

FIGURE 1.4
Road network in the southern part of Hanover, Germany.

To obtain more reliable results, we considered the travel frequency based on statistics from a survey [38] in Germany (2012) of drivers aged over 14 years, as shown in Figure 1.5.

The system was implemented and simulated in the *R* package. The agents were trained using the available real-world data set ($n = 6500$). The data were normalised in the interval [0,1]. No data transformation was performed because experiments showed that they provide no additional quality improvements in the current models. During kernel-based regression, we used bandwidths h calculated by the formula (1.13), which are listed in Table 1.2.

We combined the linear and kernel-based regression results using two additional estimates. The first estimate was the optimum, where we supposed an 'oracle' model helped to select the best forecast (with a low forecasting error) from every forecast. The oracle model represented the best forecast that could be achieved using the linear and kernel regressions.

TABLE 1.1

Factors Used for Travelling Time Forecasting

Variable	Description
Y	Travelling time (min)
$X^{<1>}$	Route length (km)
$X^{<2>}$	Average speed in system (km/h)
$X^{<3>}$	Average number of stops (units/min)
$X^{<4>}$	Congestion level (vehicles/h)
$X^{<5>}$	Number of traffic lights on the route (units)
$X^{<6>}$	Number of left turns on the route (units)

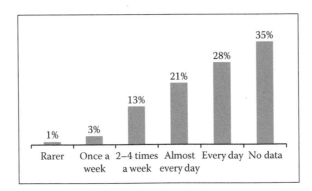

FIGURE 1.5
Frequency of driving according to the survey. (Adapted from Frequency: Travelling with passenger cars, *Statista*, 2012. [Online]. Available: http://de.statista.com/statistik/daten/studie/176124/umfrage/haeufigkeit-pkw-fahren/.)

The second 'average' estimate was the average of the kernel and linear estimates. There was a strong positive correlation between the linear and kernel estimates (about 0.8), but we demonstrated that the average estimate was often better than the kernel or linear estimates.

We conducted experiments using three types of models: a centralised architecture with an authority, a decentralised uncoordinated architecture and a decentralised coordinated architecture. The linear and kernel-based regression models were implemented in each model, as well as their combinations (oracle and average estimates).

We compared the results by analysing the average forecasting errors (1.7), the relative forecasting errors (1.8) and coefficients of determination R^2 (1.4 and 1.19). These characteristics are well-known measures of the effectiveness in predicting the future outcomes using regression models and they can be used with parametric and nonparametric models [39].

1.5.2 Centralised Architecture

This model assumes that agents transmit their observations to a central authority, which requires the transmission of a large amount of data, so it is very expensive. However, its forecast precision is the best of all the models considered because this method is based on all data sets.

First, we considered the linear regression model described in Section 1.3.2. This model was constructed using 2000 observations, which were selected randomly from the available data set of 6500. The parameter estimates are shown in Table 1.3. All of the parameters satisfied the significance test, which was a measure of the effect on the target variable, that is, the travelling time. These parameters could be used as the 'true' parameter values in linear models with decentralised architectures.

Second, we also considered a kernel-based regression model for the same data.

TABLE 1.2

Factors and Corresponding Bandwidth Values h

Variable	$X^{<1>}$	$X^{<2>}$	$X^{<3>}$	$X^{<4>}$	$X^{<5>}$	$X^{<6>}$
Bandwidth	h_1	h_2	h_3	h_4	h_5	h_6
Bandwidth value	0.01	0.03	0.2	0.2	0.1	0.05

TABLE 1.3

Factors and Corresponding Parameters

Variable	$X^{<1>}$	$X^{<2>}$	$X^{<3>}$	$X^{<4>}$	$X^{<5>}$	$X^{<6>}$
Coefficient estimate	b_1	b_2	b_3	b_4	b_5	b_6
Estimated value	0.222	−0.032	0.003	0.056	0.086	−0.017

TABLE 1.4

Values of the Coefficient of Determination R^2 Using the Centralised Approach

	Linear	Kernel
R^2 value	0.829	0.913

To test the efficiency of the models by cross-validation, we used 500 observations that were not used in the construction of the model. The corresponding R^2 values are provided in Table 1.4, which shows that kernel regression gave more reliable results with $R^2 = 0.913$.

Figure 1.6 shows the average forecasting errors for the linear, kernel-based, oracle and average models. Clearly, the kernel model yielded better results compared with the linear model (c. 15% smaller errors). The kernel model and average model yielded the same results after 300 time units, except that the average model had a 5% smaller standard deviation of errors. Before 300 time units, the average model was slightly better. In addition, the oracle model gave the best results, that is, 25% better than the kernel model. The best method for constructing the oracle model remains unclear.

1.5.3 Decentralised Uncoordinated Architecture

In this section, we consider an estimation model with fully decentralised uncoordinated architecture. There were no transmission costs in this case, although it was assumed that each agent was equipped with a special device for making forecasts. The main problem was that each agent constructed its own local model using a relatively small amount of data. This led to major forecasting errors, which could not be corrected due to the lack of coordination.

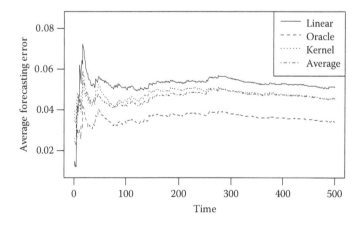

FIGURE 1.6
Average forecasting errors using different models.

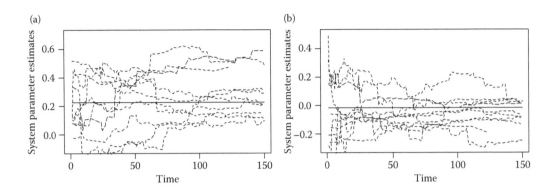

FIGURE 1.7
System parameter estimates b_1 (a) and b_6 (b) for a linear model using different agents.

We simulated 20 agents (vehicles) with an initial experience of 20 observations. The simulation experiments ran for 150 time units. The data used in the simulations were homogeneous. However, the LSE function (1.2) used for parameter estimation in the linear model had a number of local optima. If the parameters **b** of the agent fell into these local optima, they could remain there for a long time. A typical situation is shown in Figure 1.7.

Figure 1.7 shows the dynamics when changing the parameter values over time with several agents. The straight horizontal line shows the 'true' value of the corresponding parameter estimated in the centralised model. It can be seen that some agents had parameters that did not converge. Moreover, 'clusters' of agents were observed that converged to different values. Clearly, this had a negative effect on the quality of the forecasts.

The quality of the agent models was checked by cross-validation. In this case, the cross-validation checked the i-th agent's model based on observations of the other agents as test data. This procedure generated 20 coefficients of determination R^2 (one for each agent) values for the kernel-based and linear models. Figure 1.8 shows the corresponding histograms. The kernel model was better with a higher number of agents, although its performance degraded for some agents.

As with the centralised architecture, we compared the average forecasting errors for the linear, kernel-based, oracle and average models (Figure 1.9). The linear and kernel model results were approximately of the same quality (the kernel model produced c. 2% smaller

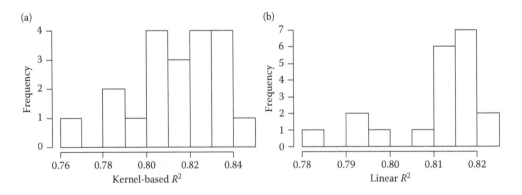

FIGURE 1.8
R^2 values for the overall system based on kernel (a) and linear (b) models.

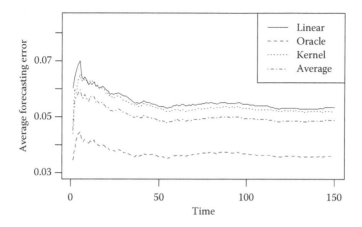

FIGURE 1.9
Average forecasting errors using the kernel, linear and combined models.

errors). However, the average model yielded considerably better results (6% smaller errors compared with the kernel model). The oracle model yielded the best results, that is, c. 35% better than the kernel model.

1.5.4 Decentralised Coordinated Architecture

This section considers an estimation model with decentralised coordinated architecture. The agents made a local estimation of the system parameters and used cooperative mechanisms to adjust their parameters (linear model) or observations (kernel-based model) according to those of other agents. The amount of information transmitted was lower than that in the centralised model, because only locally estimated parameters or selected observations were transmitted rather than global data.

We used the same simulation parameters as those used in the uncoordinated architecture, that is, 20 agents with the initial experience of 20 observations. Most simulation experiments ran for 100 time units. The agents were equipped with a special device for making forecast calculations and communicating. In the linear model, the experiences of agents were used as the weights (reliability level) for their parameters during the adjustment process. In the kernel-based model, the agents transmitted the two best observations.

First, we show the system dynamics for one randomly selected agent (Figure 1.10). Large errors disappeared over time and the number of communication events also decreased.

We analysed the number of communication events in greater detail. Figure 1.11 shows that the number of communication events with the linear and kernel-based models depended on the model parameters p and bp, which regulated the communication conditions. There was a significant decrease in the number of communication events with the kernel-based model. With the linear model, there was a significant difference in the number of communication events that depended on parameters that did not decrease over time. The number of communication events also decreased slightly with time.

We analysed the dynamics of the parameter changes in the linear model (Figure 1.12). The straight horizontal line shows the 'true' value of the corresponding parameter, which was estimated by the centralised architecture. There was good convergence of the parameters to the 'true' values. Further, there were no clusters of agents with different parameters, so the quality of forecasts was expected to be good.

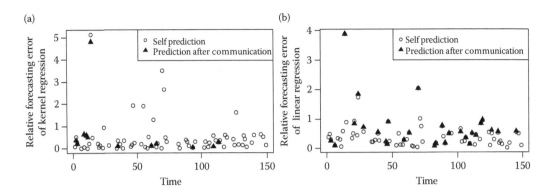

FIGURE 1.10
System dynamics of a single agent using the kernel (a) and linear (b) models, $bp = 1.5$, $p = 0.85$.

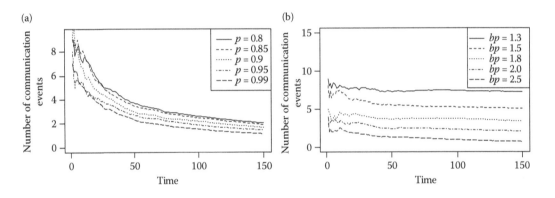

FIGURE 1.11
Number of communication events using the kernel (a) and linear (b) models.

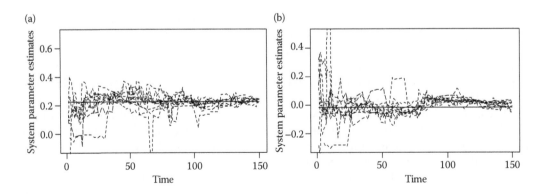

FIGURE 1.12
System parameter estimates b_1 (a) and b_6 (b) for the linear model with different agents, $bp = 1.5$, $p = 0.85$.

Cooperative Regression-Based Forecasting in Distributed Traffic Networks

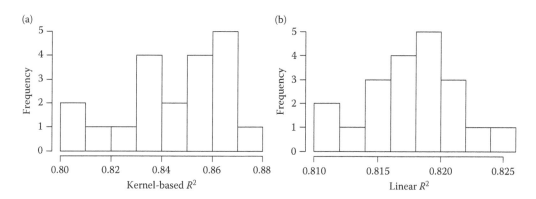

FIGURE 1.13
R^2 for the whole system for kernel (a) and linear (b) models, $bp = 1.5$, $p = 0.85$.

The quality of the agent models was also checked by cross-validation technique, as with the previous model. Figure 1.13 shows that the kernel model yielded better results than the previous uncoordinated architecture. The linear model did not yield much better results, but the 'bad' agents disappeared so that all agents had equal R^2 values. Thus, it is necessary to make a trade-off between system accuracy (represented by R^2) and the number of necessary communication events. The trade-off depended on the communication and accuracy costs.

As previously discussed and shown in Figure 1.11, changes in the model parameter values p and bp affected the number of communication events. Fewer communication events meant less precision. Figure 1.14 shows how the dynamics of the average forecasting errors changed, depending on the system parameters. With the kernel model, a change in parameter p from 0.8 to 0.99 decreased the errors by ≥12% while the number of communication events increased by 35%. Changes in the linear model parameter bp from 1.3 to 2.5 yielded approximately the same increase in quality of about 10%; however, the number of communication events increased by 300%. This demonstrated the importance of making a trade-off between an amount of communication and accuracy.

We compared the average forecasting errors using different models (Figure 1.15), as with the centralised and uncoordinated architecture. The results were similar to the centralised architecture results, that is, the kernel model was c. 5% better than the linear model.

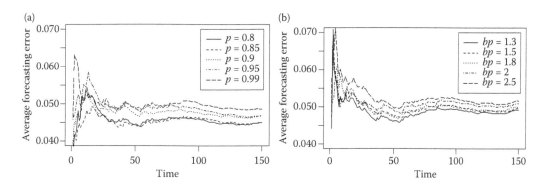

FIGURE 1.14
Average forecasting errors using the kernel-based (a) and linear (b) models.

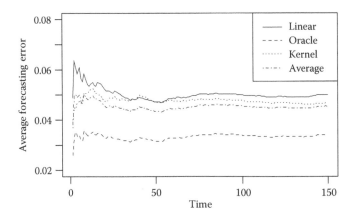

FIGURE 1.15
Average forecasting errors using the kernel, linear and combined approaches, $bp = 1.5$, $p = 0.85$.

The average estimate was similar to the kernel estimate, but it was slightly better (2%). The oracle model is considerably better than the other estimates (c. 40% better than the kernel model).

In the coordinated architecture, the agents with more experience (historical data set) helped the agents with less experience to improve their forecasts. This cooperation helped to improve the quality of forecasts, especially those of less-experienced agents. As a result, the average quality of the system was increased compared with the uncoordinated architecture (by c. 15% for the kernel-based, c. 8% for the linear, c. 6% for the average and c. 9% for the oracle estimate). There was almost no difference from the centralised architecture, although the amount of communication was smaller and the communication events occurred at a local level.

1.6 Discussion

Table 1.5 summarises the results of the average forecasting errors and goodness-of-fit criteria R^2 using the different forecasting models (linear, kernel, average and oracle) with various architectures (centralised, decentralised uncoordinated/coordinated). This demonstrated the advantages and disadvantages of the algorithms we implemented.

The linear regression model used all the data points for forecasting. This was a positive aspect if there were no historical observations close to the requested point. However, the equal effect of the distant and nearby data points, and the linear dependency, led to inaccurate forecasts. Furthermore, the coordination of the parameters always increased the squared error calculated using the agent's data because the current parameters were calculated based on the squared error minimisation. This meant that the straight regression line was being adjusted continuously via coordination and the new data points, so it could not be fitted well for all data points. Thus, linear regression gave the worst forecasts: $R^2 = 0.83$ for the centralised architecture and $R^2 = 0.82$ for the coordinated architecture. The centralised architecture was worse because a single line could not make good forecasts for 2000 data points. Figure 1.6 shows that there was a decrease in the average forecasting

TABLE 1.5

Average Forecasting Errors and Goodness-of-Fit Criteria R^2 for the Different Forecasting Models, $bp = 1.5$, $p = 0.85$

Model	Average Forecasting Errors	R^2 (After Cross-Validation)
Linear		
Centralised architecture	0.051	0.829
Uncoordinated architecture	0.054	0.811
Coordinated architecture	0.050	0.819
Kernel-Based		
Centralised architecture	0.045	0.913
Uncoordinated architecture	0.053	0.814
Coordinated architecture	0.047	0.859
Aggregated (Average)		
Centralised architecture	0.046	—
Uncoordinated architecture	0.049	—
Coordinated architecture	0.046	—
Aggregated (Oracle)		
Centralised architecture	0.034	—
Uncoordinated architecture	0.037	—
Coordinated architecture	0.034	—

errors with about 100 points, followed by an increase. The linear model within the uncoordinated architecture was worse due to the convergence of the parameters of several agents to local optima instead of global optima.

The kernel-based regression model used only neighbouring points for forecasting. In contrast to the linear model, the absence of nearby points prevented good forecasting. However, there was no strong linear dependency and the model fitted well to the data points with nonlinear dependencies. Coordination greatly improved the quality of forecasting because new neighbouring data points were provided. The centralised architecture made the best forecasts because there were many nearby points for most requested points ($R^2 = 0.91$). However, the absence of experience (uncoordinated architecture) produced bad forecasts ($R^2 = 0.81$), which were comparable with the results of the linear model. Coordination improved the R^2 to 0.86 (with the linear model only to 0.82). For large amounts of data, the kernel-based model required numerous computations (all data points needed to be checked to ensure nearby points for each forecast).

With the average estimator, that is, the mean of the kernel-based and linear estimators, R^2 could not be calculated because of the nature of the model. However, analysis of the relative forecasting errors showed that the average estimator in coordinated and uncoordinated architectures for relatively small amounts of data produced relatively better results than kernel-based or linear estimators. This was not explained by a negative correlation between the linear and kernel-based forecasting errors, because there was a strong positive correlation of about 0.8); instead, it was due to the structure of these errors. As shown in Figure 1.10, the kernel model was more accurate on average, but it sometimes yielded highly inaccurate forecasts (if only one neighbouring observation was used for forecasting). The linear model was less accurate, but it did not produce such big outliers.

The average method avoided big outliers of the kernel model and it provided more accurate forecasts. However, this advantage was lost when the kernel model had sufficient data to avoid outliers (in Figure 1.6, this occurred after c. 300 observations).

The oracle-based estimator, which combined the kernel-based and linear estimators, provided a possible lower bound for the average forecasting error. This algorithm could improve forecasts by c. 25%. However, it is still unclear how to choose between kernel-based and linear estimates. It is important to identify the factors or characteristics of estimators, which allow a comparison of linear and kernel-based regression estimates for one actual point.

1.7 Conclusions

This study considered the problem of intelligent data processing and mining in decentralised multi-agent networks. We focused on cooperative regression-based travelling time forecasting in distributed traffic networks.

We proposed a multi-agent architecture and structure for intelligent autonomous agents and paid special attention to the data processing and mining modules. This approach allows input data processing and makes travelling time forecasting possible, while adjusting the agent models if necessary.

We analysed the linear and kernel-based regression models, which can be applied efficiently for travelling time forecasting. Regression models were tested with special attention to their implementation for streaming data, which is important for intensive data flows in traffic systems. We proposed regression-based coordination learning algorithms based on the confidence intervals of estimates, which allowed agents to improve the quality of their forecasts. Each model was illustrated with simple examples in a tutorial style, which facilitated understanding of the algorithms.

We demonstrated the practical application of the proposed approaches using real-world data from Hanover, Germany. We proposed the structure of regression models by selecting significant factors. Three types of architectures were compared: distributed, centralised and decentralised coordinated/uncoordinated.

For each of the proposed architectures, we tested the accuracy of the linear and kernel-based regression estimators of the travelling time, as well as combinations of both. We used the relative error and determination coefficient as goodness-of-fit criteria. The quality of each agent model was checked by cross-validation.

The results demonstrated the appropriate goodness-of-fit of all the models considered ($R^2 > 0.8$). The kernel model usually produces better results compared with the linear model, although it was more sensitive to outliers. A simple combination of the linear and kernel-based estimates (average) reduced the effect of outliers and provided better estimates with small amounts of data (<300). However, a more effective algorithm for choosing between the kernel-based and linear estimation method could theoretically yield 25–30% better results, which was demonstrated by the oracle model.

The centralised architecture provided the best quality of forecasts. However, it required a large amount of communication and computation, which was frequently impossible to implement. In decentralised architectures, the cooperation between agents and the corresponding model adjustment improved the forecasts considerably. It was possible to obtain almost the same average forecasting errors as in centralised architectures, especially when we aggregated the kernel-based and linear estimates. This showed that a decentralised architecture with communication could be a good alternative to the centralised architecture, although it dispenses with the expensive authority and uses lesser communication and computation, which are mostly local and executed in parallel.

Our future work will be continued in three directions: (a) construction of the distributed model of multiple multivariate regression, which allows forecasting of several response variables simultaneously from the same set of explanatory variables (factors); (b) application of other regression model types and different methods of combination of their estimates; (c) modification of the parameter adjustment algorithm (new strategies for the calculation of the reliability level of agents, resampling approach [40], etc.) in order to be more robust to outliers.

Acknowledgements

The research leading to these results has received funding from the European Union Seventh Framework Programme (FP7/2007-2013) under grant agreement no. PIEF-GA-2010-274881. We also thank Professor J.P. Müller for his useful and valuable ideas during the preparation of this chapter.

References

1. H. Kargupta and P. Chan, Eds., *Advances in Distributed and Parallel Knowledge Discovery*, CA, USA: AAAI Press/MIT Press, 2000.
2. J. C. da Silva, C. Giannella, R. Bhargava, H. Kargupta and M. Klusch, Distributed data mining and agents, *Eng. Appl. Artif. Intell.*, 18(7), 791–801, 2005.
3. H. Lin, R. Zito and M. Taylor, A review of traveltime prediction in transport and logistics, *Proceedings of the Eastern Asia Society for Transportation Studies*, 5, 1433–1448, 2005.
4. M. Fiosins, J. Fiosina, J. Müller and J. Görmer, Agent-based integrated decision making for autonomous vehicles in urban traffic, *Advances on Practical Applications of Agents and Multiagent Systems (Advances in Intelligent and Soft Computing)*, 88, 173–178, 2011.
5. M. Fiosins, J. Fiosina, J. P. Müller and J. Görmer, Reconciling strategic and tactical decision making in agent-oriented simulation of vehicles in urban traffic, in *Proceedings of 4th International. ICST Conf. on Simulation Tools and Techniques (SimuTools'2011)*, 2011.
6. H. A. Simon, *The Sciences of the Artificial*, Cambridge, MA: The MIT Press, 1996.
7. N. R. Jennings, An agent-based approach for building complex software systems, *Communications of the ACM*, 44(4), 35–41, 2001.
8. M. Wooldridge, *An Introduction to Multi-Agent Systems*, Chichester, UK: John Wiley & Sons, 2002.
9. P. A. Mitkas, D. Kehagias, A. L. Symeonidis and I. N. Athanasiadis, A Framework for constructing multi-agent applications and training intelligent agents, *Agent-Oriented Software Engineering IV (Lecture Notes in Computer Science)* 2935, 96–109, 2004.
10. A. Bazzan, Opportunities for multiagent systems and multiagent reinforcement learning in traffic control, *Autonomous Agents and Multiagent Systems*, 18(3), 342–375, 2009.
11. M. Rieser, K. Nagel, U. Beuck, M. Balmer and J. Rümenapp, Truly agent-oriented coupling of an activity-based demand generation with a multi-agent traffic simulation, *Transportation Research Record*, 2021, 10–17, 2007.

12. P. Davidsson, L. Henesey, L. Ramstedt, J. Törnquist and F. Wernstedt, An analysis of agent-based approaches to transport logistics, *Transportation Research Part C: Emerging Technologies*, 13(14), 255–271, 2005.
13. A. Freitas, *Data Mining and Knowledge Discovery with Evolutionary Algorithms*, Berlin, Heidelberg: Springer-Verlag, 2002, 264pp.
14. A. L. Symeonidis and P. A. Mitkas, Agent intelligence through data mining, in *Multiagent Systems, Artificial Societies, and Simulated Organizations*, New York: Springer-Verlag, 2005.
15. L. Cao, D. Luo and C. Zhang, Ubiquitous intelligence in agent mining, *Agents and Data Mining Interaction (Lecture Notes in Computer Science)*, 5680, 23–35, 2009.
16. M. Klusch, S. Lodi and G. Moro, Agent-based distributed data mining: The KDEC scheme, *Proceedings of International Conference on Intelligent Information Agents–The AgentLink Perspective; Lecture Notes in Computer Science*, 2586, 104–122, 2003.
17. C. Guestrin, P. Bodik, R. Thibaux, M. Paskin and S. Madden, Distributed regression: An efficient framework for modeling sensor network data, in *Proceedings of the 3rd International Symposium on Information Processing in Sensor Networks*, Berkeley, CA, USA, 2004.
18. S. S. Stankovic, M. S. Stankovic and D. M. Stipanovic, Decentralized parameter estimation by consensus based stochastic approximation, *IEEE Trans. Automatic Control*, 56, 1535–1540, 2009.
19. A. Bazzan, J. Wahle and F. Kluegl, Agents in traffic modelling–from reactive to social behaviour, *Advances in Artificial Intelligence (Lecture Notes in Artificial Intelligence)*, 1701, 509–523, 1999.
20. G. Malnati, C. Barberis and C. Cuva, Gossip: Estimating actual travelling time using vehicle to vehicle communication, in *Proceedings of Fourth International Workshop on Intelligent Transportation*, Hamburg, 2007.
21. W. Lee, S. Tseng and W. Shieh, Collaborative real-time traffic information generation and sharing framework for the intelligent transportation system, *Information Sciences*, 180, 62–70, 2010.
22. J. Ehmke, M. Fiosins, J. Görmer, D. Schmidt, H. Schumacher and H. Tchouankem, Decision support for dynamic city traffic management using vehicular communication, in *Proceedings of 1st International Conference on Simulation and Modeling Methodologies, Technologies and Applications (SIMULTECH 2011)*, pp. 327–332, Noordwijkerhout, The Netherlands, 29–31 July, 2011.
23. R. Claes and T. Holvoet, Ad hoc link traversal time prediction, in *Proceedings of the 14th International IEEE conference on Intelligent Transportation Systems*, pp. 1803–1808, Washington, USA, 5–7 October, 2011.
24. J. Fiosina, Decentralised regression model for intelligent forecasting in multi-agent traffic networks, *Distributed Computing and Artificial Intelligence (Advances in Intelligent and Soft Computing)*, 151, 255–263, 2012.
25. J. Fiosina and M. Fiosins, Cooperative Kernel-based forecasting in decentralized multi-agent systems for urban traffic networks, in *Proceedings of Ubiquitous Data Mining (UDM) Workshop at the 20th European Conference on Artificial Intelligence*, Montpellier, France, 27–31 August, 2012 (CEUR-WS.org, Aachen, Vol. 960, pp. 3–7).
26. B. L. Smith, B. L. Williams and R. K. Oswald, Comparison of parametric and nonparametric models for traffic flow forecasting, *Transportation Research Part C*, 10, 303–321, 2002.
27. C. McKnight, H. S. Levinson, C. Kamga and R. Paaswell, Impact of traffic congestion on bus travel time in northern New Jersey, *Transportation Research Record Journal*, 1884, 27–35, 2004.
28. N. Draper and H. Smith, *Applied Regression Analysis*, New York: John Wiley and Sons, 1986.
29. A. Albert, *Regression and the Moor-Penrose Pseudoinverse*, New York: Academic Press, 1972.
30. A. Andronov, A. Kiselenko and E. Mostivenko, *Forecasting of the Development of Regional Transport System*, Syktyvkar: KNZ UrO RAN (in Russian), 1991.
31. W. Härdle, *Applied Nonparametric Regression*, Cambridge: Cambridge University Press, 2002.
32. W. Härdle and M. Müller, Multivariate and semiparametric kernel regression, in *Smoothing and Regression*, M. Schimek, ed., New York: Wiley, 2000, pp. 357–391.
33. B. W. Silverman, *Density Estimation for Statistics and Data Analysis*, London: Chapman and Hall, 1986.
34. D. W. Scott, *Multivariate Density Estimation: Theory, Practice, and Visualization*, New York: John Wiley and Sons, 1992.

35. P. Hall, Large sample optimality of least squares cross-validation in density estimation, *The Annals of Statistics*, 11(4), 1156–1174, 1983.
36. W. Härdle, M. Müller, S. Sperlich and A. Werwatz, *Nonparametric and Semiparametric Models*, Berlin: Springer, 2004.
37. J. Gentle, W. Härdle and Y. Mori, Eds., *Handbook of Computational Statistics: Concepts and Methods*, Berlin: Springer, 2004.
38. Frequency: Travelling with passenger cars, *Statista*, 2012. [Online]. Available: http://de.statista.com/statistik/daten/studie/176124/umfrage/haeufigkeit-pkw-fahren/.
39. J. S. Racine, Consistent significance testing for nonparametric regression, *Journal of Business and Economic Statistics*, 15, 369–379, 1997.
40. H. Afanasyeva and A. Andronov, On robustness of resampling estimators for linear regression models, *Communications in Dependability and Quality Management: An international Journal*, 9(1), 5–11, 2006.

2

A Sensor Data Aggregation System Using Mobile Agents

Tomoki Yoshihisa, Yuto Hamaguchi, Yoshimasa Ishi, Yuuichi Teranishi, Takahiro Hara and Shojiro Nishio

CONTENTS

- 2.1 Introduction ... 40
- 2.2 Related Work ... 42
  - 2.2.1 C/S-Type Integrated Sensor Networks ... 42
  - 2.2.2 P2P-Type Integrated Sensor Networks ... 42
  - 2.2.3 Spatiotemporal Query Descriptions ... 43
  - 2.2.4 MA Systems ... 43
- 2.3 Integrated Sensor Networks ... 43
  - 2.3.1 System Architecture ... 43
  - 2.3.2 Sensor Network Sites ... 43
  - 2.3.3 Gateways ... 45
  - 2.3.4 Portal Sites/Centre Server ... 45
- 2.4 Query Description for Integrated Sensor Networks ... 45
  - 2.4.1 Issues of Conventional Query Description ... 45
  - 2.4.2 Design of Query Description ... 45
    - 2.4.2.1 Designation of Spatiotemporal Ranges and Units ... 46
    - 2.4.2.2 Designation of Temporal Units ... 47
    - 2.4.2.3 Designation of Spatial Units ... 48
    - 2.4.2.4 Data Selection of Spatiotemporal Ranges ... 50
    - 2.4.2.5 Sensor Data Extraction from Spatiotemporal Ranges ... 51
    - 2.4.2.6 Loop Operations ... 51
  - 2.4.3 Implementation of Query Description ... 51
  - 2.4.4 Example of a Query Description ... 51
  - 2.4.5 Discussion ... 53
- 2.5 Sensor Data Aggregation Systems Using MAs ... 56
  - 2.5.1 Design for Sensor Data Aggregation Systems Using MAs ... 56
    - 2.5.1.1 Compatibility of Sensor Network Sites ... 56
    - 2.5.1.2 Management of MAs ... 56
    - 2.5.1.3 Database Access by MAs ... 57
    - 2.5.1.4 Behaviour Determination of MAs ... 57
  - 2.5.2 Implementation ... 58
    - 2.5.2.1 Definition of Sensor Network Information ... 58
    - 2.5.2.2 Use of the MA System ... 59
    - 2.5.2.3 MAs Generation ... 59
    - 2.5.2.4 MA Migration ... 60

	2.5.2.5	Web Browser Interface...60
2.5.3	System Evaluation...61
	2.5.3.1	Evaluation Environments 61
	2.5.3.2	Evaluation Results...61
	2.5.3.3	Discussion 62

2.6 Conclusion 62
    2.6.1 Summary...62
    2.6.2 Future Work..63
Acknowledgement...63
References..63

2.1 Introduction

Owing to the recent development of sensing technologies, sensor networks, which are information networks consisting of sensors such as weather sensors or cameras, have been managed by many organisations. For example, the Kansei project described in Ref. [1] manages a sensor network that includes 96 sensors and opens the sensor network to researchers. A sensor network typically has a sink node, which collects the sensor data generated from the sensor nodes connected to the sensor network. Sensor networks have a restriction on the number of sensor nodes since the communication or the computational resources for sink nodes have upper limits. Therefore, we need to integrate some sensor networks to construct large-scale sensor networks that have many sensor nodes. We call such sensor networks *integrated sensor networks* (more details are provided in Section 2.3). With integrated sensor networks, users can aggregate sensor data obtained from some sensor networks. Aggregation means collecting the sensor data that have been executed in operations such as averaging or finding the maximum. The following are examples of aggregating sensor data from integrated sensor networks. There are three sensor networks deployed at Osaka, Kyoto and Nara, which are notable Japanese cities in the Kansai area that are located near one another.

- For weather forecasting, a user calculates the average temperature values of these cities every 1 h and every 1 km. To calculate them, the client aggregates the temperature sensor data obtained from all the sensor networks.
- For environmental observation, a user finds the hottest city among the three cities. To find the city that has the maximum temperature value, the client compares the temperature sensor data obtained from all the sensor networks.
- To find the broken temperature sensors, a user finds the sensors generating abnormal values by analysing the sensor data obtained from all the sensor networks.

As shown in Figure 2.1, in integrated sensor networks, sink nodes are connected to gateways and gateways are connected to a centre server through a wide area network. The centre server aggregates sensor data according to the users' submitted queries. In this chapter, we call sensor networks connected to a gateway and constructing integrated sensor networks *sensor network sites*. Examples of integrated sensor networks are X-Sensor or Live E! [2,3]. Integrated sensor networks can enhance sensing accuracy, sensing type and sensing area.

A Sensor Data Aggregation System Using Mobile Agents

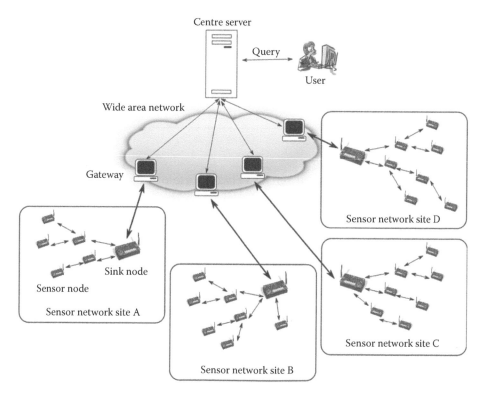

FIGURE 2.1
Example of integrated sensor networks.

To aggregate sensor data from integrated sensor networks, several data aggregation systems have been proposed [2–8]. However, sensor network sites in integrated sensor networks are often managed by some organisations and have different sensor data aggregation architecture.

Therefore, to realise the above examples, the users must collect their necessary sensor data from all sensor network sites for their clients. After this, the clients execute aggregation operations, such as averaging and analysis, on the collected sensor data. However, such processes to aggregate sensor data cause heavy network traffic in collecting data from all the sensor network sites. To solve this problem, data aggregation operations should be performed only on necessary data at each sensor network site. For example, to calculate the average temperature for weather forecasting, the clients do not need to collect all the sensor data, but only the average temperature and the number of sensors of each sensor network.

Hence, in this chapter, we describe a sensor data aggregation system using mobile agents (MAs) for integrated sensor networks. MAs are computer programs that migrate among sensor network sites. In the system, the clients generate MAs. MAs execute the programs migrating among sensor network sites, and finally return to the clients. The executed programs are automatically generated based on the users' described queries. Since MAs migrate while executing the aggregation operations on the sensor data, the clients do not need to collect and aggregate the sensor data from all the sensor networks, and thus the network traffic can be reduced.

The rest of this chapter is organised as follows. We introduce related work in Section 2.2. In Section 2.3, we explain integrated sensor networks. We explain a query description for integrated sensor networks in Section 2.4 and a sensor data aggregation system using MAs in Section 2.5. Finally, in Section 2.6, we conclude the chapter.

2.2 Related Work

System architectures for integrated sensor networks are of two types: the client–server (C/S) type and the peer-to-peer (P2P) type.

2.2.1 C/S-Type Integrated Sensor Networks

In the C/S-type integrated sensor networks, the users aggregate sensor data stored in the gateways via the centre server as shown in Figure 2.1 [2,3,7–12]. The queries submitted to the centre server by the users are extracted for each sensor network site. The queries for each sensor network site include the necessary queries to execute the submitted queries. The extracted queries are sent to each sensor network site and the centre server receives the results. The centre server aggregates the collected sensor data. The users receive the aggregated results from the centre server. It is easy to manage information for sensor network sites such as sensing type, sensing interval and so on by managing them intensively at the centre server. However, the response time lengthens when the network load increases since the load for collecting or operating sensor data concentrates on the centre server.

LiveE! [2], which is one of the integrated sensor network systems, collects the sensor data observed by digital instrument shelters. The shelters are managed by 11 decentralised servers, and have temperature, humidity, pressure, wind-direction, wind-speed and rainfall sensors. Since the servers are decentralised, LiveE! can relieve the servers' load. However, the users must access all the servers to execute the queries.

X-Sensor 1.0 [3] is a sensor network test-bed. X-Sensor 1.0 has several sensor networks that are deployed in major Japanese universities. By registering a sensor network with the X-Sensor 1.0 test-bed server, the sensor network can be managed by the centralised server. Users can collect the sensor data from the X-Sensor 1.0 system via the centralised server.

2.2.2 P2P-Type Integrated Sensor Networks

In the P2P-type integrated sensor networks, the gateways are connected with each other using P2P technology [4,5,13,14]. The users submit their queries to the gateway (peer) that their clients are connected to. The queries are sent to the gateways that have the necessary sensor data to execute the queries. Compared with the C/S type, the centre server's load can be distributed to the gateways. However, it takes a long time to collect information for the sensor network sites since they are not managed intensively at the centre server.

IrisNet [6] stores the sensor data in the distributed sensor databases. In IrisNet, sensing agents (SAs) collect the sensor data from several sensors. SAs aggregate the collected sensor data for nearby organising agents (OAs). OAs select the most relevant database in which to store the aggregated data, from the viewpoint of the overall system performance, and transfer their data to this database.

Environmental Monitoring 2.0 described in Ref. [15] is a data-sharing and visualisation system using Sensormap and a global sensor network (GSN) [4,5,14,16,17]. Sensormap provides data-sharing services and GSN provides data-management services. In contrast to the three systems above, Environmental Monitoring 2.0 focusses on the visualisation system. It can show the sensor type, sensor data, and so on visually.

2.2.3 Spatiotemporal Query Descriptions

Some query descriptions can describe the sensor data aggregation from spatiotemporal ranges [18–25]. In Ref. [22], the users can define spatiotemporal ranges and after that, they can be used in the queries. In Ref. [24], the spatiotemporal ranges are declared by using INSIDE phrases. However, they can only designate the spatiotemporal ranges and cannot aggregate sensor data to every spatiotemporal range.

2.2.4 MA Systems

Some systems designed to manage MAs have been proposed [26–31]. AgentTeamwork described in Ref. [29] uses MAs for grid computing. MAs migrate to the computers that have the necessary data, and execute the required tasks there. PIAX [30] is a P2P-based MA system. However, AgentTeamwork and PIAX do not focus on sensor data aggregation. MADSN [31], on the other hand, does not focus on integrated sensor networks. An MA system customised for integrated sensor networks offers users easy and efficient aggregation of sensor data generated from such networks.

2.3 Integrated Sensor Networks

In this section, we explain integrated sensor networks, which is a large-scale sensor network integrating some sensor networks.

2.3.1 System Architecture

Figure 2.2 shows the system architecture for integrated sensor networks. In integrated sensor networks, each sensor network is connected with each other via gateways. Sensor networks consist of some sensor nodes. As we explained in Section 2.1, sensor network sites mean sensor networks and gateways managed by the same organisation. Sensor network sites include sensor networks more than or equal to one. Integrated sensor networks consist of sensor network sites, gateways, the portal site and the centre server.

2.3.2 Sensor Network Sites

Each sensor network site is managed by different organisations. These can be integrated by commonalising their sensor network architecture. However, each organisation often adopts a different sensor network architecture. Changing the sensor network architecture is expensive. Therefore, we do not discuss the sensor network integration by commonalising their sensor network architecture but by exploiting the sensor data collected by the gateways. Since the sensor data are collected by the sink node and the sink node is

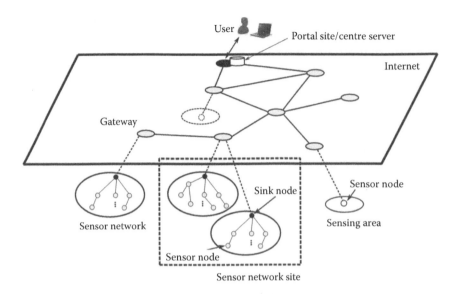

FIGURE 2.2
System architecture for integrated sensor networks.

connected to the gateway, the integrated sensor network can access the sensor data at the gateways. Sensor nodes consisting of sensor networks are the temperature sensor, humidity sensor and so on. They have sensor information listed in Table 2.1. The sensor information is used to select sensor data based on the users' described queries. Integrated sensor networks can identify the sensor network by using the organisation name and the sensor network name. The data type is the data type for the sensor data such as integer or float. The accuracy represents the error of the sensor data and the resolution is the highest resolution that the sensor node can provide. For example, the resolution is 0.5°C when the temperature sensor observes temperature of every 0.5°C. The sensor type is the sensor type of the sensor node such as the temperature sensor or humidity sensor.

TABLE 2.1

Sensor Information

Name	Content
Sensor ID	The identifier for the sensor nodes in each sensor network
Sensor model	The manufacturing model of the sensor node
Organisation name	The name of the sensor network organiser
Sensor network name	The name of the sensor network
Data type	The data type for the sensor data
Accuracy	The accuracy of the sensor data
Resolution	The resolution of the sensor data
Unit	The unit of the sensor data
Coordinate	The coordinate of the sensor node
Sensor type	The sensor type of the sensor node
Interval	The interval of sensing the data

2.3.3 Gateways

Sensor data collected by the sink node are stored in the database that the gateway has. The gateways have enough capacity of storage and are equipped with external powers. The sink nodes can also serve as the gateways when they have enough computational resources, including the storage.

2.3.4 Portal Sites/Centre Server

Integrated sensor networks have portal sites and centre servers. The users describe the queries for sensor data aggregation and submit them to the centre server via the portal site. They can obtain the results from the portal sites. The centre server grasps the sensor networks connected to the integrated sensor network and manages their sensor information.

The integrated sensor networks described above are often used in the sensor network research field and are realistic.

2.4 Query Description for Integrated Sensor Networks

In this section, we explain a query description for integrated sensor networks. The sensor data aggregation system using MAs uses this description.

2.4.1 Issues of Conventional Query Description

In conventional query descriptions for integrated sensor networks such as those used in IrisNet or GSN, the clients must collect the sensor data needed to execute the queries. For example, to realise the first example in Section 2.1, the client must collect the sensor data that are obtained by the sensor networks in Osaka city. After this, the client averages the collected sensor data every 1 h and every 1 km using other programming languages such as C language. Moreover, the user must submit many queries to collect the sensor data every 1 h and every 1 km. The query descriptions for such spatiotemporal sensor data aggregation become very complex in conventional query descriptions.

By describing temporal ranges for aggregating sensor data with a certain interval and area ranges for aggregating sensor data with a certain area interval, the complexity of the query description can be relieved. Also, describing aggregation operations such as averaging, finding the maximum and so on can save the trouble of describing many queries.

2.4.2 Design of Query Description

To make the descriptions of the sensor data aggregation easy, the query description explained here uses declarative description. The users declare their desirable aggregated sensor data. The integrated sensor network executes the queries describing the declarations. The aggregated sensor data are declared by SELECT syntax. SELECT syntax consists of the phrases to designate spatiotemporal ranges, to select the sensor data from them. SELECT syntax is based on the often-used SQL query description. Different from ordinary SQL, our explained query description can designate spatiotemporal ranges.

2.4.2.1 Designation of Spatiotemporal Ranges and Units

In this chapter, we use the words *temporal ranges* as the user designated temporal ranges, *spatial ranges* as the designated spatial ranges, and *spatiotemporal ranges* as both of them. In the first example in Section 2.1, the temporal range is an entire day. The spatial range is the Osaka, Kyoto and Nara cities. The temporal ranges are designated by specifying the start time and the end time. The spatial ranges are designated by specifying the rectangles. Since rectangles make it easy to combine some ranges and it is easy to specify rectangles, we use the shape of rectangles. The rectangles are specified by describing the coordinate of the left-bottom corner and the lengths of each edge. In our explained query description, the spatiotemporal ranges are designated by the FROM clauses. Here, clauses mean sets of phrases.

AREA phrases in FROM clauses designate the spatial ranges. An example is shown by *rect* in Figure 2.3. The horizontal axis is the longitude, the depth axis is the latitude and the vertical axis is the time. The sensor data series are denoted by cylinders. There are eight sensor data series in the figure. To enable the sensor data series to be easily seen, we denote them in the form of cylinders, though they are actually vertical lines. TIME phrases in FROM clauses designate the temporal ranges. An example is shown by *span* in Figure 2.3. START and END phrases appearing after TIME phrases specify the start time and the end time. WINDOW phrases are used to specify the current time as the temporal ranges. By specifying *window size* after WINDOW phrases, the following temporal range *timespan* is designated. Here, t is the current time.

$$0 < t - timespan < window\ size$$

SENSOR phrases designate the sensor types such as the temperature sensors or humidity sensors.

The spatiotemporal ranges are designated by using AREA, TIME and SENSOR phrases in FROM clauses. By describing some FROM clauses, the users can designate multiple

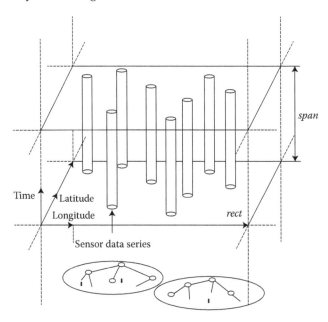

FIGURE 2.3
Designation of spatiotemporal ranges.

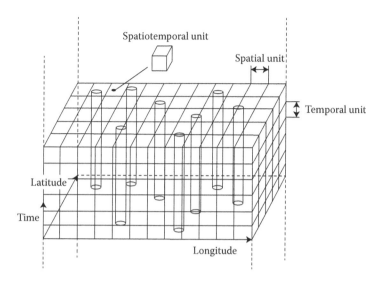

FIGURE 2.4
Spatiotemporal units in a spatiotemporal range.

spatiotemporal ranges. In this case, AS clauses name each spatiotemporal range to identify them. FROM clauses are summarised as the following:

FROM {(AREA *rect*, TIME *span*, SENSOR *type*) AS *alias* [,...]}

The query description does not need to be changed when a new sensor network is added to the integrated sensor network since the ranges to aggregate sensor data are designated by spatiotemporal ranges, not the sensor network themselves. The centre server extracts the appropriate sensor networks from the spatiotemporal ranges using the sensor network information.

When queries operate sensor data at every spatiotemporal interval, the users need to designate their intervals. For example, the interval of the spatial range is 1 km^2, and that of the temporal range is 1 h. Here, we denote the interval of the spatial ranges as *spatial units* and that of the temporal ranges *temporal units* and both of them as *spatiotemporal units*. Figure 2.4 shows examples of temporal units, spatial units and spatiotemporal units. There are eight sensor data series in the figure. Sometimes, there are no sensor data in a spatiotemporal unit or there are some sensor data in a spatiotemporal unit. For example, when there are no sensor data in a 1 km^2 range in Osaka city, the query substitutes the sensor data with the sensor data of neighbouring units. Or, when there are multiple sensor data in a 1 km^2 range, the query substitutes the sensor data with the maximum sensor data in the unit. The representative value of each spatiotemporal unit is denoted by *representative sensor data*. We explain temporal units and spatial units, respectively, from the next section.

2.4.2.2 Designation of Temporal Units

TIMESPAN clauses designate temporal units.

TIMESPAN *slot* USING *aggmethod*
[ELSE {INTERP USING *method*|CONTINUE()|ABORT()}]

TABLE 2.2

Aggregation Methods Used for Spatiotemporal Units

Method	Content
AVG()	Average the sensor data in the unit
MAX()	Find the maximum value in the unit
MIN()	Find the minimum value in the unit
COUNT()	The number of the sensor data in the unit
SUM()	Sum up the sensor data in the unit
UDEF()	Uses the user-defined method

TABLE 2.3

Interpolation Methods for Temporal Units

Method	Content
PREV()	Uses the sensor data of the previous temporal unit
LERP()	Interpolates the sensor data by linear function using the sensor data in the previous and the next temporal units
UDEF()	Uses the user-defined method

Temporal units are specified by a *slot*. *span* specified in the TIME clause is divided into some temporal units of every *slot*. When ALL specifier is specified in a *slot*, the temporal unit is the same as the temporal range. When there are multiple sensor data in the temporal unit, the aggregation method is specified by *aggmethod* in USING clauses. Table 2.2 shows the aggregation methods that can be used in *aggmethod*. For example, the users describe MAX(), which finds the maximum value in the temporal unit, to *aggmethod*. Also, the users can use their defined method by using UDEF(). When there are no sensor data in the temporal unit, the aggregation method is specified in the phrases after the ELSE phrase. When INTERP USING phrase is described after the ELSE phrase, the users specify the interpolation method by *method*. Table 2.3 shows the methods that can be specified in *method*. For example, the sensor data of the temporal unit that has no sensors are interpolated by the sensor data of the previous temporal unit by using the PREV method. When CONTINUE() is used, the aggregation process skips the current temporal unit when there are no sensors in the temporal unit. When ABORT() is used, the aggregation process finishes. When the ELSE phrase is abbreviated, the aggregation process continues to the next unit.

2.4.2.3 Designation of Spatial Units

AREASPAN clauses designate spatial units.

<p align="center">AREASPAN side USING aggmethod
[ELSE {INTERP USING <i>method</i>|CONTINUE()|ABORT()}]</p>

Spatial units are specified by *side*. The shapes of the spatial units are square and *side* is the edge length of the spatial units. Figure 2.5 shows spatial units. The horizontal axis is the longitude and the vertical axis is the latitude. The sensor data in the spatial range are represented by circles. The spatial range is designated by the left-bottom corner of the spatial range (*la, lo*) and the longitude and latitude offsets. The spatial units are designated by square shapes and the length of the edge is *side*. As shown in the figure, the spatial range

A Sensor Data Aggregation System Using Mobile Agents

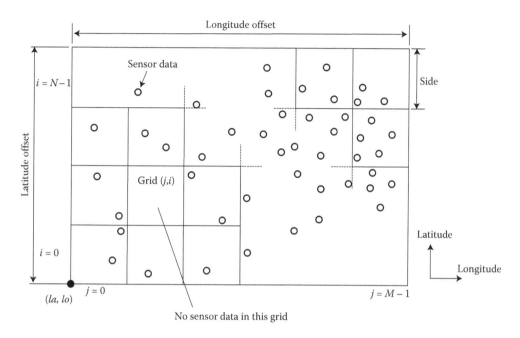

FIGURE 2.5
Spatial units in a spatial range.

designated by the AREA phrase is divided into the spatial units on every *side*. We denote each spatial unit by *grid(j,i)* ($0 <= i <= M - 1, 0 <= j <= N - 1$). Here, M and N are the number of i and j. When ALL specifier is specified in a *side*, the spatial unit is the same as the spatial range. When there are multiple sensors in the spatial unit, the aggregation method is specified by *aggmethod* in USING clauses. Similar to the TIMESPAN phrases, the aggregation methods shown in Table 2.2 can be used in *aggmethod*. Also, the users can use their defined method by using UDEF(). When there are no sensors in the spatial unit such as *grid(1,1)*, the aggregation method is designated after the ELSE phrase. When the INTERP USING phrase is described after the ELSE phrase, the users specify the interpolation method by *method*. Table 2.4 shows the methods that can be used in *method*. For example, the sensor data are interpolated using the maximum value or the average value of the neighbouring spatial units when there are no sensors in the spatial unit. Also, the users can use their defined method by using UDEF(). When ABORT() is used, the aggregation process finishes when the unit has no sensors. When the ELSE phrase is abbreviated, the aggregation process continues to the next spatial unit.

TABLE 2.4

Interpolation Methods for Spatial Units

Method	Content
E_AVG()	Uses the average sensor data in the neighbouring spatial unit
E_MAX()	Uses the maximum sensor data in the neighbouring spatial unit
E_MIN()	Uses the minimum sensor data in the neighbouring spatial unit
UDEF	Uses the user-defined method

2.4.2.4 Data Selection of Spatiotemporal Ranges

The query description designates the sensor data for aggregation in SELECT phrases. Table 2.5 shows the aggregation methods for SELECT phrases. The identifiers specified in AS phrases for spatiotemporal ranges can be used for the arguments of these methods. These methods are applied for the representative data. Also, we can use four arithmetic operations. The details are as follows:

$$\text{SELECT SUM } (c.value)$$

SUM method is the aggregation operation that calculates the sum of the data series specified by the arguments. In the above description, the query calculates the sum of the sensor data for each spatial unit. Here, *value* is one of the attributes for spatiotemporal ranges and specifies all sensor data in the spatiotemporal range c.

$$\text{SELECT TOPK } (10, c.value, order)$$

TOPK method is the aggregation operation that extracts the top K sensor data from the spatiotemporal ranges. The above description aggregates the top 10 temperature sensor data from the spatiotemporal range c. If *order* is ASC, the result is an ascending order. If *order* is DESC, the result is a descending order.

$$\text{SELECT MAX } (c1.value, ..., cn.value)$$

MAX method is the aggregation operation that finds the sensor data that have the maximum value. The above description aggregates the maximum sensor data from the spatiotemporal ranges $c1$ to cn.

$$\text{SELECT CORR } (c1.value, c2.value)$$

CORR method is the aggregation operation that calculates the correlation of two spatiotemporal ranges. The above description calculates the correlation between the spatiotemporal ranges $c1$ and $c2$.

$$\text{SELECT UDEF } (c1, ..., cn)$$

TABLE 2.5
Aggregation Methods Used for SELECT Phrase

Method	Arguments	Content
AVG	Spatiotemporal ranges	Average the sensor data in the spatiotemporal ranges
MAX	Spatiotemporal ranges	Find the maximum sensor data in the spatiotemporal ranges
MIN	Spatiotemporal ranges	Find the minimum sensor data in the spatiotemporal ranges
COUNT	Spatiotemporal ranges	The number of the sensor data in the spatiotemporal ranges
SUM	Spatiotemporal ranges	Sum up the sensor data in the spatiotemporal ranges
TOPK	K, spatiotemporal ranges, order	Find the top K sensor data in the spatiotemporal ranges
ABS	Spatiotemporal ranges	Absolute values of the sensor data in the spatiotemporal ranges
CORR	Two spatiotemporal ranges	Calculate the correlation of the spatiotemporal ranges
UDEF	Spatiotemporal ranges	Uses the user-defined method

The UDEF method aggregates the sensor data based on the user-described method. Therefore, the users can program their desirable aggregation operation by using the UDEF method. The operation for the UDEF method can be programmed by other programming languages such as C language.

$$\text{SELECT } c1.value - c2.value$$

The above description calculates the difference between two spatiotemporal ranges.

2.4.2.5 Sensor Data Extraction from Spatiotemporal Ranges

WHERE phrases extract the sensor data from the spatiotemporal ranges. Similar to SELECT phrases, WHERE phrases can use the identifiers of spatiotemporal ranges and their attributes. An example description follows:

$$\text{WHERE } c.value > a$$

The above description extracts the sensor data of that value is larger than a from the spatiotemporal range c. The operation also includes four arithmetic operations and logical operations such as AND and OR.

2.4.2.6 Loop Operations

The aggregation operations can be repeated while a condition is satisfied by using WHILE phrases.

$$\text{WHILE } c.value >= b$$

The above description repeats the aggregation operations until the value of the spatiotemporal range c becomes larger than or equal to b.

2.4.3 Implementation of Query Description

Figure 2.6 shows the summary of SELECT syntax. We explain the SETAGENT phrases in Section 2.5. This is related to the migration of MAs. Figure 2.7 shows the syntax definition using Backus–Naur form (BNF) description. Also, the syntax definitions for FROM, AREASPAN, TIMESPAN, WHERE and WHILE phrases are shown in Figures 2.8 through 2.10. The spatial units are operated from the left-bottom unit to the right-top unit sequentially along the longitudinal axis. The temporal units are operated from the oldest unit to the newest unit sequentially. The operations are executed after the interpolation. Table 2.6 shows the attributes of spatiotemporal ranges. These can be used for the arguments of aggregation operations.

2.4.4 Example of a Query Description

Figure 2.11 shows an example of a query description for sensor data aggregation. This is a query description to calculate the correlation of the temperature between Osaka and Kyoto for environmental monitoring. The monitoring period starts at 12:00 hours and finishes at 17:00 hours. The query calculates the average temperature every 1 km^2 and

```
SELECT [ ALL | DISTINCT ] { * | expression   [, ...] }
    FROM { (
        AREA INSIDE(rectangle),
        TIME { {START start END end} | {WINDOW wsize [, slide] } },
        SENSOR {sensor_type}
    ) AS alias } [, ...]
    [ AREASPAN { {side | ALL } USING aggmethod }
        [ ELSE { INTERP USING method | CONTINUE() | ABORT() } ] ]
    [ TIMESPAN { {slot | ALL } USING aggmethod }
        [ ELSE { INTERP USING method | CONTINUE() | ABORT() } ] ]
    [ WHERE condition ]
    [ WHILE condition ]
    [ SETAGENT { TRAVERSE | PARALLEL } ]
```

FIGURE 2.6
Summary of SELECT syntax.

```
<specification> ::= SELECT [ <set quantifier> ]
        <select list> <spatio temporal expression>
<set quantifier> ::= DISTINCT | ALL
<spatio temporal expression> ::= <from clause>
        [ <areaspan clause> ] [ <timespan clause> ] [ <where clause> ]
        [ <while clause> ] [ <setagent clause> ]
```

FIGURE 2.7
BNF description of SELECT syntax.

```
<from clause> ::=
        FROM <spatio temporal extraction>
        [ { ',' <spatio temporal extraction> }... ]
<spatio temporal extraction> ::=
        '(' <area clause> ',' <time clause> ',' <sensor clause> ')'
        AS <{block name}>
<area clause> ::=
        AREA INSIDE '(' <longitude value> ',' <longitude offset value>   ','
        <latitude value> ',' <latitude offset value>   ')'
<latitude value> ::= <signed numeric value>
<longitude value> ::= <signed numeric value>
<longitude offset value> ::= <unsigned numeric value>
<latitude offset value> ::= <unsigned numeric value>
<time clause> ::=
        TIME { {START '" <timestamp string literal> '"
        END '" <timestamp string literal> '" }
        | { WINDOW <window size> [ ',' <slide size> ] } }
<sensor clause> ::=
        SENSOR <sensor type string literal>
<window size> ::= <time unit>
<slide size> ::= <time unit>
<time unit> ::= <unsigned numeric value> { SEC | MIN | HOUR }
```

FIGURE 2.8
BNF description of FROM clauses.

10 min. For the correlation calculation, the area and the timespan specified by the TIME and AREA phrases are similar to both cities. If there are no sensors in a spatial range, the representative value is interpolated by the neighbouring values. If there are no sensors in a temporal range, the previous value is used as the representative value. Figure 2.12 shows an example of a continuous query. This is a query description to repeatedly find the maximum temperature every 0.0001° geometric square and every 10 min to find emergency situations.

A Sensor Data Aggregation System Using Mobile Agents

```
<areaspan clause> ::=
        AREASPAN <side> <using aggmethod> [ <areaelse clause> ]
<timespan clause> ::=
        TIMESPAN <slot> <using aggmethod> [ <timeelse clause> ]
<areaelse clause> ::= ELSE
        { INTERP { <areainterp method> '(' ')' }
        | CONTINUE '(' <signed numeric value>   ')' | ABORT '(' ')' }
<timeelse clause> ::= ELSE
        { INTERP { <timeinterp method> '(' ')' }
        | CONTINUE '(' <signed numeric value>   ')' | ABORT '(' ')' }
<using aggmethod> ::=
        USING { <function type> '(' ')' }
<side> ::= ALL | <unsigned numeric value>
        | DIST '(' <unsigned numeric value> <dist unit> ')'
<slot> ::= ALL | <time unit>
<dist unit> ::= KM | M
<function type> ::=
        AVG | MAX | MIN | SUM | COUNT | UDEF
<areainterp method> ::=
        E_AVG | E_MAX | E_MIN | E_SUM | E_COUNT | UDEF
<timeinterp method> ::= PREV | LERP | UDEF
```

FIGURE 2.9
BNF description of AREASPAN and TIMESPAN clauses.

```
<select list> ::= '*' | <term> [ { ',' <term> }... ]
<term> ::= <factor> | <set function>
<factor> ::= <column> | <factor> '+' <column> | <factor> '-' <column>
<column> ::=   <column reference>
        | <column> '*' <column reference> | <column> '/' <column reference>
<set function> ::= COUNT '(*)'
        | TOPK '(' <unsigned integer> ',' <arguments> ',' <ordering> ')'
        | CORR '(' <factor> ',' <factor> ')'
        | ABS '(' <factor> ')'
        | <general function> | <user function>
<general function> ::= <function type> <arguments>
<arguments> ::= '(' <factor> [ ':' <factor> ] ')'
<user function> ::= UDEF '(' <block name> [ { ',' <block name> }...] ')'
<column reference> ::= [ <block name> '.' ] <column name>
<ordering> ::= ASC | DESC
<where clause> ::= WHERE <condition>
<while clause> ::= WHILE <condition>
<setagent clause> ::= TRAVERSE | PARALLEL
<condition> ::= <boolean term> | <condition> OR <boolean term>
<boolean term> ::= <boolean factor> | <boolean term> AND <boolean factor>
<boolean factor> ::= <predicate> | '(' <condition> ')'
<predicate> ::= <comparison predicate> | <factor> <null predicate>
<comparison predicate> ::= <factor> <operator literal> <value expression>
<value expression> ::= <factor> | <signed numeric value> | <string value>
<string value> ::= <unsigned literal> | <column reference>
<unsigned literal> ::= <string literal> | <timestamp literal>
<null predicate> ::= IS [ NOT ] NULL
```

FIGURE 2.10
BNF description of WHERE and WHILE phrases.

2.4.5 Discussion

We compare the character length of query descriptions. Our explained query description is compared with conventional SQL and a query description that can describe spatiotemporal ranges called PLACE. The detail is described in Refs. [24,25]. The result is shown in Table 2.7. x is the character length of the procedural descriptions. To compare only the declarative descriptions, we show them by x. The queries used for the comparison

TABLE 2.6

Attributes of Spatiotemporal Ranges

Attribute	Content
value	The value of the sensor data
sensor_id	The sensor ID of the sensor
sensor_model	The sensor model of the sensor
organization_name	The organisation name of the sensor network including the sensor
sensornetwork_name	The sensor network name
data_type	The data type of the sensor data
accuracy	The accuracy of the sensor data
resolution	The resolution of the sensor data
unit	The unit of the sensor data
longitude	The longitude of the sensor
latitude	The latitude of the sensor
sensor_name	The sensor name
interval	The interval of the sensor data

```
SELECT CORR(c1.value,c2.value)
    FROM (
        AREA INSIDE(135.519876,0.009335,34.816411,0.007486),
        TIME START '2012-01-10 12:00:00' END '2012-01-10 17:00:00',
        SENSOR temperature
    ) AS c1, (
        AREA INSIDE(135.772455,0.009335,35.018829,0.007486),
        TIME START '2012-01-10 10:00:00' END '2012-01-10 15:00:00',
        SENSOR temperature
    ) AS c2
    AREASPAN DIST(1 km) USING AVG() ELSE INTERP USING E_AVG()
    TIMESPAN 10 MIN USING AVG() ELSE INTERP USING PREV()
    WHERE c1.value > 0 AND c2.value > 0
```

FIGURE 2.11
Query to calculate the correlation of different spatiotemporal ranges.

```
SELECT c.value
    FROM (
        AREA INSIDE(135.519876,0.009335,34.816411,0.007486),
        TIME   WINDOW 30 MIN,
        SENSOR temperature
    ) AS c
    AREASPAN 0.000100 USING MAX() ELSE INTERP USING CONTINUE(0)
    TIMESPAN 10 MIN USING MAX() ELSE INTERP USING ABORT()
    WHILE c.value < 100
```

FIGURE 2.12
Example of continuous query.

are shown in Figures 2.13 through 2.15. The query description *A* does not use any procedural descriptions and does not specify the spatiotemporal units. The spatial units for the query descriptions *B* and *C* are every 1 km and the temporal units are every 10 min. The query description *B* finds the top 10 temperature data and the query description *C* calculates the correlation.

A Sensor Data Aggregation System Using Mobile Agents

TABLE 2.7

Comparison of Description

Query Description	SQL	PLACE	Our Description
A	31	20	10
B	$36 + x$	$25 + x$	27
C	$65 + x$	$46 + x$	47

```
SELECT MAX(c.value)
    FROM (
        AREA INSIDE(135.519876,0.009335,34.816411,0.007486),
        TIME   START '2012-01-10 12:00:00' END '2012-01-10 17:00:00',
        SENSOR temperature
    ) AS c
```

FIGURE 2.13
Query description A.

```
SELECT TOPK(10,c.value,ASC)
    FROM (
        AREA INSIDE(135.519876,0.009335,34.816411,0.007486),
        TIME   START '2012-01-10 12:00:00' END '2012-01-10 17:00:00',
        SENSOR temperature
    ) AS c
      AREASPAN DIST(1 km) USING MAX()
      TIMESPAN 10 MIN USING MAX()
```

FIGURE 2.14
Query description B.

```
SELECT CORR(c1.value,c2.value)
    FROM (
        AREA INSIDE(135.519876,0.009335,34.816411,0.007486),
        TIME   START '2012-01-10 12:00:00' END '2012-01-10 17:00:00',
        SENSOR temperature
    ) AS c1,(
        AREA INSIDE(135.772455,0.009335,35.018829,0.007486),
        TIME   START '2012-01-10 10:00:00' END '2012-01-10 15:00:00',
        SENSOR humidity
    ) AS c2
      AREASPAN DIST(1 km) USING AVG()
         ELSE INTERP USING E_AVG()
      TIMESPAN 10 MIN USING AVG()
         ELSE INTERP USING PREV()
```

FIGURE 2.15
Query description C.

The conventional descriptions cannot flexibly describe the aggregation operations for spatiotemporal units. Although GROUP BY phrases of SQL is similar to the operations for spatiotemporal units, GROUP BY only groups the same values and does not flexibly describe the units. The character lengths of the conventional descriptions are longer than our explained description since they have to write some queries to realise spatio-temporal queries.

2.5 Sensor Data Aggregation Systems Using MAs

As noted in Section 2.1, the network traffic required to execute the query explained in the previous section by using conventional architecture is large since the clients must often collect all the sensor data from all the sensor network sites. By executing the aggregation operations on the gateways that have the sensor data needed for the aggregation and aggregating only the necessary aggregated results, the network traffic can be reduced. MA systems are suitable for executing operations on respective sensor networks, since they can migrate among the sensor networks, executing operations. Therefore, we explain a sensor data aggregation system for integrated sensor networks using MAs. The detail is explained in Ref. [32].

2.5.1 Design for Sensor Data Aggregation Systems Using MAs

In this section, we describe the design for sensor data aggregation systems using MAs for integrated sensor networks. For this, we have to consider the following items.

2.5.1.1 Compatibility of Sensor Network Sites

There is a high possibility that different sensor network sites use a different sensor network architecture. One of the main differences is the attribute names for the databases. For example, the attribute name of the time stamps for the sensor data can be 'time', 'time stamp' and so on. To make these attributes compatible, the sensor data aggregation systems use unified attribute names. By associating the attribute names of the sensor network sites with the unified attribute names, the queries can use the unified attribute names. For example, the system associates the attribute names that mean time stamps such as 'time' or 'time stamp' with the unified attribute name 'timestamp'. Also, the centre server should grasp the sensor network information described in Table 2.1 to execute the queries.

2.5.1.2 Management of MAs

To enable MAs to aggregate the sensor data, we need to manage MAs. That is, the system generates MAs and controls them according to the users' operations. For example, the centre server selects the sensor network sites that have the necessary sensor data to execute the query. After this, the centre server generates MAs to execute the extracted queries and makes them migrate to the sensor network sites.

The query description explained in Section 2.4 can describe spatiotemporal ranges. Therefore, by migrating MAs to each spatiotemporal range, the system can easily aggregate sensor data from the integrated sensor networks. Hence, the system generates MAs for each spatiotemporal range. They migrate to their associated spatiotemporal range and collect the sensor data from the gateways.

Figure 2.16 shows our designed architecture for MAs to aggregate sensor data. The syntax parser parses the queries described using the query description explained in Section 2.4. The query is extracted for some queries for each sensor network site. MAs migrate among the gateways and execute the query. The results are returned to the centre server by JavaScript Object Notation (JSON) or text formats. Here, MA_0 is the MA that manages the sensor data aggregation of the user-described query. $MA_1, \ldots, MA_n$ are MAs for each spatiotemporal unit. The sensor data aggregation methods in each spatiotemporal range

A Sensor Data Aggregation System Using Mobile Agents

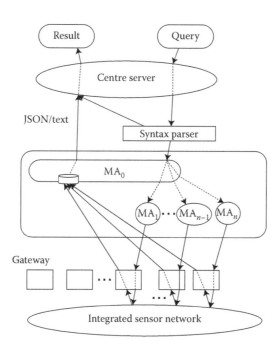

FIGURE 2.16
Architecture of sensor data aggregation by MAs.

are described in the query. Finally, their aggregated sensor data are gathered by a mobile agent (MA_0). The results are returned to the centre server and shown to the user.

2.5.1.3 Database Access by MAs

The syntax parser generates SQL queries by using the sensor network information and the unified attributes. By the user query extraction of the syntax parser, MAs can obtain the SQL queries. At each gateway, the MA that knows the way to access the database is required. We call them the *gateway agents*. The gateway agents reside at the gateways. They receive the SQL query from the MA that the system migrates (*user agents*) to and submit the query to the databases that the gateways have. Figure 2.17 shows our designed architecture to access databases from MAs.

2.5.1.4 Behaviour Determination of MAs

The sensor data collected from some gateways are sometimes aggregated to a gateway. For this, the behaviour of MAs to collect sensor data has two policies, which are shown in Figure 2.18.

In the figure, the black circles show the gateways and the lines show their connections. The trajectories of MAs are shown by dotted lines. MAs finally migrate to the centre server.

In the policy for MAs to collect sensor data by only one MA (TRAVERSE), an MA traverses the gateways described in the spatiotemporal ranges of the queries. In this policy, the MA has sensor data needed to execute the query and it is easy to execute the query. However, the time to aggregate the sensor data increases as the number of gateways needed to migrate increases.

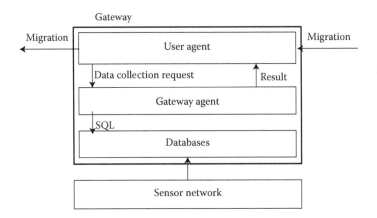

FIGURE 2.17
Database access architecture.

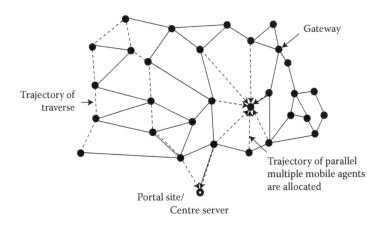

FIGURE 2.18
Routes for MAs to aggregate sensor data.

On the other hand, in the policy for MAs to collect sensor data by multiple MAs (PARALLEL), some MAs migrate to the gateways that have necessary sensor data for executing the query. Since some MAs aggregate the sensor data in parallel, it can reduce the time to aggregate sensor data. However, the number of MAs increases as the number of the gateways needed to collect sensor data increases. In the query description explained in Section 2.4, SETAGENT phrases designate the MA route.

2.5.2 Implementation

We explain our implementation of the sensor data aggregation system based on the previous section.

2.5.2.1 Definition of Sensor Network Information

To ensure the compatibility of sensor network sites, we make the centre server grasp the attribute names for the sensor databases of each sensor network site associated with the unified attribute names. The associations are written in XML format. MAs read the XML documents and change the attribute names for the gateways to the unified attribute names.

2.5.2.2 Use of the MA System

Various systems to manage MAs have been developed. Among these, P2P-based systems are suitable for integrated sensor networks. By using a P2P-based MA system, the system load is not concentrated on the server, and thus it can run effectively on MAs. Although our system has a centre server, the load is low since the clients communicate with it only when they start running the system.

We exploit PIAX [30] for managing MAs. PIAX is a P2P-based MA system, suitable for aggregating data from integrated sensor networks. The MAs are implemented by Java. In PIAX, MAs can migrate between PIAX peers. Figure 2.19 shows the implementation of PIAX to the gateways. The observed sensor data by the sensor networks are stored in the databases in the gateways. We use PostgreSQL for the databases. Java virtual machine (JVM) is required to run PIAX, and JDBC is the library to access the databases from Java. PIAX peers run on JDBC and PIAX libraries, and manage MAs.

2.5.2.3 MAs Generation

After the syntax parser parses the submitted query, the centre server generates the user agent to aggregate the sensor data. The original user agent traverses the sensor network sites or generates other user agents based on the SETAGENT phrase in the query. The user agents have the SQL queries generated by the syntax parser and collect the sensor data by communicating with the gateway agents. After they collect the sensor data, they execute the aggregation operations to the sensor data. The aggregation operations are extracted from the user-submitted query by the centre server based on the spatiotemporal ranges. Each user agent has the sensor data needed to execute the aggregation operations. After that, they return the aggregated sensor data for each spatiotemporal range to the original user agent. The user agent aggregates the collected sensor data and the users can obtain desirable results.

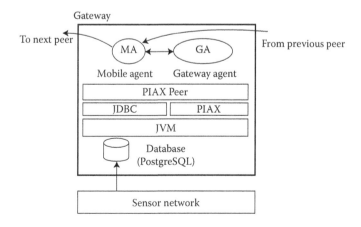

FIGURE 2.19
Implementation of gateways.

2.5.2.4 MA Migration

As described in Section 2.5.1.4, MAs have two policies. To realise them, MAs need to know the network address for each gateway. PIAX, which we used for the MA management system, provides the MA migration functions and the finding of the network address is realised by PIAX. The detail is described in Ref. [30].

2.5.2.5 Web Browser Interface

To provide visual interfaces, we use web browsers. This is because users do not need to instal new software to visualise the state of sensor data aggregation, since most of the clients have web browsers. In addition, we can employ various web applications such as maps and databases through the Internet, and exploit these, through web browsers, for the purposes of visualisation.

The interface is shown in Figure 2.20. When the mouse cursor points at the indicator, the detailed information for the sensor network pops up. If there are a number of sensor networks in a narrow area, it may be difficult to recognise each indicator. In this case, the indicators are bundled up in one square indicator. By zooming the map, the bundled indicator is divided into the respective squares. To display the information for each sensor network when users access the website, the web server accesses the sensor network information from the centre server.

To aggregate the data from integrated sensor networks, users first select the respective sensor networks. The selected sensor networks are listed in the left column. When users click the selected sensor network name in the list, the location is shown on the map. By double-clicking the sensor network name, the sensor network is removed from the list.

After selecting the sensor networks, users push the 'Create Query' button to create queries. Here, 'query' means a query to aggregate sensor data, including the program for MAs. To execute the queries, the system generates the user agents. The query creation dialogue appears when users push the button. With the 'AgentRoute' tab in the dialogue, users enter

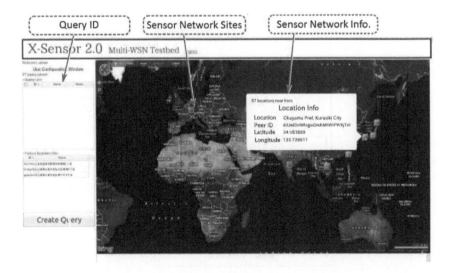

FIGURE 2.20
Web interface.

A Sensor Data Aggregation System Using Mobile Agents

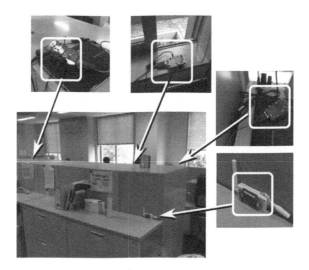

FIGURE 2.21
Sensors in our laboratory.

the query name. Then, users select the agent mode. When users finish creating the query, they push the 'OK' button. Then, the generated user agent begins to migrate among the sensor networks. The created queries are listed in the 'QueryList' located on the top left of the web page. The users can check the results of the data aggregation by double-clicking the query name listed in the 'QueryList'. In addition, since MAs can return the results at runtime, users can check the intermediate results when needed.

Figure 2.21 shows the physical sensors that compose the integrated sensor network. We deployed more than 100 sensor nodes in our laboratory.

2.5.3 System Evaluation

We evaluated our implemented sensor data aggregation system using MAs.

2.5.3.1 Evaluation Environments

We measured the network traffic and the response time for the sensor data aggregation using MAs. In this evaluation, we used three sensor network sites. We described the query assuming that the user checks the trip destination is rainy or not by using camera images. The query calculates the feature amounts that the images taken by the cameras in each spatiotemporal unit suggest rain. Then, the query calculates the average of the feature amounts for each sensor network site and finally obtains the probability that the trip destination is rainy. For comparison, we realise the query by our implemented query description and conventional SQL descriptions by C/S systems. The data size of MAs is 7700 bytes and the SQL query size becomes 110 bytes. The data size of the camera image is 1800 Kbytes.

2.5.3.2 Evaluation Results

Table 2.8 shows the evaluation results. In our implemented sensor data aggregation system using MAs, MAs can calculate the feature amount at each sensor network site. Therefore, MA systems do not need to transmit the image data itself, but only transmit the feature amounts. On the other hand, the C/S system cannot calculate the feature amount at each

TABLE 2.8

Evaluation Results

Number of Sensor Network Sites	Traffic (Kbytes)		Response Time (s)	
	C/S	MA	C/S	MA
1	1844.468	796.240	12.535	8.261
2	3689.302	1584.726	23.432	18.984
3	5534.238	2373.619	27.821	32.192

sensor network site and has to transmit the image data to the centre server. So, the traffic for C/S is larger than that for MA.

Regarding the response time, when the number of the sensor network sites is small, the response time of the MA system is smaller than that of the C/S system. This is because the MA system does not need to transmit the image data. However, as the number of the sensor network sites increases, the response time increases and that of the MA system becomes longer than that of the C/S system. This is because the time to traverse MA increases as the number of sensor network sites increases.

2.5.3.3 Discussion

We can see that the MA systems can reduce the network traffic compared with the C/S systems. The C/S systems require more network traffic than the MA systems as the sensor data size increases. However, when MAs cannot aggregate sensor data effectively, the network traffic increases. Therefore, the effectiveness of using MAs depends on the application of data aggregation.

Our implemented system is suitable for sensor data aggregation using MAs, since the system is designed to make MAs aggregate the sensor data. Other MA systems such as AgentTeamwork or the original PIAX are designed for, and can be applied to, various MA applications. However, an MA system customised for integrated sensor networks offers users easy and efficient aggregation of the sensor data generated from integrated sensor networks. A further merit may be seen in the ability to visualise the locations of sensor networks and MAs. Therefore, it is easy to determine the status of MAs. Regarding the fault tolerance, by programming MAs so that they can find unusual response from sensors, we can make the system tolerant to troubles.

On the other hand, one of the demerits of our implemented system is that it is hard for nonexpert users to write the agent programs. We can solve this problem by improving the script language for programming MAs, and by supporting more intuitive GUI. In addition, since the users have flexibility in writing MA programs, they can create malicious agents such as computer viruses. However, we can detect these by checking the user-written MA programs at the centre server.

2.6 Conclusion

2.6.1 Summary

In this chapter, we have explained the query description and the sensor data aggregation systems from the integrated sensor networks. Our explained query description can declaratively describe the data aggregation operations for spatiotemporal units. The users can easily describe the queries that include operations for every spatial and temporal unit such as averaging, similarity calculations, and so on. Therefore, the users do not need to submit multiple queries to each sensor network site that has different sensor network architecture. Moreover, the query description is used for our implemented sensor data aggregation system using MAs. MAs are computer programs that migrate among sensor network sites. MAs can decrease the network traffic when the data size for the aggregated sensor data is small by migrating the MAs to the sensor network sites that have necessary sensor data for the aggregation.

Our evaluation results show that the character length of the query is small compared with other descriptions when the query includes the aggregation operations for spatiotemporal units. Also, the traffic and the response time are short when the number of sensor network sites is small.

2.6.2 Future Work

The current sensor data aggregation system using MAs for integrated sensor networks assumes that the sensor nodes do not move. However, recently prevailing smart phones or laptop computers have many sensors and we can exploit them for sensor networks. In that case, the sensor nodes move and the system must collect the sensor data from the moving sensor nodes. When the sensor nodes move, their connected gateways change. Therefore, MAs should chase the gateways that the necessary sensor nodes to execute the queries connect to.

Also, when the users submit multiple queries, a part of the aggregation operations can be shared with other queries. For example, averaging temperature data to every spatio-temporal unit can be used for many queries. In that case, by sharing the result of the aggregation operations, the network traffic and the response time can be reduced. We will consider the network traffic reduction by sharing the aggregation results.

Acknowledgement

This research was supported by Grants-in-Aid for Scientific Research (S) numbered 21220002 and collaborative research of NICT and Osaka University (research and development, validation of integrated management technique for heterogeneous, wide-area sensor networks).

References

1. A. Arora, E. Ertin, R. Ramnath, M. Nesterenko and W. Leal, Kansei: A high-fidelity sensing testbed, *IEEE Internet Computing*, 10(2), 35–47, March 2006.
2. H. Esaki and H. Sunahara, Live E! Project: Sensing the earth with Internet weather stations, *Proceedings of the International Symposium on Applications and the Internet (SAINT 2007)*, Hiroshima, Japan, No. 67, January 2007.

3. A. Kanzaki, T. Hara, Y. Ishi, T. Yoshihisa, Y. Teranishi and S. Shimojo, X-Sensor: Wireless sensor network testbed integrating multiple networks, In: *Wireless Sensor Network Technologies for the Information Explosion Era, Studies in Computational Intelligence*, Vol. 278, Springer-Verlag, Berlin, Heidelberg, pp. 249–271, November 2010.
4. K. Aberer, M. Hauswirth and A. Salehi, Global sensor networks, Technical Report LSIR-REPORT-2006-001, September 2006.
5. K. Aberer, M. Hauswirth and A. Salehi, Infrastructure for data processing in large-scale interconnected sensor networks, *Proceedings of the International Conference on Mobile Data Management (MDM 2007)*, Mannheim, Germany, pp. 198–205, May 2007.
6. P.B. Gibbons, B. Karp, Y. Ke, S. Nath and S. Srinivasan, IrisNet: An architecture for a worldwide sensor web, *IEEE Pervasive Computing*, 2(4), 22–33, December 2003.
7. J. Campbell, P.B. Gibbons, S. Nath, P. Pillai, S. Seshan and R. Sukthankar, IrisNet: An Internet-scale architecture for multimedia sensors, *Proceedings of the ACM Multimedia Conference*, Singapore, pp. 81–88, November 2005.
8. M. Nakayama, S. Matsuura, H. Esaki and H. Sunahara, Live E! Project: Sensing the earth, *Lecture Notes in Computer Science, Technologies for Advanced Heterogeneous Networks II*, Vol. 4311, pp. 61–74, December 2006.
9. M. Balazinska, A. Deshpande, M.J. Franklin, P.B. Gibbons, J. Gray, M. Hansen, M. Liebhold, S. Nath, A. Szalay and V. Tao, Data management in the world-wide sensor web, *IEEE Pervasive Computing*, 6(2), 30–40, April 2007.
10. K. Chang, N. Yau, M. Hansen and D. Estrin, A centralized repository to slog sensor network data, *Proceedings of the International Conference on Distributed Computing in Sensor Systems (DCOSS 2006)*, San Francisco, USA, June 2006.
11. C.L. Fok, G.C. Roman and C. Lu, Towards a flexible global sensing infrastructure, *Proceedings of the ACM SIGBED Review, Special Issue on the Workshop on Wireless Sensor Network Architecture (WSNA 2007)*, 4(3), 1–6, July 2007.
12. A. Kansal, S. Nath, J. Liu and F. Zhao, SenseWeb: An infrastructure for shared sensing, *IEEE Multimedia*, 14(4), 8–13, October 2007.
13. Y. Hamaguchi, T. Yoshihisa, Y. Ishi, Y. Teranishi, T. Hara and S. Nishio, A data aggregation system using mobile agents on integrated sensor networks, *Proceedings of the International Conference on Advances in P2P Systems (AP2PS 2011)*, Lisbon, Portugal, Vol. 3, pp. 33–38, November 2011.
14. S. Michel, A. Salehi, L. Luo, N. Dawes, K. Aberer, G. Barrenetxea, M. Bavay, A. Kansal, K.A. Kumar and S. Nath, Environmental monitoring 2.0, *Proceedings of the IEEE International Conference on Data Engineering (ICDE 2009)*, Shanghai, China, pp. 1507–1510, March 2009.
15. S. Nath, J. Liu, J. Miller, Z. Feng and A. Santanche, SensorMap: A web site for sensors worldwide, *Proceedings of the International Conference on Embedded Networked Sensor Systems (Sensys 2006)*, Boulder, USA, pp. 373–374, October 2006.
16. N. Dawes, K.A. Kumar, S. Michel, K. Aberer and M. Lehning, Sensor metadata management and its application in collaborative environmental research, *Proceedings of the IEEE International Conference on e-Science*, Indianapolis, USA, pp. 143–150, December 2008.
17. H. Jeung, S. Sarni, I. Paparrizos, S. Sathe, K. Aberer, N. Dawes, T.G. Papaioannou and M. Lehning, Effective metadata management in federated sensor networks, *Proceedings of the IEEE International Conference on Sensor Networks, Ubiquitous, and Trustworthy Computing (SUTC 2010)*, Newport Beach, USA, pp. 107–114, June 2010.
18. C.X. Chen and C. Zaniolo, SQLST: A spatio-temporal data model and query language, *Proceedings of the International Conference on Conceptual Modeling (ER 2000)*, Salt Lake City, USA, pp. 96–111, October 2000.
19. M. Egenhofer, Spatial SQL: A query and presentation language, *IEEE Transactions on Knowledge and Data Engineering*, 6(1), 86–95, February 1994.
20. H.G. Elmongui, Query optimization for spatio-temporal data stream management systems, *SIGSPATIAL Special 2009*, 1(1), 21–26, March 2009.
21. M. Erwig, Design of spatio-temporal query languages, *Proceedings of the NSF/BDEI Workshop on Spatio-Temporal Data Models of Biogeophysical Fields for Ecological Forecasting*, April 2002.

22. M. Erwig, R.H. Guting, M. Schneider and M. Varzigiannis, Spatio-temporal data types: An approach to modeling and querying moving objects in databases, *Proceedings of the Chorochronos Intensive Workshop on Spatio-Temporal Database Systems*, 3(3), 269–296, September 1999.
23. M. Erwig and M. Schneider, Developments in spatio-temporal query languages, *Proceedings of the IEEE International Conference on Spatio-Temporal Data Models and Languages (STDML 1999)*, Florence, Italy, pp. 441–449, August 1999.
24. M.F. Mokbel, X. Xiong, W.G. Aref, S. Hambrusch, S. Prabhakar and M. Hammad, PLACE: A query processor for handling real-time spatio-temporal data streams (demo), *Proceedings of the International Conference on Very Large Data Bases (VLDB 2004)*, Toronto, Canada, Vol. 9, pp. 343–365, December 2004.
25. M.F. Mokbel, X. Xiong and W.G. Aref, SINA: Scalable incremental processing of continuous queries in spatio-temporal databases, *Proceedings of the ACM SIGMOD International Conference on Management of Data*, 1398(2), 623–634, June 2004.
26. M. Chen, T. Kwon, Y. Yuan, Y. Choi and V.C.M. Leung, Mobile agent based directed diffusion in wireless sensor networks, *EURASIP Journal on Advances in Signal Processing*, 2007(1), 1–14, January 2007.
27. A. Knoll and J. Meinkoehn, Data fusion using large multi-agent networks: An analysis of network structure and performance, *Proceedings of the IEEE International Conference on Multisensor Fusion and Integration for Intelligent Systems (MFI 1994)*, Las Vegas, USA, pp. 2–5, October 1994.
28. G. Lu, J. Lu and J. Yip, Mobile agent based data fusion for wireless sensor networks with a XML framework, *Proceedings of the International Conference on Internet Computing (ICOMP 2008)*, pp. 298–303, July 2008.
29. M. Fukuda, C. Ngo, E. Mak and J. Morisaki, Resource management and monitoring in AgentTeamwork grid computing middleware, *Proceedings of the IEEE Pacific Rim Conference on Communications, Computers and Signal Processing (PacRim 2007)*, pp. 145–148, August 2007.
30. Y. Teranishi, PIAX: Toward a framework for sensor overlay network, *Proceedings of the International IEEE Consumer Communications and Networking Conference Workshop on Dependable and Sustainable Peer-to-Peer Systems (CCNC 2009)*, pp. 1–5, June 2009.
31. H. Qi, S.S. Iyengar and K. Chakrabarty, Multi-resolution data integration using mobile agents in distributed sensor networks, *IEEE Transactions on System, Man and Cybernetics (Part C)*, 31(3), 383–391, August 2001.
32. Y. Hamaguchi, T. Yoshihisa, Y. Ishi, Y. Teranishi, T. Hara and S. Nishio, A data aggregation system using mobile agents on integrated sensor networks, *International Conference on Advances in P2P Systems (AP2PS 2011)*, November 2011.

3

Underlay-Aware Distributed Service Discovery Architecture with Intelligent Message Routing

Haja M. Saleem, Seyed M. Buhari, Mohammad Fadzil Hassan and Vijanth S. Asirvadam

CONTENTS

3.1 Introduction 68
3.2 Underlay Routing versus Overlay Routing 68
3.3 Mismatch between the Overlay and the Underlay Topology 69
3.4 Related Work 69
    3.4.1 Underlay Awareness in P2P and SD Approaches 69
    3.4.2 Message Level Intelligence 71
3.5 P2P-Based DSD Routing Architecture 72
    3.5.1 Layered Design of the Current P2P-Based DSD Systems 72
    3.5.2 Layered Design of the Proposed P2P-Based DSD Systems 73
    3.5.3 ISP Level Perspective of the Proposed System 75
    3.5.4 Design of an Underlay Layer Query Routing Algorithm 75
    3.5.5 Underlay (AON) Query Routing Process 77
    3.5.6 Design of AON Router 79
    3.5.7 AON-Based Query Message Structure 80
3.6 Performance Comparison of Underlay versus Overlay Routing 81
    3.6.1 Hypothesis 1 81
        3.6.1.1 Estimation of the Number of Routers Crossed in the Overlay 82
        3.6.1.2 Estimation of the Number of Routers Crossed in the Underlay 82
        3.6.1.3 Complexity Analysis 83
    3.6.2 Hypothesis 2 83
        3.6.2.1 Estimation of the Number of Links Utilised in the Overlay 83
        3.6.2.2 Estimation of the Number of Links Utilised in the Underlay 83
    3.6.3 Hypothesis 3 85
    3.6.4 Hypothesis 4 85
        3.6.4.1 Inter-ISP Traffic in the Overlay Routing 85
        3.6.4.2 Inter-ISP Traffic in the Underlay Routing 86
    3.6.5 Hypothesis 5 86
3.7 Limitations of the Study 87
3.8 Conclusion 87
References 88

3.1 Introduction

Distributed service (or resource) discovery (DSD) is becoming an important research area in service-oriented computing (SOC) because many software applications are now developed with services from different vendors. Most of the current DSD approaches follow the techniques used in peer-to-peer (P2P) applications. It is estimated that 70% of the Internet traffic today is consumed by these P2P applications. Also, the volume of P2P traffic is on the rise. P2P and DSD applications function by forming an overlay on top of the existing Internet protocol (IP) layer (underlay layer). The query routing mechanism of the current P2P applications functions purely on the overlay without incorporating the topological and routing knowledge of the underlay. Consequently, Internet service providers (ISPs) are tested to their limits due to underlay-ignorant query forwarding that is employed by the overlay applications such as P2P and DSD. This underlay-ignorant query forwarding leads to inefficient usage of the underlying links because the peers in the overlay are ignorant of the location of their neighbours in the underlay. As a result, ISPs need to handle large volumes of traffic as transiting nodes, which results in increased inter-ISP traffic. Moreover, the neighbours that could be reached through the same path in the underlay are not known in the overlay, and therefore the traffic is redundantly sent in the underlying links, consuming their bandwidth and causing high interlayer communication overhead.

To alleviate these problems that are caused by underlay-ignorant query forwarding in the overlay, this research proposes to move the query routing process from the overlay layer to the IP layer. This research focuses on providing the IP layer with both the service location information and message level knowledge to minimise the stand-off between the ISPs and P2P-based DSD applications. For this purpose, it is proposed to employ an intelligent message routing (IMR) by implementing application-oriented networking (AON) at the routing layer of the systems that are based on DSD architecture. A suitable query routing algorithm was developed and tested with the DSD architecture.

The proposed DSD architecture and the query routing algorithm are hypothetically analysed and the improvements obtained in terms of query retrieval time, inter-ISP traffic and underlay links utilisation are demonstrated. The remainder of the chapter is organised as follows. Section 3.2 describes the overlay versus underlay routing, Section 3.3 discusses the cause for IP-oblivious query forwarding in DSD systems and Section 3.4 covers the related work. The proposed DSD model is elaborated in Section 3.5 and it is analysed hypothetically in Section 3.6; Section 3.7 concludes the chapter with issues and future work.

3.2 Underlay Routing versus Overlay Routing

Overlay routing is the approach of forming a network of nodes (peers) on top of an existing network, usually the IP network. The neighbours formed in the overlay are independent of the IP network. The IP network is called the underlay layer. The simplest form of an overlay network is the virtual private network (VPN) in which a VPN packet is encapsulated over the IP packet. The concept of overlay formation has been used by many applications and this chapter, in particular, focuses on the DSD application.

3.3 Mismatch between the Overlay and the Underlay Topology

One of the major reasons for the poor performance of DSD applications is the mismatch between the overlay neighbour formation by the peers and the underlay topology. To understand this, let us inspect Figure 3.1, shows a schematic view of the peers super registry (SR) in the overlay layer. The continuous line in the overlay depicts the neighbour relationship between the peers. In the underlying IP layer, a continuous line depicts a physical neighbour connected via the routers. The dotted lines in the figure map the peers in the overlay to that of their physical location in the underlay. In the overlay, peers A and B, which appear to be neighbours, are in fact four hops away in the underlay. When a query routing mechanism is implemented in the overlay without considering the underlay's physical distance, then peers A and B are considered to be neighbours. Also, there could be peers C and D that are neighbours in the overlay as well as in the underlying topology. The overlay query mechanism cannot differentiate between the pair A–B and the pair C–D. Both pairs are considered to be neighbours, which results in inefficient query forwarding because the number of hops (routers) between them in the underlay is not taken into account. The following sections show the layered approach of the current P2P systems and how improvements can be made.

3.4 Related Work

3.4.1 Underlay Awareness in P2P and SD Approaches

As per the records in 2010, P2P applications contributed to about 70% of the total traffic on the Internet [1]. It was also predicted to have 2.8 megabytes of traffic per month that

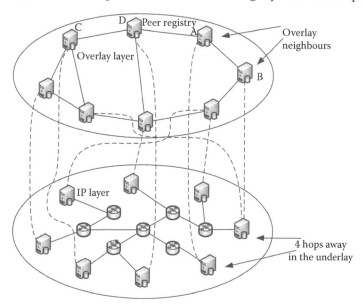

FIGURE 3.1
Mismatch between the overlay and the underlay topology.

would be consumed by P2P applications by 2011. It is quite clear as this chapter is being read that the P2P traffic is much more than that predicted in 2011. This situation clearly demands that something must be done to minimise that traffic for the following major reasons:

1. P2P traffic redundantly uses different ISPs as transit nodes without any kind of optimisation.
2. ISPs need to spend lots of effort on traffic management.
3. This can lead to huge congestion and affects the performance of the network as a whole.
4. All of the above tremendously increase the cost of ISPs.

To this end, there has been a considerable amount of effort made by researchers around the globe. This section will review those works and how this research work is different from those works will be elaborated.

PIPPON [2] attempts to bring the underlay awareness to the overlay by clustering the peers based on proximity. The proximity of the peers is measured with two parameters:

1. Longest prefix IP matching
2. Round trip time (RTT)

Apart from cluster formation, it forms a dynamic 'key tree' for routing the query messages on the overlay. This tree is employed to route the queries based on the IP prefixes. However, PIPPON does not take into account the similarity of queries in the service discovery, which is a key factor in improving the efficiency. Also, it does not solve the problem of redundant traffic in the underlay.

$$d(j,k) = \sum_{v=0}^{31} |j-k| \cdot 2^v \qquad (3.1)$$

TOPLUS [3] has adopted a slightly different approach to PIPPON in terms of underlay awareness. The cluster formation in TOPLUS involves three tier hierarchies such as groups, super-groups and hyper-groups so that the queries can be easily forwarded with a minimal number of hops. In contrast to the longest prefix match of the IP addresses, TOPLUS makes use of the Exclusive-Or (XOR) metric for identifying the proximity. The distance $d(j,k)$ between the two nodes j and k can be expressed in the XOR metric as shown in Equation 3.1. The range for the summation in Equation 3.1 is the number of bits in the IP address, which is 32(0–31). In each tier, the routing of the query messages is performed with the XOR metric. TOPLUS claims that the routing performance it provides is very close to IP routing. TOPLUS does not cover the aspects of inter-ISP, redundant traffic and interlayer communication overhead.

In general, there are two approaches with respect to finding the locality information of the underlay:

1. Dynamic (on the fly calculation with RTT and longest prefix match [LPM])
2. Static (prior knowledge of the Internet such as autonomous system [AS] numbers)

Plethora [4] follows the second approach, that is static, whereas TOPLUS and PIPPON follow the first approach. Plethora makes use of *cSpace*, which makes use of a local broker, and *gSpace*, which makes use of a global broker. Hence, it ends up with a two-layer architecture. If a request is generated, it checks *cSpace* for answers; if *cSpace* is able to answer it, then the query is not forwarded to *gSpace*. It only forwards it to *gSpace* if the query is not answered by *cSpace*. Here again, it does not cover the problems that are addressed in this research work.

P4P [5] addresses the locality awareness problem by two interfaces, *iTrackers* and *appTrackers*. The purpose of *iTrackers* is to provide the locality parameters of a particular node to the querying node. *iTrackers* are designed to reside in each of the ISPs. *appTrackers* are designed to reside globally so that they have a complete picture of the P2P applications. *appTrackers* decides whether a particular peer a could establish a neighbour relation with another peer b based on the information from *iTrackers*.

Internet engineering task force (IETF) formed an application layer traffic optimisation (ALTO) group in 2009 and has been tasked with identifying the means of optimising the traffic in the underlay that are generated by the ever-increasing P2P applications. Seedorf et al. [6] elaborate on the effort of the ALTO group, which intends to provide the ALTO service to the P2P applications. As per the ALTO approach, any P2P application that needs to know the topological and proximity information could obtain the same from the ALTO server. Information such as topology, operational cost and policies from the ISPs could be provided to the querying peers. In this approach, letting the peers be aware of these network-related parameters could be misused by any malicious user.

SLUP [7] is another approach that forms clusters based on semantics and RTT. It forms a three-tier clustering: normal peer, level-2 super peer and level-1 super peer. Redundancy in the underlay and ISP considerations are not in their scope.

Bindal et al. [8] have provided a strategy for improving the traffic locality in the BitTorrent network. They highlight that ISPs often end up throttling the P2P traffic as BitTorrent-like systems do not take ISPs' cost into consideration. Their work focuses on biased neighbour selection in which a peer chooses the majority of its neighbours from the same ISP so as to reduce the cross-ISP traffic. The authors point out here that if there are N users within the ISP wanting to download a particular file, then in the current BitTorrent system, the file is downloaded N times, which heavily increases inter-ISP traffic. In their work, they have demonstrated that biased neighbour selection could improve this situation. Here, their application is specific to the BitTorrent system that uses the rarest first-piece algorithm at the client side. Our work differs from Bindal et al. [8] by providing a generic solution for the service discovery mechanism with improved performance.

To relieve the tension between the P2P applications and ISPs, Shen et al. [9] have provided an HTTP-based approach for P2P traffic caching. As ISPs already have cache proxies for web traffic, they propose to encapsulate the P2P traffic into the HTTP so that P2P traffic can also be cached by the same set of proxies. Their focus is on just providing a caching mechanism for the P2P application and providing an architecture infrastructure, for SD is out of their scope.

P4P Pastry [10] is an approach to bring the locality features to the PASTRY [11] structured system.

3.4.2 Message Level Intelligence

The ever-increasing demand for Internet business and entertainment applications has warranted the need for moving some of the processes from the end nodes to the networking components such as routers and switches [12]. IMR is needed even in the core routing process for the success of the next-generation Internet [13]. CISCO responded by introducing the concept of AON in 2005. AON can enable IMR and switching. CISCO has introduced the AON module into its component, which can differentiate between normal and AON-specific packets [14]. AON-specific packets are checked for their message contents and routing or switching decisions are made as per the AON module.

The approach in this research is to perform content-based routing of queries with the help of AON [14]. Since the introduction of the AON concept by CISCO, it has been a widely researched area in terms of its applicability in various areas of networking. AON is applied in the DSD process so that query routing is performed in the IP layer independent of the overlay layer.

Cheng et al. [15] have provided a model based on a holistic approach with an AON for autonomic service composition. Their focus is on combining the service oriented architecture (SOA), AON and autonomic service computing so that the problems of scalability and time to market products could be addressed collectively, which are handled individually otherwise. In this work, the authors have provided a scheme; however, they have not provided a concrete implementation model.

Xiaohua et al. [16] have improved the definition of AON introduced by CISCO and have given a more generalised approach called 'Application Oriented Studies'. Part of our research comes under this category of studies as it has explored the possibility of employing AON in the SD routing process in the underlay. In this work, the authors have provided a multicast approach that can replace the overlay-based multicast that is usually poor in performance and scalability. However, their approach does not provide a concrete implementation model for DSD.

Yao et al. [17] have provided a mechanism for AON-based switching that can perform effective message scheduling at data centres.

Zille Huma et al. [18] discussed the service location problem (SLP) on how a particular service could be placed in the network so that AON could perform with better time efficiency. Here, their focus is on assisting AON with SLP and not exactly on AON itself.

The application of AON in the area of DSD by providing a query routing process in the underlay (IP layer) that is capable of mapping the service classes to the geographical location on the target registries is the focus of this research work, and it is the first effort of its kind.

3.5 P2P-Based DSD Routing Architecture

3.5.1 Layered Design of the Current P2P-Based DSD Systems

In general, the current P2P system can be abstractly divided into three layers. The network (IP) layer is completely isolated from the P2P system. The IP layer does not have a clue about what kind of message it is forwarding. Therefore, the IP layer does not consider the packet content of the P2P applications.

Also, layers 2 and 3 of the P2P systems function completely independently of the IP layer and the physical topology. Layer 2 is where the core P2P protocols such as Chord [19], Kademlia [20] and so on are implemented. Finally, layer 3 is where application- specific

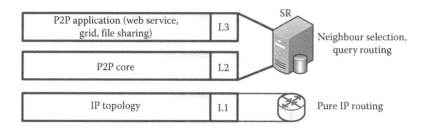

FIGURE 3.2
Layered approach—current P2P systems.

modules such as semantic matching and ontology matching are implemented. In Figure 3.2, the L2 and L3 modules are implemented in the peers (service registry). Henceforth, both the neighbour selection process and the query routing mechanism are implemented in the overlay, making it IP oblivious.

3.5.2 Layered Design of the Proposed P2P-Based DSD Systems

As the issues with the current approach to P2P systems are explained in Section 3.4, the goals for the architecture design of the proposed system have emerged clearly:

1. *Enhanced security*: As it was highlighted in the literature review, the present topologically aware query routing mechanism obtains the network-related parameter such as AS number, geographical location and so on from the ISP and passes it to the overlay peers, enlightening them with topological awareness. However, malicious users or compromised peers could misuse the system to launch attacks and destabilise the system. To prevent this, it is proposed that the network-aware query routing must be delegated to the underlying layer and kept transparent to the overlay layer [21,22].

2. *Reduced inter-ISP traffic*: Most of the P2P traffic on the Internet today uses intermediate nodes as the transit node, which belongs to a particular ISP. These ISPs may have to process a lot of intermediate traffic that would stress them to the limit and cause issues such as bandwidth throttling and high cost. In view of this, the second design focus of this research is to let the query-forwarding traffic enter a particular ISP domain only if it hosts the targeted service registry so that transit traffic could be minimised [22–24].

3. *Minimised overlay process overhead*: Every time a peer receives a packet and a routing decision is made in the overlay, the packet goes until it reaches the application layer. The third design goal is to minimise this overhead by letting the query reach the overlay layer only if it is a targeted peer. Other intermediate forwarding decisions are made in the underlay itself. This approach would eliminate the involvement of intermediate registries (peers) in query processing and, in turn, the processing overhead [23].

4. *Interoperability*: The Internet is very huge and it cannot be guaranteed that the underlay-based queries always pass through AON routers. Our final design goal is to integrate seamlessly with non-AON routers, if encountered, along the path of query forwarding.

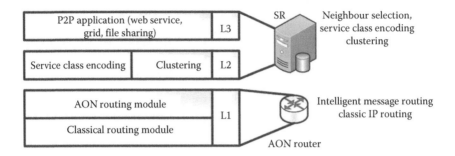

FIGURE 3.3
Layered approach—proposed P2P system.

Considering the goals listed above, a novel service discovery architecture has been designed as shown in Figure 3.3. The major modifications to the existing P2P systems are at layers 2 and 1.

As per the design goals, interoperability is one of the major concerns for the success of AON routing. Moreover, we propose AON routing as an additional service offered by the router on top of the classical routing. Consequently, at layer 1, pure IP routing has been split into two sublayers, namely, the AON routing module and the classical routing module.

- *Classical routing*: The classical routing module carries out classic (legacy) IP level routing based on network layer parameters.
- *AON routing*: The AON routing module is an additional module that is capable of carrying out IMR. IMR is a process in which a router can make routing decisions based on the content of the message instead of the IP headers. Making use of this IMR feature of the architecture is leveraged by the design at layer 2.

Layer 2 of the architecture performs two important functions, namely, service class encoding and peer clustering.

- *Service class encoding*: The services published on the registries by various participating entities such as users, organisations and so on are classified into different categories as per the Global Industry Classification Standards [24] as shown in Table 3.1. These codes in the messages are used for classifying the services. There could be a semantic encoder running at layer 2, which maps the service description to class codes. At this layer, the output needed for the approach is given a query; the service class encoder function should take the query as an input and

TABLE 3.1
Partial List of Service Classification

Industries	Class Codes
Air freight and logistics	20301010
Airlines	20302010
Hotels, resorts and cruise	25301020
Leisure facilities	25301030
Restaurants	25301040

provide a set of classes in which the query belongs as shown in Equation 3.2. The details of this semantic encoder are out of the scope of this research and from now on, it will be assumed that such an encoder exists. This assumption is valid as it will not affect the outcome of this research.

$$f_{(query)} = \{C_0, C_1 \ldots C_n\} \tag{3.2}$$

- *Clustering*: As pure unstructured P2P systems do not scale well, it has been adapted to use a hybrid structure by means of a clustering mechanism. As one of the design goals is the reduction of inter-ISP traffic, our clustering is ISP based. Peers belonging to a particular ISP are clustered together by electing a super peer. In this approach, the first service registry registering with the ISP becomes the super peer, otherwise called the SR. There are many other approaches [25] available for selecting the super peer, which can also be employed in this process. The way super peer is selected is also not included in the scope and will not affect the outcome of this research.

Layer 3 is similar to the current P2P system, where application-specific modules such as semantic matching and ontology matching are implemented.

The main features of the proposed architecture are

- The query routing mechanism for service discovery has been moved down to the IP layer. With the advent of service class encoding, the AON module at the IP layer is capable of bridging the information gap regarding the proximity of the nodes between the overlay and the underlying topology.
- The neighbour selection process is delegated to the AON module, where the neighbours are learnt dynamically.

The aspects regarding how this architecture can be made to fit with the current Internet infrastructure and the role of ISP in the proposed architectures are discussed in the following section.

3.5.3 ISP Level Perspective of the Proposed System

ISPs need to play an important role in underlay query processing as they are the key players in providing the service.

Any organisation, company or DSD system that needs to use underlay query processing would have to register themselves with the system. ISPs use the registration list to cluster those service registries in the organisation with the SR. The ISP needs to make sure that SR always uses the AON router when sending out query messages for discovery. Figure 3.4 depicts SR within an AS under the control of an ISP. Also, Figure 3.4 clearly shows the overlay links among the overlay neighbours and the physical links in the underlay. The SR in a particular AS is responsible for accepting and forwarding queries on behalf of users present in its AS.

3.5.4 Design of an Underlay Layer Query Routing Algorithm

AON routers are capable of making routing decisions based on application level messages. This feature fits quite nicely into the DSD. Any query generated from an AS is forwarded

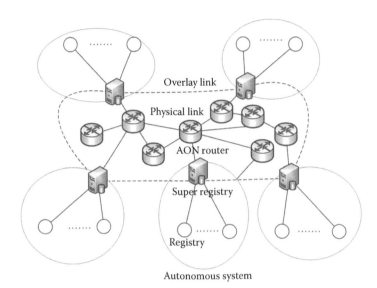

FIGURE 3.4
ISP level view of DSD architecture.

to the SR in its AS, which classifies the query into one or more of its service classes. Then, the query is encapsulated with its service class and forwarded to its nearest AON router. The packet structure in the IP layer is designed to record the interface(s) of the intermediate routers through which a particular router has received its query request or reply along with the intended source and destination IP addresses. The AON router uses these recorded fields in the header to inspect and update its AON-specific routing table, which is used for query routing. The query routing algorithm is shown in Figure 3.5.

It is important to note that the algorithm in this research does not assume that all the packets entering the router are AON packets. Instead, it checks that the incoming packet is indeed an AON packet or not. The motivation behind this implementation is due to the

```
Algorithm: Routing
Input: packet
If→packet is AON
    If→packet is DSD
        If→packet is a request
            If→AON routing table entry exists
                Route as per the entry
            Else
                Forward to all egress ports
        Else
            Update the AON routing table
            Forward the packet in the return path
    Else
    Forward to the appropriate AON module
Else
Perform classical routing
```

FIGURE 3.5
Query routing algorithm.

fact that the existing services on the Internet are not to be interrupted due to the AON service. The AON service has been implemented as an additional service on the existing infrastructure of the Internet.

There are two stages in the query routing process, namely, the learning state and steady state. During the learning state, the AON routing table would be empty, and hence all incoming queries are forwarded to all egress interfaces of the router, which is in contrast to classical routers where the packets are dropped if there is no route to a particular destination. During the reply stage of this routing process, routers would eventually learn the routes for the service classes by inspecting the packet for the preceding node through which the query reply has been received. It is worth noting that a single query can span one or more classes depending on the user request. Once the routes are learned, the router enters the steady state and any further queries are forwarded in a single (common) path without duplication to a certain router from which the queries are distributed to multiple targets. This particular router is termed as a proxy router that forwards the query on behalf of the query initiator.

3.5.5 Underlay (AON) Query Routing Process

The current IP layer packet structure is designed to record the interface(s) of the intermediate routers through which a particular router has received its query and reply along with the intended source and destination IP addresses. This feature can be easily incorporated with the help of extension headers in the IPv6 format or within the payload in the IPv4 format. The AON router uses this feature to inspect and update these fields and its AON-specific routing table that is used for selective query forwarding to multiple SRs. This approach, which is multicast in nature, does not maintain any group memberships and forwarding is purely based on the AON-specific routing table. Our design is in such a way that the whole system, including the AON router, is interoperable with the existing classical routers in the Internet. There is every chance that the AON router will encounter non-AON packets as well. For this reason, every packet is inspected for its type as soon as it enters the router. The possible scenarios that could be encountered during query forwarding are depicted in Table 3.2.

Figure 3.6 illustrates a scenario of four ASs, each with their own SRs, connected via AON routers. A query forwarded from an AS is received by the border routers of the ASs, in this case, router R1. Figure 3.6 demonstrates a sample query forwarding from SR1 to SR2, SR3 and SR4 during the learning state. In the initial state, the AON routes are not yet learnt, and therefore the query is forwarded through all interfaces of the router except for the incoming interface. It is worth noting here that when queries are forwarded from one peer to another, a connectionless protocol such as UDP is used. There is no need to maintain the connection as the queries are independent of each other. The looping of the query is

TABLE 3.2

Possible Scenarios Encountered in Query Forwarding

Router	Packet	Remark
AON	AON	Routing based on extension headers (message level intelligence)
AON	Non-AON	Classical routing based on IP header
Non-AON	AON	Classical routing based on IP header
Non-AON	Non-AON	Classical routing based on IP header

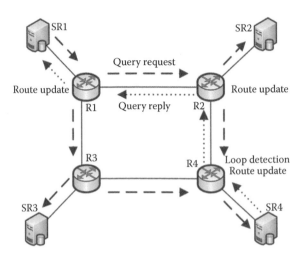

FIGURE 3.6
Query routing in learning state.

prevented by providing a unique ID for each query so that the router does not forward the query if it has already done so. During this state, the AON routing table is updated along the path of the reply, as shown in Equation 3.3, where r refers to the existing routes of the routing table RT and r_{new} refers to the new routes learnt through the query reply Q_{reply}. Figure 3.7 demonstrates a typical scenario of the steady state in query forwarding. SR1 forwards the query to its border router R1. AON router R1 inspects the query and finds that this query should be forwarded to R2 as per its AON routing table and this process continues until the query reaches its destination. The same process is used if a query could be answered by multiple SRs, in which case queries are forwarded to more than one SR.

$$\text{routing\_update} = RT \cup Q_{reply} \mid r \in RT, r_{new} \in Q_{reply} \tag{3.3}$$

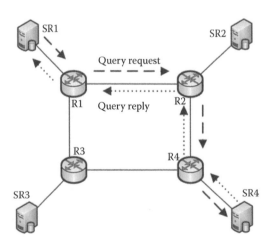

FIGURE 3.7
Query routing in steady state.

TABLE 3.3
Routing Analysis for the Given Scenario

Query Propagation Method	Number of Peers Involved	Number of Links Involved
Flooding (selects all 3 neighbours)	4 (SR1, SR2, SR3 and SR4)	13 (SR1 → R1 × 3, R1 → R2 × 2, R1 → R3 × 2, R2 → R4, R3 → SR3, R4 → SR4 × 2, R3 → R4 and R2 → SR2)
Random/probabilistic (selects 2/3 neighbours)	3 (SR1, SR2 and SR4)	7 (SR1 → R1, R1 → R2 × 2, R2 → SR × 2, R2 → R4 and R4 → SR4)
AON based	2 (SR1, SR4)	4 (SR1 → R1, R1 → R2, R2 → R4 and R4 → SR4)

In pure overlay-based routing, for instance, Gnutella-like systems, considering the worst-case scenario (flooding), a query from SR1 to SR2, SR3 and SR4 would generate traffic along the paths SR1 → SR2, SR1 → SR3 and SR1 → SR4. Particularly, the link R1 → R2 and R1 → R3 would carry the same request twice as the routing is performed in the overlay. However, if AON routing is employed, the traffic generated is just along the path SR1 → SR4. Even during the learning stage, the redundant queries are eliminated. This clearly illustrates that ineffective query propagation could effectively be overcome by AON to improve the efficiency of the search mechanism. The same can be visualised in terms of inter-ISP traffic as well. In our illustration, the only inter-ISP traffic is from the source to the AS in which the target SR resides, as the intermediary peers are not involved in query forwarding, whereas in overlay routing, all the four peers belonging to different ISPs are involved in query processing. The performance can also be improved in the case of the other query-forwarding heuristics such as random or probabilistic selection of neighbours that is summarised in Table 3.3. In this and the previous sections, the role of an AON router and what it has to accomplish has been elaborated. The design of the AON router follows next.

3.5.6 Design of AON Router

Figure 3.8 shows the structure of the AON router inspired from Ref. [14]. It has been developed for the purpose of message level routing. The main additional components to the

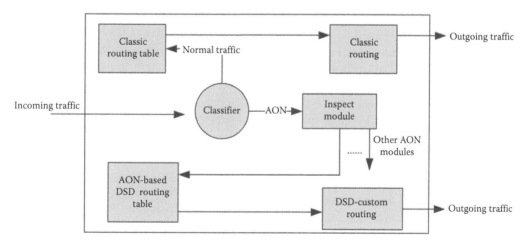

FIGURE 3.8
Model of an AON router.

normal router are the classifier, AON routing module and AON routing table. In the AON router, all incoming packets are inspected for their type as shown in Equation 3.4.

The classifier first inspects the incoming packet to determine if it is normal or AON based; if it is normal, the packet is diverted to the classic routing module. If the packet is AON, it needs to be further checked for what specific service needs to be offered to AON. The specific service in our case is DSD; however, in future, there may be more AON services that could be offered for the application layer.

$$R_{p_{in}} = \begin{cases} \text{Routing}_{\text{Classic}} \text{ if } (p_{in} \neq \text{AON}) \\ \text{Routing}_{\text{AON}} \text{ if } (p_{in} = \text{AON}) \end{cases} \quad (3.4)$$

3.5.7 AON-Based Query Message Structure

So far, the AON router design and the algorithm design for the query routing have been seen. Now, let us have a closer look at the AON message structure that is used in the query routing process. The AON-based query routing message has been designed as shown in Figure 3.9. This message body is sent by encapsulating the IP header. Once the router has identified the AON packet, it makes use of the other fields in the message for IMR as per the algorithm in Figure 3.5.

- *Request/reply*: The query routing algorithm makes use of this field to distinguish between the request and the reply message. During the reply messages, the AON-based routing table is updated.
- *Timestamp*: This field is used to calculate the RTT.
- *Query message*: This field is used to carry the actual query message that needs to be searched. This contains the service class information that is used by the AON router.
- *Result*: This field is used to store and send the result of the search back to the source.
- *Is AON*: This field is used to identify the AON packet. For the sake of this simulation, the AON field has been built inside the AON message itself. However, in real-time implementation, this could be applied into IPv6 extension headers.
- *Previous node*: This field is used to learn the previous node that the router has passed through. This information, along with the service class, is used for updating the AON routing table during the reply path.
- *Hop count*: This field is used for the purpose of data collection regarding the number of hops crossed by each query. In real-time mode, this field would not be necessary.
- *Query ID*: This field is used for preventing loops in the query forwarding.

Request/Reply Integer: 0/1	Time stamp, double	Query message, String	Result, String	Is AON, Boolean	Previous node, String	Hop count, Integer	Query ID
32 bits	64 bits	n x 16 bits	n x 16 bits	1 bit	32 bits	32 bits	32 bits

FIGURE 3.9
AON query structure.

3.6 Performance Comparison of Underlay versus Overlay Routing

3.6.1 Hypothesis 1

The routing complexity in the underlay is always less than that of the overlay.

Proof 1:

Figures 3.10 and 3.11 depict the connection of SRs in the underlay via routers. The structure can be modelled to be a tree of order 2 with n levels for the sake of this analysis.

Analysis scenario: Let us consider that a user issues a query in the DSD system. The SR in the AS distributes the query into the system. Let us also assume that there are eight neighbours to the SR and that the topology is as shown in Figure 3.10.

Let us consider the worst-case scenario that all the neighbours are well distributed in the underlay as shown in Figure 3.10 for the overlay routing and Figure 3.11 for the underlay routing.

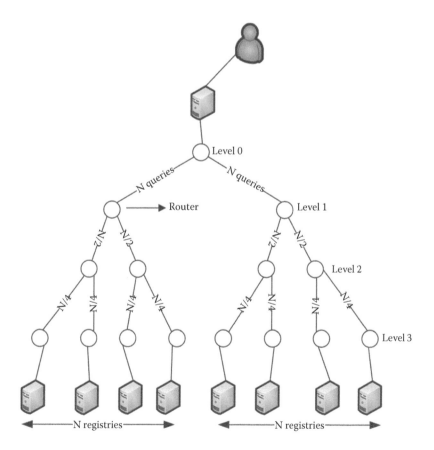

FIGURE 3.10
Overlay routing scenario.

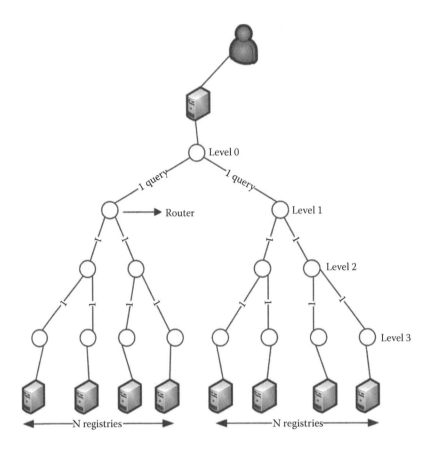

FIGURE 3.11
Underlay (AON) routing scenario.

3.6.1.1 Estimation of the Number of Routers Crossed in the Overlay

Let n be the number of levels in the tree with a degree of 2. Then, the number of leaf nodes in the tree can be written as 2^{n-1}. This value also corresponds to the number of neighbours. The number of routers passed by overlay routing is $n \cdot 2^{n-1}$.

It is worth noting here that though the number of routers in each level decreases when moving from the leaf nodes to the root, the number of query messages passing through each level remains the same, which is shown in Figure 3.10. This is due to the redundant transmission that is caused by the underlay's ignorance of overlay routing.

3.6.1.2 Estimation of the Number of Routers Crossed in the Underlay

As shown in Figure 3.11, there is no duplicate query transmission in the links in the underlay when employing AON-based routing. This is due to the fact that the query messages are forwarded as per the AON routing table in the AON router and it is clearly aware of which interface the query needs to be forwarded to. The SR submits the query just once to its connected AON router; thereupon, the AON routers handle query routing. In contrast to this, in overlay routing, query routing is carried out by the end peers

(SRs). Keeping these factors in mind, now let us estimate the number of routers crossed in underlay routing.

- The number of leaf nodes in this scenario would be the same, which is 2^{n-1}.
- The number of routers crossed in each level of the tree is 0, 2, 4 and 8 for the given topology.
- The number of routers crossed can be generalised as $2^{n-1} + 2^{n-2} + 2^{n-3} + \cdots + 2^0$.

3.6.1.3 Complexity Analysis

The numbers of routers crossed have been estimated both in overlay and in underlay routing. To analyse the complexity of query processing, the complexity of routing a packet in the router needs to be found.

The complexity of routing a packet is given as $O(1)$ [26].

Knowing that if the complexity of routing a packet in a router is $O(1)$, then the complexity analysis of the routing process solely depends on the number of routers crossed.

Therefore, in the overlay, the complexity would be $O(n \cdot 2^{n-1})$, whereas that of underlay routing is $O(2^{n-1} + 2^{n-2} + 2^{n-3} + \cdots + 2^0) \approx O(2^{n-1})$ [26,27].

$O(2^{n-1}) < O(n \cdot 2^{n-1})$, and hence the proof.

3.6.2 Hypothesis 2

The number of links utilised in underlay routing is always less than that in overlay routing.

Proof 2:

Let us consider that a query needs to be forwarded from registry r_0 in AS0 to r_2 in AS2. Let LST be the number of links between peer registries r_0 and r_2.

3.6.2.1 Estimation of the Number of Links Utilised in the Overlay

As shown in Figure 3.12, if a peer has n neighbours, then the total number of links nl utilised at the overlay is given by

$$nl = LST_1 + LST_2 + LST_3 + \cdots + LST_n = \sum_{i=1}^{n} LST_i \tag{3.5}$$

3.6.2.2 Estimation of the Number of Links Utilised in the Underlay

In AON-based query routing in the IP layer, LST could be divided into LSP and LPT as shown in Figure 3.13, where LSP is the number of links between the source and the proxy and LPT is the number of links between the proxy and the given target.

If there is more than one neighbour in the overlay for a node, the queries are sent redundantly in the underlay layer. To prevent the redundant query propagation in the underlay, this AON-based routing algorithm identifies a particular router until which the query need not be duplicated by way of updating the AON routing entries. In these scenarios, router C is termed as the proxy router from which the fanning out of the queries to the multiple registries starts. In general, during the steady state of AON routing,

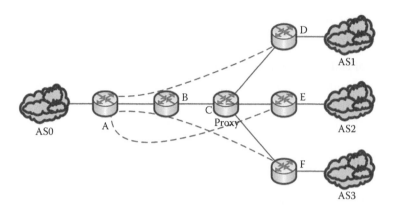

FIGURE 3.12
Links utilised—overlay routing.

proxies from all directions are learnt by means of updating the AON routing table during the return path. The proxies are learnt dynamically depending on the current state of the system.

Thus, the total number of links utilised in the AON-based IP layer is

$$nl = LSP + LPT_1 + LPT_2 + \cdots + LPT_n = LSP + \sum_{i=1}^{n} LPT_i \qquad (3.6)$$

The percentage of the savings in terms of link utilisation for AON-based query processing to that of the overlay is

$$\frac{LSP + \sum_{i=1}^{n} LPT_i}{\sum_{i=1}^{n} LST_i} \qquad (3.7)$$

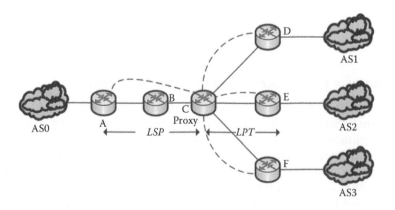

FIGURE 3.13
Links utilised—AON routing, learning state.

Underlay-Aware Distributed Service Discovery Architecture

From Equation 3.7, the complexity of query forwarding can be inferred as

$$\frac{O(1) + O(n)}{O(n)} \quad (3.8)$$

On the basis of Equation 3.8, it can be observed that a higher percentage of saving can be obtained if nl_{pt} is much smaller compared to nl_{st}, so that the complexity of query forwarding could be brought closer to

$$\frac{O(1) + O(1)}{O(n)} \approx \frac{O(1)}{O(n)} \quad (3.9)$$

To achieve a high percentage of reduction in the underlying traffic, the proxy should be as close as possible to the target registries, which would result in shorter nl_{pt} compared to that of nl_{sp}. This is achieved by the current AON-based implementation as it detects the closest possible proxy to the target.

3.6.3 Hypothesis 3

Redundant traffic caused in the overlay is eliminated by the underlay query routing process.

Proof 3:

Let us consider Figures 3.12 and 3.13 for this proof.

If a query message is sent from the source (AS0) to different destinations (AS1, AS2 and AS3), then the redundant traffic can be measured by identifying the common path between the source and the destinations.

From Equation 3.5, it can be inferred that the overlay routing cannot differentiate between *LSP* and *LPT*, and therefore redundancy is caused by sending the same query messages, a number of times that is equal to the number of destinations.

However, underlay query routing is capable of differentiating between *LSP* and *LPT* and does not duplicate query messages along the path of *LSP*.

3.6.4 Hypothesis 4

The inter-ISP traffic is always reduced in underlay query routing.

Proof 4:

Let us consider Figures 3.12 through 3.14 for proving this.

Let us assume that AS0, AS1, AS2 and AS3 belong to different ISPs. The inter-ISP traffic could be measured by ASs that receive the query messages.

3.6.4.1 Inter-ISP Traffic in Overlay Routing

In Figure 3.12, routers B and C that are in between the source and the destination would be in different ISPs and therefore handle the transit traffic of query routing.

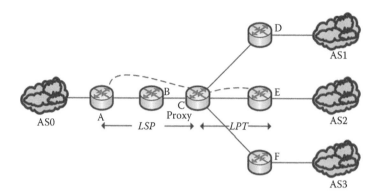

FIGURE 3.14
Links utilised—AON routing, steady state.

Overlay routing is not capable of detecting the transit ISPs, and therefore the inter-ISP traffic is proportional to the number of target registries m and the number of routers n along the path, which can be expressed as

$$m \cdot n \tag{3.10}$$

3.6.4.2 Inter-ISP Traffic in Underlay Routing

Let x be a factor by which the proxy node C is closer to the target registry. In other words, x represents *LPT* in Figure 3.13. Therefore, with reference to Figure 3.13, $n - x$ corresponds to the portion of the topology *LSP* and $m \cdot x$ corresponds to the portion of the topology *LPT*. Hence, the inter-ISP traffic for the underlay routing is as shown below in Equation 3.11, which is reduced to a factor of x

$$(n - x) + m \cdot x \tag{3.11}$$

If $x = 0$ (negligible) when compared to the distance between the source and the target, then there is a higher percentage of reduction in the inter-ISP traffic.

Moreover, routers D, E and F are the destination routers that receive the query message. Among D, E and F, only a subset, for example, {D,E}, {E,F}, {D,F}, {D}, {E} or {F} and in the worst case, {D,E,F}, would have the results for the queries that are received. Inter-ISP traffic is reduced in this approach as well as depicted in Figure 3.17 in the case of underlay routing.

3.6.5 Hypothesis 5

The time taken for query routing in the underlay is less than that of the overlay.

Proof 5:

Let μ be the processing time taken by the overlay registry, α the processing time taken by the routers and β the time taken for each of the link propagations.

Then, a query to be sent to m neighbours by a source registry in overlay query routing is given by the following equation:

$$m[2\mu + (n + 1)\beta + n\alpha] \tag{3.12}$$

where n is the number of routers in between the source and the targets.

Let us now calculate the time taken for a query to be sent to m neighbours in the underlay. The nodes in the underlay can be divided into two sections: as *LSP* and *LPT* as per Equation 3.6.

A query to be sent from a source to a proxy node is given by

$$\mu + (n - x + 1)\beta + (n - x)\alpha$$

A query to be sent from the proxy to m neighbours is given by

$$m[\mu + (x + 1)\beta + x\alpha]$$

Therefore, in underlay query routing, a query to be sent from a source registry to all its neighbours is given by the following equation:

$$\mu + (n - x + 1)\beta + (n - x)\alpha + m[\mu + (x + 1)\beta + x\alpha] \tag{3.13}$$

Here again, it can be clearly observed that the time taken is considerably reduced depending on the location of the proxy. In Equation 3.13, x is the fraction that determines the factor of the proximity of the proxy router to the target registries.

3.7 Limitations of the Study

Router performance: The AON router model was designed to have additional modules such as a packet classifier, an AON-specific routing table and an AON-specific inspection module. These additional modules could cause overhead to the routers and would demand higher processing and memory capacity from the routers. However, it has been argued that with the tremendous increase in processing power and memory capacity of the current routers, this issue could be resolved.

Involvement of ISPs: The design and implementation of the AON-based DSD relies on installing AON routers on the Internet backbone, which could only be provided by service providers and not by consumers. Therefore, ISPs play an important role in providing AON services to its customers. There could be additional costs incurred by the ISPs for providing the AON infrastructure to the customers. If ISPs are not willing to provide the AON services, then the success of the system cannot be guaranteed in real-time implementations.

Involvement of the participating organisations: The routing algorithm employed depends on looking into the contents of query messages. The AON routers would inspect every AON packet that passes through. Therefore, there needs to be a good understanding in terms of security and reliability between the service provider and the clients. A security mechanism has been provided for AON implementation to keep the messages from the prying eyes of intruders. However, the router itself needs to look into the messages, which is only possible with clients willing to participate in the DSD process.

3.8 Conclusion

This chapter provided a novel implementation of underlay query routing for the DSD process. It elaborated how AON is implemented for incorporating IMR. It also provided a novel query routing algorithm that runs in the underlay. The benefits obtained by the proposed architecture and the query routing algorithm were proved hypothetically.

The following are the possible future works that need attention for the successful implementation of the proposed system in real time and also to study the possibility of the emerging AON technology in other domains such as computational grids and cloud computing.

1. There is a need to study how ISPs could be encouraged to provide AON services. The reduction of the cost due to the reduced inter-ISP traffic could be one of the incentives.
2. The focus in this research has been on the DSD process. Its applicability in file sharing and downloading systems such as BitTorrent needs to be studied.
3. In this research, the clustering mechanism is based only on ISPs. The performance of the system while clustering the peers purely based on service classes needs to be studied.
4. The effect of the clustering of the registries across ISPs also needs to be studied.
5. In this research, the system built for testing employs hybrid architecture with an unstructured backbone. However, the possibility of employing IMR in structured distributed systems needs to be studied.

References

1. S. James et al., IMP: ISP-managed P2P, in *IEEE 10th International Conference on Peer-to-Peer Computing (P2P)*, Delft, Netherlands, 2010.
2. D. B. Hoang, H. Le and A. Simmonds, PIPPON: A physical infrastructure-aware peer-to-peer overlay network, presented at *TENCON 2005, IEEE Region* 10, 2005.
3. L. Garcés-Erice et al., Topology-centric look-up service, in *Group Communications and Charges. Technology and Business Models*. Vol. 2816, B. Stiller et al., eds., Springer, Berlin, 2003, pp. 58–69.
4. R. Ferreira, A. Grama and S. Jagannathan, Plethora: An efficient wide-area storage system, in *High Performance Computing—HiPC 2004*, Vol. 3296, *Lecture Notes in Computer Science*, L. Bougé and V. Prasanna, eds., Springer, Berlin, 2005, pp. 305–310.
5. H. Xie et al., P4P: Provider portal for applications, presented at the *Proceedings of the ACM SIGCOMM 2008 Conference on Data Communication*, Seattle, WA, USA, 2008.
6. J. Seedorf et al., Traffic localization for P2P-applications: The ALTO approach, in *P2P '09. IEEE 9th International Conference on Peer-to-Peer Computing*, Seattle, USA, 2009, pp. 171–177.
7. S. Xin et al., SLUP: A semantic-based and location-aware unstructured P2P network, in *10th IEEE International Conference on High Performance Computing and Communications, HPCC '08*, Dalian, China, 2008, pp. 288–295.
8. R. Bindal et al., Improving traffic locality in BitTorrent via biased neighbor selection, in *26th IEEE International Conference on Distributed Computing Systems, ICDCS 2006*, 2006, pp. 66–76.
9. G. Shen et al., HPTP: Relieving the tension between ISPs and P2P, in *USENIX IPTPS*, 2007.
10. G. Zhengwei et al., P4P Pastry: A novel P4P-based Pastry routing algorithm in peer to peer network, in *The 2nd IEEE International Conference on Information Management and Engineering (ICIME)*, Chengdu, China, 2010, pp. 209–213.
11. A. Rowstron and P. Druschel, Pastry: Scalable, decentralized object location, and routing for large-scale peer-to-peer systems middleware, In *Lecture Notes in Computer Science*, Vol. 2218, R. Guerraoui, ed., Springer, Berlin, 2001, pp. 329–350.

12. A. A. Bare, A. P. Jayasumana and N. M. Piratla, On growth of parallelism within routers and its impact on packet reordering, presented at *15th IEEE Workshop on Local & Metropolitan Area Networks, LANMAN 2007*, Princeton, NJ, 2007.
13. Cisco-white-paper, The evolution of high-end router architectures: Basic scalability and performance considerations for evaluating largescale router designs, Available: www.cisco.com/en/US/products/hw/routers/ps167/products_white_paper09186a0080091fdf.shtm.
14. CISCO, Cisco AON: A network embedded intelligent message routing system, Available: www.cisco.com/en/US/prod/collateral/modules/ps6438/prod_bulletin0900aecd802c201b.html.
15. Y. Cheng et al., Toward an autonomic service management framework: A holistic vision of SOA, AON, and autonomic computing, *IEEE Communications Magazine*, 46, 138–146, 2008.
16. T. Xiaohua et al., Multicast with an application-oriented networking (AON) approach, in *IEEE International Conference on Communications, ICC '08*, 2008, Beijing, China, pp. 5646–5651.
17. J. Yao et al., Intelligent message scheduling in application oriented networking systems, in *IEEE International Conference on Communications, ICC '08*, 2008, Beijing, China, pp. 5641–5645.
18. K. Zille Huma, A.-F. Ala and G. Ajay, A service location problem with QoS constraints, in *Proceedings of the 2007 International Conference on Wireless Communications and Mobile Computing%@ 978-1-59593-695-0*. Honolulu, Hawaii, USA: ACM, 2007, pp. 641–646.
19. S. Ion et al., Chord: A scalable peer-to-peer lookup service for Internet applications, *SIGCOMM Computing Communications* 31, 149–160, 2001.
20. P. Maymounkov and D. Mazières, *Kademlia*: A peer-to-peer information system based on the XOR metric peer-to-peer systems. Vol. 2429, P. Druschel et al., eds., Springer, Berlin, 2002, pp. 53–65.
21. J. Seedorf et al., Traffic localization for P2P-applications: The ALTO approach, in *P2P '09. IEEE 9th International Conference on Peer-to-Peer Computing*, Seattle, USA, 2009, pp. 171–177.
22. T. Karagiannis et al., Should Internet service providers fear peer-assisted content distribution?, in *Proceedings of the 5th ACM SIGCOMM Conference on Internet Measurement*, USENIX Association, Berkeley, CA, 2005, pp. 6–7.
23. E. Meshkova et al., A survey on resource discovery mechanisms, peer-to-peer and service discovery frameworks, *Computer Networks*, 52, 2097–2128, 2008.
24. MSCI, Global industry classification standard, Available: www.msci.com/products/indices/sector/gics/.
25. C. Mastroianni et al., A super-peer model for building resource discovery services in grids: Design and simulation analysis, in *Advances in Grid Computing–EGC 2005*, Vol. 3470, P. M. A. Sloot et al., eds., Springer, Berlin, 2005, pp. 132–143.
26. A beginner's guide to big Oh notation, Available: rob-bell.net/2009/06/a-beginners-guide-to-big-o-notation/.
27. Some rules for big-Oh notation, Available: www.augustana.ualberta.ca/~hackw/csc210/exhibit/chap04/bigOhRules.html.

4

System-Level Performance Simulation of Distributed Embedded Systems via ABSOLUT

Subayal Khan, Jukka Saastamoinen, Jyrki Huusko, Juha Korpi, Juha-Pekka Soininen and Jari Nurmi

CONTENTS

4.1 Introduction 92
4.2 Introduction to ABSOLUT 94
    4.2.1 Structure of Application Workload Models 95
    4.2.2 Application Workload Modelling Techniques 96
    4.2.3 Platform Modelling 96
    4.2.4 Mapping and Co-Simulation 96
4.3 Requirements for SLPE of Distributed Applications 97
    4.3.1 Requirements for Performance Evaluation of Distributed Systems 97
    4.3.2 Enhancements Needed for ABSOLUT 99
4.4 Modelling Multithreaded Support 100
    4.4.1 System Synchronisation 101
    4.4.2 Inter-Process Communication and System Synchronisation Model 102
4.5 Modelling ABSOLUT OS_Services 103
    4.5.1 Deriving New OS_Services 103
    4.5.2 Registering OS_Services to OS Models 104
    4.5.3 Accessing OS Services 104
    4.5.4 Data Link and Transport Layer Services 106
    4.5.5 Modelling Guidelines and Lessons Learnt 106
4.6 Physical Layer Models 108
    4.6.1 Analysing Accuracy of Bit Error Rate Calculation 108
    4.6.2 Analysing Accuracy of Frame Error Rate Calculation 109
    4.6.3 Analysing Accuracy of Packet Error Rate 109
4.7 Workload Modelling via Run-Time Performance Statistics 111
    4.7.1 Workload Extraction via CORINNA 111
    4.7.2 Comparing ABSINTH, ABSINTH-2 and CORINNA 113
4.8 Conducted Case Studies 114
4.9 Conclusion and Future Work 115
References 116

4.1 Introduction

An embedded system can be defined as a special-purpose computing system (meant for information processing), which is closely integrated into the environment. An embedded system is generally dedicated to a particular application domain. Therefore, the embedding into a technical environment and the constraints that are a consequence of their application domain mostly result in implementations that are both heterogeneous and distributed. In such cases, the systems comprise hardware components that communicate by means of an interconnection network [1].

Owing to the dedication to a particular application domain, heterogeneous distributed implementations are common. In such implementations, each node specialises by incorporating communication protocols and other functionalities that facilitate optimum and reliable performance in its local environment [1]. For example, in automotive applications, each network node (usually called embedded control units) contains a communication controller, a central processing unit (CPU), memory and I/O interfaces [1]. But as per functionality, a particular node in the network might contain additional hardware resources such as digital signal processors, CPUs and a different memory capacity [1].

Distributed embedded systems can be classified as real-time and non-real-time distributed systems. Real-time systems are required to complete their tasks or deliver their services within a certain time frame. In other words, the real-time systems have strict timing requirements that must be met. Digital control, signal processing and telecommunication systems [2] are usually distributed real-time systems. On the other hand, personal computers (PCs) and workstations that run non-real-time applications, such as our email clients, text editors and network browsers, are common examples of distributed non-real-time systems.

Also, the networking revolution is driving an efflorescence of new distributed systems for new application domains, for example, telesurgery, smart cars, unmanned air vehicles and autonomous underwater vehicles. The components of these systems are distributed over the Internet or wireless local area networks (LANs) [2]. Owing to these technological advancements, the spatial limitations seem to be progressively fading away, which has given rise to new paradigms such as mobile computing. These technologies have enabled us to connect to the Internet while we are on the move via pocket-sized, battery-powered embedded devices, for example, personal digital assistants (PDAs) and cellular phones, which communicate over a wireless channel. The applications of computing devices have also been changing in response to ever-improving hardware and software technologies [2]. Nowadays, we routinely use a variety of multimedia-based services instead of text-based ones by using nomadic hand-held devices such as high-end mobile phones.

The burgeoning market for information, entertainment and other content-rich services can be seen as a consequence of the rising popularity of mobile devices with high-quality multimedia capabilities [3]. These services should not just adapt to a continuously changing computing environment but also meet the different requirements of individual users. In such cases, we need to consider additional real-time and embedded non-functional properties of multimedia applications, for example, the maximum allowable time for each delivered packet and battery life. Therefore, the challenge from the system design perspective is to reduce the form factor and energy consumption of the mobile nomadic devices, thus increasing the portability and durability of these devices. This will enable these devices to be used by many customers for everyday use by maintaining low power consumption, which is important due to the limited power available from the battery. Since Moore's law predicts that the computing power will continue to increase, the energy constraints demand that we shall sacrifice the performance in portable devices in return for a longer operation time. Generally speaking, the focus of recent research has been topics such as efficient usage of storage space, various I/O devices [4–6] and processing elements available from the device platforms. In distributed embedded systems, the end–end communications between applications is provided by transport protocols or layer 4 of the open systems interconnection (OSI) model. It provides different services to the applications such as stream support, reliability, flowcontrol and multiplexing. The most widely used transport protocols are the transmission control protocol (TCP) and user datagram protocol (UDP). TCP is used for connection-oriented transmissions, whereas UDP is connectionless and is used for simpler messaging transmissions. TCP is a more complex protocol due to its state-based design. It incorporates reliable transmission and data stream services.

Wireless sensor networks (WSNs) are also an example of distributed real-time embedded systems that are composed of a cooperative network of nodes [7]. Owing to the small form factor of the network nodes, each consists of limited processing capability (for example, microcontrollers, CPUs or digital signal processor (DSP) chips) and memory (program, data and flash memories) resources. Each node has a radio frequency (RF) transceiver, a power source (e.g., batteries and solar cells) and contains sensors and/or actuators. The nodes communicate wirelessly and have the ability to self-organise after ad hoc deployment. WSNs of 1000s or even 10,000 nodes are anticipated and are perceived to revolutionise the way we live and work. Since WSNs are distributed real-time systems that are rapidly evolving technologically, an important question is to know how many existing solutions (transport protocols and data link protocols, etc.) for existing distributed and real-time systems can be used in these systems. It has become obvious that many protocols that were developed beforehand will not perform well in the domain of WSNs. The reason is that WSNs do not employ many of the assumptions underlying the previous networks, for example, medium access control (MAC) protocols.

An MAC protocol is employed by the network nodes for the coordination of actions over a shared channel. The most commonly used MAC protocols are contention-based. One generally used distributed contention-based strategy is that a node that has a message to transmit tests the channel to see if it is busy; if the channel is not busy, it transmits, and if it is busy, it waits and tries again later. In most cases, MAC protocols are optimised for the general cases and arbitrary communication patterns and workloads. Contrarily, WSNs have more specific requirements that include a local unicast or broadcast. The traffic flow is usually from many nodes towards one or a few sinks (most traffic is thus directed in one direction). The individual nodes have periodic or rare communication and must consider energy consumption as a major factor. An effective MAC protocol for WSNs must have reduced power consumption, shall avoid collisions, should be implemented with a small code size and memory requirements, be efficient for a single application and be tolerant to changing radio frequency and networking conditions [2]. That is why many WSNs employ highly efficient MAC protocols for the transfer of frames over the wireless channels, for example, NANO MAC [8] and BMAC [9].

Many modern distributed systems contain networked embedded devices that contain multicore processors. For some of these devices, it might be necessary to remain responsive to inputs. Multithreading is a programming model that allows multiple threads to exist within the context of a single process. This allows a multithreaded program to operate faster on a device that has multiple or multicore CPUs. In some systems, the system functionality dictates that the application has to remain responsive to inputs. One convenient way to resolve this issue is to allow one or more threads to monitor the inputs and execute in parallel with the main execution thread of the application. This enables the application to retain responsiveness to input(s) while executing tasks in the background.

Owing to the various communication technologies (operating at the transport and data link layers of the OSI model) and multitude of distributed multithreaded applications supported by distributed systems, the overall complexity of the distributed embedded systems in many application domains is huge. To simplify the design of these complex systems, the methodology used for architectural exploration must provide the models that will enable efficient design space exploration by validating the non-functional properties of the system. The methodology should also report the contribution of protocols at different layers of the OSI model in non-functional properties of the system such as end–end communication delays. Therefore, the early-phase performance simulation of the distributed computer systems must use functionally accurate models of MAC and transport protocols to provide estimates of non-functional properties that are accurate enough for taking design decisions at early phases of the system development and architectural exploration.

The rest of the section is organised as follows. Section 4.2 provides an overview of the ABSOLUT SLPE approach. Section 4.3 describes the requirements that must be fulfilled by a performance evaluation approach to be applied for the performance evaluation of distributed systems. The section also describes the models and tools that must be integrated to ABSOLUT to extend it to the SLPE of distributed systems. The tools and models already provided by ABSOLUT are listed while the tools and models that the ABSOLUT approach lacks are identified. These missing tools and models (which were not provided by ABSOLUT), identified in Section 4.3, are described in Sections 4.4 through 4.7. Section 4.8 provides an overview of the case studies conducted for validating the approach. Conclusions and references are provided in the end. The research articles that provide a detailed description of the extensions made to ABSOLUT are mentioned in Section 4.9.

4.2 Introduction to ABSOLUT

ABSOLUT uses the Y-chart approach [10] for SLPE and consists of the application workload model and execution platform as shown in Figure 4.1.

The complete performance model is formed by mapping the application workload models to the execution platform model, which is simulated to obtain the performance

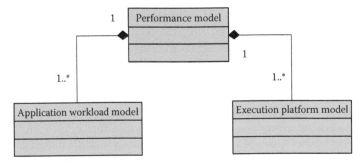

FIGURE 4.1
Main parts of an ABSOLUT performance model.

numbers. The performance numbers are analysed by the system designer. If the results do not meet the design constraints, the platform models or application models or both are changed in the next iteration.

4.2.1 Structure of Application Workload Models

The workload models consist of three layers, that is, main workload, application workload and function workload, as shown in Figure 4.2.

The topmost layer consists of the main workload that is composed of one or more application workloads, each of which corresponds to an application supported by the system:

$$W = \{Ca, A1, A2...An\}, \qquad (4.1)$$

where $A1, A2...An$ represent different application workload models and Ca is the control. In the second layer, each application workload is refined to one or more (platform-level) services or process workload models. Each of these (service or process) workloads is denoted by Pi:

$$Ai = \{Cp, P1, P2...Pn\}, \qquad (4.2)$$

where Cp is the control and $P1, P2...Pn$ show service or process workload models. In the third layer, each process or service workload is represented as a composition of one or more function workloads:

$$Pi = \{Cf, F1, F2...Fn\}, \qquad (4.3)$$

where Cf is the control between function workload models, that is, $F1, F2...Fn$. The ABSOLUT operating system (OS) model of the platform handles the scheduling of workloads at the process level. The function workloads are control flow graphs:

$$Fi = (V, G), \qquad (4.4)$$

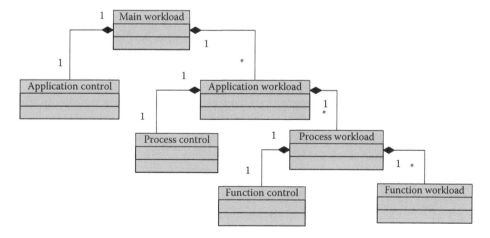

FIGURE 4.2
The ABSOLUT application workload model layers.

where the nodes $v_i \in V$ represent the basic blocks and $g_i \in G$ represent the branches. Each basic block is an ordered set of load primitives used for load characterisation.

4.2.2 Application Workload Modelling Techniques

ABSOLUT application workload models can be generated via three methods, that is, analytical, measurement-based, trace-based and compiler-based workload generation methods. The analytical workload generation requires moderate modelling effort and is based on the analysis or functional description of the algorithms. After analysis, the number of operations required to perform the application tasks is estimated. These operations are used to estimate the number of abstract instructions in the corresponding ABSOLUT workload models [4]. The measurement-based workload modelling technique is based on the extraction of data from the partial traces of the modelled use cases. The trace-based application workload generation method generates workloads by tracking the instructions executed by the processors while running the modelled application [4].

The compiler-based workload generation method uses a tool called abstract instruction extraction helper (ABSINTH) [4]. It is based on GNU compiler collection (GCC) version 4.5.1 with additional passes in the compiler middle end, which enable workload model generation. The workload generation takes place in three distinct phases. In the first phase, application source code is compiled via ABSINTH with profiling information extraction enabled. These data are used by ABSINTH in the second phase for statistical modelling of branch probabilities and extraction of a number of loop iterations. In the second phase, the selected use case is executed by running the application to produce the profiling data. In the last phase, the source code is recompiled to produce workload models that are based on actual true execution of the application [4]. ABSINTH generates one workload model for each function in the application source code. These models can contain calls to other function workloads without the knowledge of implementations of these workloads. Before compiling the models for simulation, they are post-processed with the ABSINTH manager. It is a Python script that detects function dependencies from a set of workload functions and modifies the files by linking them in the order in which the functions were called in the application.

4.2.3 Platform Modelling

The platform model is also layered and consists of three layers, that is, component layer, subsystem layer and the platform architecture layer, as shown in Figure 4.3.

The component layer is composed of processing, storage and interconnection elements. The subsystem layer is built on top of the component layer. This layer shows the components of the system and the way they are connected. The platform architecture layer, which is built on top of the subsystem layer, incorporates the platform software and also serves as a portal that links the workload models to the platform during the mapping process [4].

4.2.4 Mapping and Co-Simulation

The workload models are mapped to the execution platform model, which involves the selection of a part of the platform that will execute a particular workload model. This is performed during the initialisation of a workload model by passing a pointer to correct the host in each workload constructor. Mapping is done at each layer, that is, by mapping

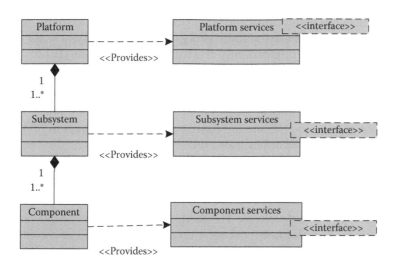

FIGURE 4.3
The ABSOLUT platform architecture model layers.

application workloads to subsystems, process workloads to OSs inside subsystems and functional workloads to processing units. The system model is built on the Linux platform via CMake [4] and the Open SystemC Initiative (OSCI) SystemC library [4]. The simulator is executed from the command line. During simulation, the progress information is printed to the standard output and after the completion of simulation, the gathered performance results are displayed.

4.3 Requirements for SLPE of Distributed Applications

After investigating the protocols and technologies employed by the distributed embedded systems in Section 4.1, we conclude that, to validate the non-functional properties (NFPs) of distributed embedded systems at an early stage, an SLPE methodology must provide the models of the communication protocols and other technologies employed by the distributed embedded systems. These models must be accurate enough so that the system designer can take design decisions during architectural exploration. We first provide a list of requirements that must be fulfilled by a performance evaluation methodology to evaluate the performance of distributed embedded systems. Afterwards, we describe the extensions needed for ABSOLUT to fulfil these requirements.

4.3.1 Requirements for Performance Evaluation of Distributed Systems

An SLPE methodology must fulfil the following requirements to perform SLPE of distributed embedded systems in different domains.

I. *Model of computation (MOC) agnostic (no domain restriction):* The methodology should not employ a specific MOC for modelling applications and platforms. Using an

MOC restricts the methodology to a particular domain of systems, for example, the methodologies that use the Kahn process network (KPN) MOC for application modelling are usually targeted only at streaming applications [11,12].

II. *Multithreaded application modelling:* For performance evaluation of multithreaded applications, the methodology must model the multithreading support [13].

III. *Physical layer models:* Physical layer models such as channel models, coding and modulation models to evaluate the contribution of these layers in non-functional properties of distributed applications [14].

IV. *Transport layer models:* Functional models of OSI data link and transport layer protocols for evaluating their contribution in non-functional properties such as end–end packet and frame delays [14].

V. *Performance evaluation of protocols:* The methodology must be capable of evaluating the performance of protocols operating on a particular layer of the OSI model in isolation just like widely used network simulators, for example, OMNeT++ and ns-2 by abstracting application workload models via traffic generators [14].

VI. *Workload model generation of user space code, external libraries and system calls:* Since, from an implementation perspective, all the application processes use user space code, external libraries, background processes and system calls, the methodology must therefore provide tools and methods for generating the workload models of not only the user space code but also the external libraries, background processes and system calls [11,14,15].

VII. *Workload generation of middleware technologies:* It must be capable of workload extraction of API functions of the various middleware technologies such as the device interconnect protocol (DIP) stack employed by network on a terminal architecture (NoTA). This will enable the methodology to span the domain of distributed streaming and context-aware applications [14].

VIII. *Detailed as well as highly abstract workload modelling:* The methodology must provide/define application workload modelling tools/techniques for generating the application workload models with varying degrees of refinement and detail. The more refined and detailed workload models result in slower simulation speed due to increased structure and control, while the less detailed workload models usually result in faster simulation speed [4,15,16] at the expense of accuracy. Once this is achieved, the system designer can freely choose the workload models that will result in more accurate or faster simulation.

IX. *Integration of application design and performance evaluation:* For early-phase evaluation of the distributed applications, the methodology must automate the workload extraction process by seamless integration of the application design and performance simulation phase. This can be achieved if the application and workload modelling phases are linked in such a way that application models act as a blueprint or starting point for the application workload models. The proposed technique must be experimented with modern service-oriented architectures (SOAs) such as the generic embedded system platform (GENESYS) and NoTA [11,17].

X. *Non-functional properties validation:* The non-functional properties of the system must be carried through the application design phase and be validated by the SLPE approach. The non-functional properties are usually modelled and elaborated in the application model views [11,17,18].

4.3.2 Enhancements Needed for ABSOLUT

ABSOLUT uses transaction-level modelling (TLM) 2.0 and SystemC for modelling platform components only; it is not restricted to a particular domain or TLM 2.0 for modelling applications. Also, ABSOLUT provides an OS model that is hosted on one or more processor models in the platform. The OS model consists of a scheduler for scheduling application processes and provides the possibility to model different OS services. A variety of services provided by the platform can be implemented by the system designer. The scheduling of the implemented services closely mimics the way services are scheduled by the widely used platforms (mostly via scheduling queues). The design and implementation of these services in SystemC (as ABSOLUT OS services) requires a thorough knowledge of the transport and data link layer technologies and event-driven simulation paradigms.

Also, ABSOLUT workload generation tools provide a high level of automation for the extraction of application workload models to test their feasibility on a variety of platforms. This is especially useful when the source code of the application is available. In the absence of a source code, the ABSOLUT workload models can be created via an analysis of the algorithmic details and the control of the application. For fast performance evaluation and architectural exploration, the ABSOLUT workload modelling phase and performance simulation phase can be integrated [18]. This reduces the time and effort involved in the performance simulation. Also, the non-functional properties must be carried through the application design phase and be validated by the performance evaluation phase [18]. This seamless integration of application design and the performance simulation phase has been demonstrated for different SOA-based application design methodologies such as GENESYS and NoTA [11,17].

In ABSOLUT, the cycle accurate platform component models are avoided, and instead, cycle approximate models are used for faster simulation speed to enable brisk iterations in the architectural exploration phase [4]. We now list the features mentioned in Section 4.3.1 that were provided by ABSOLUT beforehand, that is, before the extensions made to ABSOLUT for enabling it to be used for the performance simulation of distributed

TABLE 4.1

Features Provided by ABSOLUT Beforehand for the Performance Evaluation of Distributed Embedded Systems

	Feature	ABSOLUT Description	References
I	MOC agnostic	X^1	[4]
II	Multithreaded applications modelling	N	
III	Physical layer models	N	
IV	Transport layer models	N	
V	Performance evaluation of protocols	N	
VI	Workload model generation of user space code, external libraries and system calls	X^2	[4,16]
VII	Workload generation of middleware technologies	X^3	[4,16]
VIII	Detailed and highly abstract workload modelling	X^4	[4,16]
IX	Integration of application design and performance evaluation	N	
X	Non-functional properties validation	X^5	[4]

TABLE 4.2
Description of the Terms Shown in Table 4.1

X	Feature completely provided
N	Feature, related model or tool not provided
X^1	Restricted to the domain of non-distributed systems (single device-based systems). but not restricted to any application domain
X^2	Cannot generate the workload models of system calls, for example, Berkeley Software Distributions (BSD), API functions, etc.
X^3	Only if the middleware is implemented as an external library. It cannot generate the workload of middleware technologies if it is implemented as OS services or system calls or runs as a background process, for example, NoTA operates in Daemon mode
X^4	Only for user space code and external libraries. It cannot generate the workload for system calls
X^5	Only for non-distributed applications running on a single device. In other words, it cannot validate the non-functional properties due to MAC and transport protocols in use cases that involve two or more devices

systems. We also provide the references to research articles that describe these contributions. This information is shown in Tables 4.1 and 4.2.

From Table 4.1, it is clear that many features mentioned in Section 4.3.1 are either completely or partially absent in ABSOLUT. The modelling and integration of Feature II, that is, multithreading support, is described in Section 4.4. The modelling and integration of Features III–V are covered in Sections 4.5 and 4.6. Features VI, VII and VIII are provided by a novel methodology of workload modelling that uses run-time performance statistics. This methodology provides the ability to automatically extract the workload models of system calls. The methodology is described in Section 4.7. Features IX and X are described via a set of case studies described in Section 4.8.

4.4 Modelling Multithreaded Support

In recent years, multithreaded programming has gained popularity since general-purpose processors have evolved to multicore platforms. This has resulted in new challenges for software designers in the early stages of development of multithreaded applications (both distributed and non-distributed). In other words, designers need to make important design decisions related to load-balancing, thread management and synchronisation. This implies that even for moderately complex applications that have a few concurrent threads, the design space will be huge. The exploration of the design space will require the ability to quickly evaluate the performance of different software architectures on one or more platforms. The ABSOLUT performance simulation approach has been extended to achieve a faster simulation of multithreaded applications in the early phases of the design process. Abstract workload models are generated from the source code of the POSIX threaded applications, which are then mapped to the execution platform models for the transaction-level simulation in SystemC.

4.4.1 System Synchronisation

In case of ABSOLUT, the workload models do not contain timing information. To enhance the simulation speed, the ABSOLUT execution platform models contain cycle-approximate timing information. Thus, the platform model dictates the duration of the execution of a particular workload model. The system must ensure the correct behaviour while the concurrent processes are being executed in parallel. This demands a system synchronisation mechanism that respects the causal relations between the processes. In other words, this means that any particular execution order of processes or threads is allowed as long as their causal dependencies are respected. From the perspective of POSIX threads modelling, this can be illustrated via the example shown in Figure 4.4.

We assume that the target the application is following is the master–worker programming model. $T1$ acts as the master thread (main function) and is also a workload process. It creates two worker threads $T2$ and $T3$ by calling the 'pthread_create ()' primitive function. Apart from that, the $T3$ thread is detached ('pthread_detach ()'), that is, the creator thread ($T1$) will never block and will wait for $T3$ to terminate.

On the other hand, when $T1$ calls $T2$ to terminate ('pthread_join ()'), it will block (which is shown via the dotted line between $T12$ and $T13$) and waits for $T2$ to complete before it will continue. Both $T2$ and $T3$ are independent of each other. The order of execution within each process $T1$, $T2$, $T3$ is as follows:

$$T11 \rightarrow T12 \rightarrow T13 \rightarrow T14 \tag{4.5}$$

$$T21 \rightarrow T22 \tag{4.6}$$

$$T31 \tag{4.7}$$

From the correct system synchronisation perspective, the following are the additional constraints between the processes $T1$, $T2$, $T3$:

$$T11 \rightarrow T21 \tag{4.8}$$

$$T12 \rightarrow T31 \tag{4.9}$$

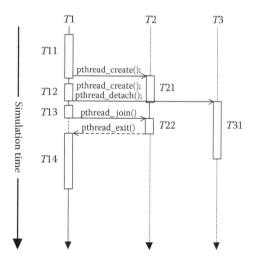

FIGURE 4.4
System synchronisation between POSIX threads.

It should be noted that the POSIX standard does not place any particular order constraint between T13 and T22, for example, the thread T22 can terminate (reach pthread_exit ()) before T13 calls it to terminate ('pthread_join ()'). In this case, T1 will not block after T13 but will proceed to T14. The following are some examples of the possible overall execution orders in this example:

$$T11 \rightarrow T21 \rightarrow T22 \rightarrow T12 \rightarrow T31 \rightarrow T13 \rightarrow T14 \tag{4.10}$$

$$T11 \rightarrow T12 \rightarrow T21 \rightarrow T31 \rightarrow T13 \rightarrow T22 \rightarrow T14 \tag{4.11}$$

Since untimed TLM does not guarantee the deterministic execution of concurrent processes, a mechanism for inter-process communication and system synchronisation must be integrated to the ABSOLUT platform model. It must also guarantee that the correct intra-process execution order is respected.

4.4.2 Inter-Process Communication and System Synchronisation Model

During the execution of the ABSOLUT performance model, the function workloads running on the execution platform model can request different software services by using the service interface called Generic_Serv_IF as shown in Figure 4.5. This interface is realised in the OS model called the Generic_Serv_OP_Sys model in Figure 4.5.

To support the POSIX service calls from the function workload models, a mechanism is needed. We modelled this mechanism in the form of a run-time library service process, that is, 'Pthread_lib' as shown in Figure 4.5. Pthread_lib acts not only as an inter-process communication mechanism but also as a thread synchronisation layer between the OS model and application workloads as illustrated in Figure 4.5. A new thread can be created by the function workload by calling the 'Use Service ()' call. The service name is used as an attribute to the call, for example, Use Service ('Serv_1') along with optional attributes.

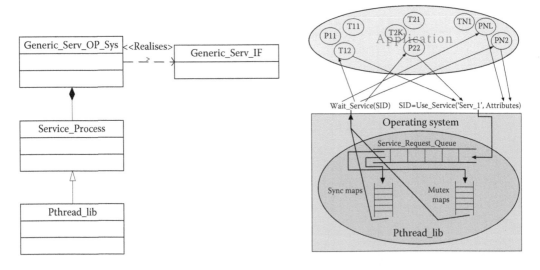

FIGURE 4.5
Application workload structure (left) and usage of Pthread_lib OS_Service by application workload models (right).

The unique service name is assigned to a particular service when it is registered to the OS model during the elaboration phase. This is explained in detail in Ref. [14]. This is a non-blocking request and the calling thread can therefore continue while the service is being processed. The Use_Service () call also returns a unique service identifier that can be given to the blocking 'Wait_Service ()' call to wait for the completion of a requested service.

The OS model relays every new service request to the Pthread_lib service process, which puts these requests in the Service_Request_Queue. As the simulation proceeds, a new service request is taken from the queue for processing. Depending on the service name, the 'Pthread_lib' object relays the service request to OS. The OS model schedules the call for execution on the platform. The relaying mechanism can be different and depends on the service type. Further details of the ABSOLUT multithreading support modelling and POSIX threads, along with a complete case study, are described in Ref. [13]. The case study describes the modelling of POSIX-based multithreaded applications on a high-level general-purpose multicore architecture-based processor [13]. A run-time thread service process that ensures correct intra-process execution of application workloads in association with a high-level OS model has been described in Ref. [13]. So far, the correctness of the inter- and intra-process execution order and compatibility with the POSIX API has been emphasised in the development. Therefore, generalisation of the methodology to support other parallel programming models like message passing will be one of the focus areas in future development of ABSOLUT methodology.

4.5 Modelling ABSOLUT OS_Services

Extension of ABSOLUT for the performance evaluation of distributed applications requires the modelling of protocols operating at different layers of the OSI model. This, in turn, requires a mechanism for instantiating new hardware (HW) and software (SW) services. These services are registered to the ABSOLUT OS model and are used by the application workload models. Furthermore, the services operating at a higher layer of the OSI model can use lower-layer services, for example, transport-level services such as TCP can use data-link-level services such as IEEE 802.11 MAC protocols for the transmissions of frames of a packet as shown in [14]. These services are instantiated by deriving them from the OS_Service base class as shown in Ref. [14]. The modelled services implement the Generic_Serv_IF as explained in Ref. [14]. ABSOLUT functional workload models request the services from the ABSOLUT OS model by using this interface. The modelling and integration of highly accurate data link- and transport-level services are explained in Ref. [14]. The relationship between OS services operating at the transport layer, data link layer, OS model, OS_Service base class and function workload models is shown in Figure 4.6.

The OS_Service base class implements the functionality related to the scheduling of requests made by processes via priority queues. After requesting the service from the OS, the requesting process goes to the sleep state. The OS informs the requesting process on service completion after which it goes back to the running state. This is shown in Figure 4.7.

4.5.1 Deriving New OS_Services

Only the service-specific functionality is implemented by the derived services, which make the modelling of services straightforward. These services are registered to the OS

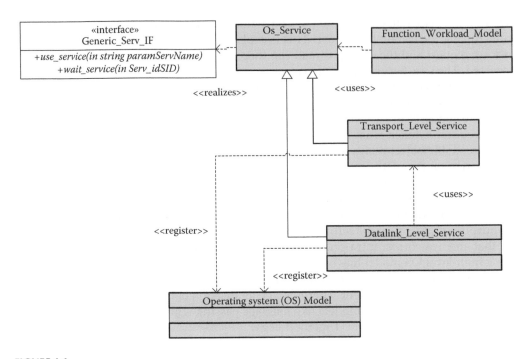

FIGURE 4.6
OS services implemented to model use cases spanning multiple devices and for modelling BSD API as OS services.

model during the elaboration phase and executed during the simulation phase when process- or application-level workload models request them from the OS. Other HW and SW services, apart from the data link and transport protocols, can also be derived from the OS_Service base class. This is shown in Figure 4.7.

4.5.2 Registering OS_Services to OS Models

The services residing at the higher layers of the OSI model use the services at the next lower layer of the OSI model in the same way as real world systems, for example, transport layer services use data link layer services for transmitting the individual frames of the packets. The services are accessed from the platform using the service name assigned during registration to the OS model. For example, the BSD socket application programming interface (API) function 'send()' can be modelled as an OS_Service and registered by a unique service name, for example, 'PktTx' to the OS as shown in Figure 4.8. It can then be accessed by the process workload models by using its unique service name via *Generic_Serv_IF* as shown in Figure 4.9.

4.5.3 Accessing OS Services

As mentioned before, the implementation of the *Generic_Serv_IF* by the *OS_Service* base class enables the function- or process-level ABSOLUT workload models to request services by their name as shown in Figure 4.9. This invokes the functionality of that service, which is implemented by the service derived from the OS_Service base class. The implementation of the OS_Service base class is described next.

System-Level Performance Simulation of Distributed Embedded Systems via ABSOLUT

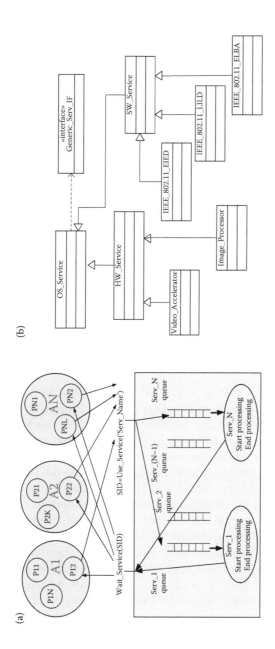

FIGURE 4.7
Processing of service requests by OS_Services base class (a) and the way new OS_Services are derived from the base class 'OS_Service' (b).

```
//The processor model
ARM_Crtx_Proc_ptr=new Scalable_MultiCore_CPU("m_ARMCortex_A9_MP_Processor");

//Creating the operating system (OS) model
m_os = new Generic_serv_op_sys ("os",ARM_Crtx_Proc_ptr->GetProcessorCores(),m_os_addr);

//Send() function of BSD API registered as "PacketTx" to OS
Pkt_Tx_Service= new Packet_Tx_Serv("Transmit_Packet",m_os);
Serv_type Msg_serv_type = { SERV_TYPE_LOCAL };
m_os->register_service(Pkt_Tx_Service,"PacketTx",Msg_serv_type);

//This service Handles transmission of single frame via IEEE 802.11 DCF registered by the
//name "FrameTx" to OS model
Frame_Tx_Service= new Frame_Tx_ServM3_Msg_Tx("Transmit_Frame", m_os);
Serv_type Frame_serv_type = { SERV_TYPE_LOCAL };
m_os->register_service(Frame_Tx_Service,"FrameTx",Frame_serv_type);
```

FIGURE 4.8
Registration of services to the OS model.

```
//Accessing an OS Service
SID=Use_Service("Service_Name");
Wait_Service(SID);
```

FIGURE 4.9
Accessing an OS_Service via Generic_Serv_IF.

4.5.4 Data Link and Transport Layer Services

The transport layer services, that is, TCP, UDP, etc., and data link services, for example, IEEE 802.11 distributed coordination function (DCF), are derived from the OS_Service base class. The transport services use the data-link-level services for the transmission of one or more frames of their packets. The IEEE 802.11 DCF operating at the data link layer can be shown in the form of a flow chart as shown in Figure 4.10.

Every frame received by the MAC layer is transmitted via the IEEE 802.11 DCF, which uses linear increase and linear decrease (LILD) or exponential increase and linear decrease (EIED) algorithms for contention resolution as shown in Figure 4.10. The transport layer simply divides a packet into frames and forwards them to the data link layer for transmission. The data link layer stores the frames in a queue and services them one by one for transmission over the channel. The recorded simulation results achieve an accuracy of over 92% when averaged after 20 simulation runs when compared with the analytical results for packet lengths of 228 and 2228 as shown in Ref. [14]. The packet loss probability achieved an accuracy of over 85% when compared with the analytical results as shown in Ref. [14]. The accuracy of these models is therefore enough for the SLPE of distributed systems.

4.5.5 Modelling Guidelines and Lessons Learnt

The following conclusions related to workload extraction and accuracy can be drawn from the case studies and literature review:

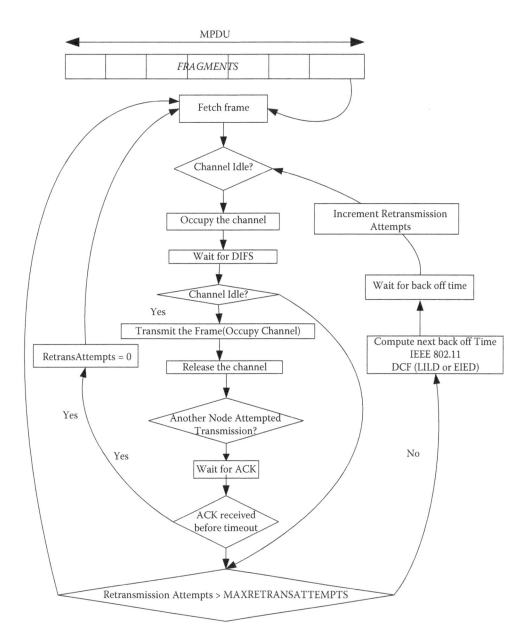

FIGURE 4.10
Flow chart of IEEE 802.11 DCF.

1. The MAC and transport layer models are functionally accurate and follow the same trend as the corresponding ns-2 and OMNeT++ models.
2. The workload models of the highly efficient and specialised MAC protocols can be obtained via ABSINTH-2 and CORINNA [15].
3. The main contribution of the MAC and transport protocols as far as non-functional properties are concerned is in end–end delays, etc., but not platform utilisation.

The following points can be taken into consideration while comparing the ABSOLUT MAC and transport models to ns-2, OMNeT++ or other network simulators:

1. The functionality related to the network layer has been abstracted out in ABSOLUT currently.
2. The alignment of traffic generators in time at the start of simulation.
3. The way system calls are modelled, that is, blocking or non-blocking.
4. Random number generators used for back-time calculation and the seeds used for randomisation can vary the simulation results significantly from other simulators, though the trend will be the same [14], and hence the models provided by these simulators are always functionally correct.
5. The queue sizes used for implementing the OS_Services must match those used in the simulators.
6. The way a collision is determined and defined in a network simulator is also very important to consider. In many network simulators, two or more nodes will collide if they transmit at exactly the same time. In the real world though, the propagation time due to distance between the nodes might also be a potential cause of collision. For example, if two nodes can sense the channel is idle, one starts to transmit and the other node(s) can still sense the channel is idle and transmit, causing collision.

The functionally correct models of MAC protocols and related OS_Services will be provided as a part of the ABSOLUT component library. Specialised probes and use cases will be provided to facilitate the system designer to modify and enhance the MAC and transport protocol models and make necessary adjustments as per use case.

4.6 Physical Layer Models

To study the MAC protocols in isolation under a particular scenario, the application workloads can be abstracted by using traffic generators. Three types of traffic generators, that is, pareto on off, exponential and constant-bit rate available in ns-2, have been integrated to ABSOLUT [14]. Two modulation techniques, that is, QPSK and BPSK have been modelled along with the multicarrier code division multiple access (MC-CDMA) technique. Two channel coding techniques, that is, convolutional and Reed Solomon codes, and two channel models, that is, a binary symmetric channel and additive white Gaussian noise (AWGN) channels, have been integrated by using models available in the itpp library [14]. The performance model is configured with a certain type of modulation scheme, coding scheme and channel model. Bit errors are computed using the functions available in the itpp library. Frame lengths can be chosen randomly or fixed to a value before simulation to analyse the MAC and transport protocols in a particular scenario.

4.6.1 Analysing Accuracy of Bit Error Rate Calculation

Different modulation schemes available in the itpp library have been used without modification. We present the results for multicode CDMA with QPSK modulation. For $1e^6$ bits, the results are over 99.8% accurate (when compared with theoretical results) as shown in Figure 4.11.

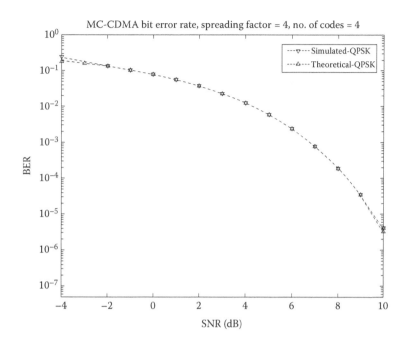

FIGURE 4.11
Theoretical versus simulation bit error rate for MC-CDMA with QPSK. Number of codes (M) = 4. Spreading factor (k) = 4. Number of bits = 100,000.

4.6.2 Analysing Accuracy of Frame Error Rate Calculation

In the absence of any encoding in IEEE 802.11, the fragment and the bit error rate are related by Equation 4.12.

$$P_e = 1 - (1 - BER)^S, \qquad (4.12)$$

where S is the fragment size, BER is the bit error rate and P_e is the probability of frame error. The bit error rates are plotted against the frame error rates for different values of frame lengths as shown in Figure 4.12. The frame and bit error rates can be recorded directly from simulation and plotted for different values of bit error rates as shown in Figure 4.12. The recorded simulation results are over 92% accurate when averaged after 20 simulation runs. The simulation results are compared with the analytical results for packet lengths of 228 and 2228 as shown in Figure 4.12.

4.6.3 Analysing Accuracy of Packet Error Rate

In case of IEEE 802.11, one MAC service data unit (MSDU) can be partitioned into a sequence of smaller MAC protocol data units (MPDUs) to increase reliability. Fragmentation is performed at each immediate transmitter. The process of recombining MPDUs into a single MSDU is called defragmentation. Defragmentation is also done at each immediate recipient. When a directed MSDU is received from the LLC with a length greater than a fragmentation threshold, the former is divided into MPDUs. Each fragment's length is smaller

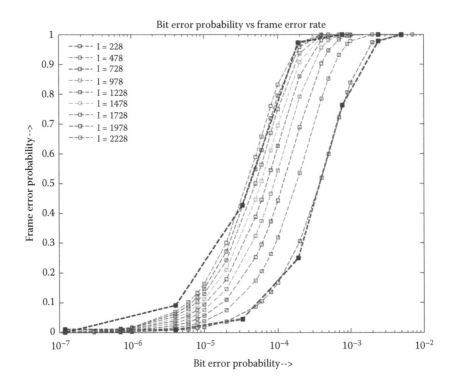

FIGURE 4.12
Frame error probability versus bit error rate. Theoretical results compared with simulation results for frame lengths 228 and 2228.

or equal to a fragmentation threshold [14]. MPDUs are sent as independent transmissions, each of which is separately acknowledged. The loss probability of transmitting a transport packet fragmented at the MAC layer into N fragments is given by the Equation 4.13 [14]

$$P_{wl} = 1 - \left(\sum_{i=1}^{1=M} P_l^{i-1}(1 - P_l) \right)^N = 1 - (1 - P_l^{M-1})^N, \tag{4.13}$$

where P_l denotes the successful transmission probability of one attempt, i denotes the retransmission attempts and M is the maximum number of retransmission attempts. Figure 4.13 shows the transport packet loss rate as a function of the MAC frame loss probability during each transmission retry for a fixed number of fragments ($N = 10$) and for different values of maximum retransmission attempts [14] ($M = 1 \to 10$). The simulation results are compared with the analytical results as shown in Figure 4.13. The values of M and N were fixed, the value of the signal-to-noise ratio (SNR) was varied and the simulation was repeated several times. The results for each value of SNR were averaged to obtain each point on the two curves. The simulation was run 20 times and the averaged results achieved an accuracy of over 85% when compared with analytical results as shown in Figure 4.13.

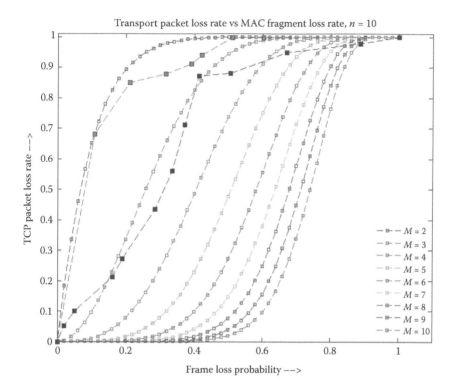

FIGURE 4.13
Theoretical versus simulation results. MAC frame loss probability versus transport packet loss rate, for maximum retransmission attempts ($M = 2$ and 3) and number of fragments ($N = 10$).

4.7 Workload Modelling via Run-Time Performance Statistics

The novel methodology for automatic workload extraction presented in this section is called configuration and workload generation via code instrumentation and performance counters (CORINNA) [15]. This methodology is completely dependent on the information read from CPU performance counters and is not compiler dependent. Furthermore, the instantiation of this methodology on a certain platform only requires re-implementation of the interface functions so as to access the CPU performance counters of that machine. The methodology can generate the application workload models of system calls, user space code and external libraries automatically. Furthermore, it does not employ additional programs like Valgrind and SAKE (abstract external library workload extractor) for the extraction of workload of external libraries as in ABSINTH-2 [16]. CORINNA is implemented as C++ classes and can be compiled in the form of a static and dynamic library.

4.7.1 Workload Extraction via CORINNA

The application workload model generation via CORINNA consists of three phases, that is, the pre-compilation, application execution and post-execution phases. In the pre-compilation phase, tags are inserted at different points in the source code automatically via a Python script called tag source-code parser written in python for CORINNA tags insertion (CORINNA-SCENT) [15]. In the execution phase, the run-time performance statistics of the application are recorded by reading performance counters for generating the function workload model primitive instructions.

After the execution phase, two CORINNA output files are obtained as shown in Figure 4.14, apart from the normal output (when the application is not compiled with the CORINNA library and no tags are inserted) of the program. In the post-execution phase, the two CORINNA output files are parsed to generate the classes for function workload models. Also, the configuration of CPU models is carried out by adjusting cache-hit and cache-miss probability, etc., according to the run-time statistics. Also, a top-level process model is generated that calls the generated function workload models in the order in which they appear in the trace information. The post-execution phase is also done via a Python script called CORINNA output parser for function workload generation and pro-

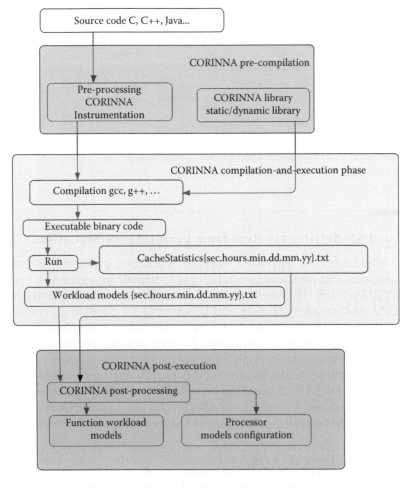

FIGURE 4.14
Pre-compilation and post-compilation steps of workload model extraction via CORINNA.

cess model configuration (CORINNA-PERFUME). The workload modelling and platform configuration via CORINNA is summarised in Figure 4.14.

As explained earlier, the workload modelling via CORINNA consists of three phases: the pre-compilation phase, the application execution phase and the post-execution phase. The pre-compilation phase inserts CORINNA tag pairs around selected source code lines in the source code. One tag in each tag pair marks the entry to that code line while the other marks the exit. The second phase is called the application execution phase. This phase ends when the application execution ends. During application execution, a separate data structure for function workload model creation is generated for each instruction between tags and a separated data structure for function workload model creation for the set of instructions encountered along the execution path to the entry of another tag pair, thus giving full coverage of the application source code. Also, the execution order is recorded by storing the names of workload models as they are generated. This trace information is used to call the function workload models in correct order, mimicking the true order of execution of the instructions during application execution. In the final step, the text files created after the application execution that contains the workload model creation data and CPU models configuration data are parsed and the function workload model classes generated and the CPU models are configured according to run-time statistics. The aforementioned three phases are described in detail in Ref. [15].

4.7.2 Comparing ABSINTH, ABSINTH-2 and CORINNA

The ABSINTH application workload generation methodology is compiler based [4], whereas CORINNA generates application workload models based on the run-time statistics gathered during the execution of an application. The salient features of ABSINTH and CORINNA are compared as follows:

1. *gcc compiler compatibility*: ABSINTH works with certain versions of gcc compiler, for example, gcc-4.3.1, whereas CORINNA is totally independent of the gcc compiler version used to compile the application source code.
2. *Workload of external libraries*: To extract workload models, for example, the current version of ABSINTH uses a tool called SAKE (abstract external library workload extractor) [16]. CORINNA does not use any other tool to extract workload models for external libraries.
3. *Workload of system calls*: ABSINTH is limited to the user space code and external libraries. Therefore, it cannot generate function workload models for the system calls. CORINNA has no such limitations and the insertion of tags around system calls will provide the required workload models.
4. *Coverage of C++ applications/g++ compiler*: ABSINTH cannot generate the workload of C++ applications. ABSINTH works as a patch of GNU gcc compiler and has certain limitations with error and exception handling what comes to C++ applications compilation. CORINNA has no such limitations and is not compiler based.
5. *Distributed applications*: Most of the distributed applications communicate via transport technologies such as TCP/IP and UDP. These transport technologies are available to the application programmer as system calls and are used for both message-based communications and also audio/video streaming in the case of

real-time multimedia client server applications. ABSINTH cannot generate function workload models for these system calls since it is limited to user space code and external libraries. CORINNA only requires the insertion of tags in the pre-compilation step via CORINNA-SCENT for the generation of function workload models for these transport API functions.

6. *Processor models configuration*: CORINNA records the cache-hits/miss statistics during the entire execution of the application, and when the execution of the use case ends, it writes the overall cache-hit and cache-miss rates of the application to the CacheStatistics {sec.hours.min.dd.mm.yy}.txt file for different cache levels. These data are used to configure the CPU models in the ABSOLUT platform model for more accurate SLPE. ABSINTH does not provide this functionality.

7. *Portability*: CORINNA is highly portable, but it might require the re-implementation of only the class member functions used to access the hardware counters in some cases. The complicated tasks of gathering the performance statistics, generation of function workload models during execution, reporting results, pre-compilation and post-execution Python scripts, that is, CORINNA-SCENT and CORINNA-PEFUME, do not need to be re-implemented and are totally machine independent.

Therefore, the application workload extraction methods previously employed by ABSOLUT, called ABSINTH and ABSINTH-2, had some shortcomings, for example, these are compiler-based workload extraction methods and cannot be used to extract the application workload models of kernel space code, for example, system calls. They lack the support for the g++ C++ compiler and cannot configure the platform processor models according to run-time statistics of the application, for example, cache-hits/misses [4]. To solve these issues, a novel method for workload generation based on run-time performance statistics called CORINNA has been developed, which is non-compiler-based. Also, CORINNA has some shortcomings; since it uses the hardware counters to generate the workload models, the latter might result in less accurate results when the ABSOLUT platform model has very different hardware architecture.

4.8 Conducted Case Studies

In Ref. [4], the performance evaluation of the transport and MAC protocols via ABSOLUT is described. The article describes the way the different MAC and transport protocols can be modelled as OS services and integrated into the ABSOLUT framework. In Refs. [17] and [18], the performance evaluation of distributed and non-distributed GENESYS applications via ABSOLUT is described. These articles describe the way the application design and performance simulation phases can be linked to reduce time and effort in the performance evaluation phase. In Ref. [11], the system-level performance of distributed NoTA systems via ABSOLUT is described. The modelling of NoTA DIP in all the three modes is explained and the approach is demonstrated via a case study. Also, in Ref. [15], we give two different case studies of how to extract ABSOLUT workload models and to show how to map the application workload models to the platform model for performance evaluation. For case studies, we used two different applications, Qt-based submarine attack game and office security application with video streaming.

4.9 Conclusion and Future Work

The complexity of the distributed embedded systems is growing rapidly in different application domains. To achieve faster deployment and a more optimal design of these systems, their feasibility must be evaluated at an early design phase. For a performance evaluation methodology to be applicable across multiple domains of distributed systems, the methodology must provide the easy instantiation and modelling of transport and data link protocol models as well as multithreaded applications.

This chapter presents the modelling and integration of the related protocols and models in ABSOLUT. New models and protocols can be added to the framework as they are developed and used in the performance models for SLPE. The functionalities that are common among these protocols are implemented in the base classes so that new models can be developed with minimal effort. The modelled protocols developed during the research presented in this section were applied in a number of case studies. The models integrated to ABSOLUT are accurate enough for SLPE and architectural exploration [14]. Table 4.3 provides an overview of the authors' contributions. It lists the research articles that describe the missing/partially provided features/models elaborated in Table 4.1. These references related to novel contributions are listed in bold font.

The network layer protocols of the OSI model were not modelled and the research efforts were mainly directed towards the SLPE of single hop distributed embedded systems. The network protocols can be instantiated by using the OS_Service base class models used for modelling the transport and data link layer protocols. The protocols at the data link and transport layer are functionally correct, and therefore the functional network layer protocols can be integrated into the framework just like the transport and data link layer protocols. Furthermore, the models/probes for estimating the energy consumption of the application, transport, middleware and data link OSI model layers are planned to be integrated to the ABSOLUT framework.

TABLE 4.3

Features Provided by ABSOLUT Beforehand for the Performance Evaluation of Distributed Embedded Systems

Feature		ABSOLUT Description	References
I	MOC agnostic	X	[4]
II	Multithreaded applications modelling	X	[13]
III	Physical layer models	X	[14]
IV	Transport layer models	X	[14]
V	Performance evaluation of protocols	X	[14]
VI	Workload model generation of user space code, external libraries and system calls	X^2	[4,15,16]
VII	Workload generation of middleware technologies	X^3	[4,15,16]
VIII	Detailed and highly abstract workload modelling	X^4	[4,15,16]
IX	Integration of application design and performance evaluation	X	[8,11,17]
X	Non-functional properties validation	X^5	[11,17,18]

References

1. S. Perathoner, E. Wandeler, L. Thiele, A. Hamann, S. Schliecker, R. Henia, R. Racu, R. Ernst and M. González Harbour, Influence of different abstractions on the performance of distributed hard real-time systems, in *Proceedings of the 7th ACM & IEEE International Conference on Embedded Software*, pp. 193–202, 2007.
2. I. Lee, J.Y-T. Leung, S.H. Son, *Handbook of Real-Time and Embedded Systems*, Chapman & Hall/CRC Computer & Information Science Series, Taylor & Francis Group, Boca Raton, pp. 207, 2008, ISBN-10: 1-58488-678-1.
3. K. Tachikawa, A perspective on the evolution of mobile communications, *IEEE Communications Magazine*, 41(10), 66–73, October 2003.
4. J. Kreku, M. Hoppari, T. Kestilä, Y. Qu, J.-P. Soininen, P. Andersson and K. Tiensyrjä, Combining UML2 application and SystemC platform modelling for performance evaluation of real-time embedded systems, *EURASIP Journal on Embedded Systems*, 2008, 712329, January 2008.
5. A. Vahdat, A. Lebeck and C.S. Ellis, Every Joule is precious: The case for revisiting OS design for energy efficiency, in *Proceedings of the 9th ACM SIGOPS European Workshop*, pp. 31–36, September 2000.
6. M. Weiser, Some computer science issues in ubiquitous computing, *Communications of the ACM*, 36(7), 75–84, July 1993.
7. J. Hill, R. Szewczyk, A. Woo, S. Hollar, D. Culler and K. Pister, System architecture directions for networked sensors, *ASPLOS*, 35(11), 93–104, November 2000.
8. J. Haapola, Nano MAC: A distributed MAC protocol for wireless ad hoc sensor networks, in *Proceedings of the XXVIII Convention on Radio Science & IV Finnish Wireless Communication Workshop*, pp. 17–20, 2003.
9. J. Polastre, J. Hill and D. Culler, Versatile low power media access for wireless sensor networks, in *SenSys '04 Proceedings of the 2nd International Conference on Embedded Networked Sensor Systems*, pp. 95–107, November 2004.
10. B. Kienhuis, E. Deprettere, K. Vissers and P. van der Wolf, Approach for quantitative analysis of application-specific dataflow architectures, in *Proceedings of the IEEE International Conference on Application- Specific Systems, Architectures and Processors (ASAP '97)*, pp. 338–349, Zurich, Switzerland, July 1997.
11. P. Lieverse, P. van der Wolf and E. Deprettere, A trace transformation technique for communication refinement, in *Proceedings of the 9th International Symposium on Hardware/Software Codesign* (CODES 2001), pp. 134–139, 2001.
12. P. Lieverse, P. van der Wolf, K. Vissers and E. Deprettere, A methodology for architecture exploration of heterogeneous signal processing systems, *Kluwer Journal of VLSI Signal Processing* 29(3): 197–207, 2001.
13. J. Saastamoinen, S. Khan, K. Tiensyrjä and T. Taipale, Multi-threading support for system-level performance simulation of multi-core architectures, in *Proceedings of the 24th International Conference on Architecture of Computing Systems 2011* (ARCS 2011), VDE Verlag Gmbh, 169–177, pp. 2011.
14. S. Khan, J. Saastamoinen, M. Majanen, J. Huusko and J. Nurmi, Analyzing transport and MAC layer in system-level performance simulation, in *Proceedings of the International Symposium on System on Chip*, SoC 2011. Tampere, Finland, 31 October–2 November 2011. IEEE Computer Society, 8 p.
15. S. Khan, J. Saastamoinen, J. Huusko, J.-P. Soininen and J. Nurmi, Application workload modelling via run-time performance statistics, *International Journal of Embedded and Real-Time Communication Systems (IJERTCS)*, IGI Global, Copyright © 2013 (In press.).
16. J. Saastamoinen and J. Kreku, Application work load model generation methodologies for system-level design exploration, in *Proceedings of the 2011 Conference on Design and Architectures for Signal and Image Processing, DASIP 2011*. Tampere, Finland, 2–4 November 2011. IEEE Computer Society 2011, pp. 254–260.

17. S. Khan, J. Saastamoinen, K. Tiensyrjä and J. Nurmi, SLPE of distributed GENESYS applications on multi-core platforms, in *Proceedings of the 9th IEEE International Symposium on Embedded Computing (EmbeddedCom 2011)*. Sydney, pp. 12–14 December 2011.
18. S. Khan, S. Pantsar-Syväniemi, J. Kreku, K. Tiensyrjä and J.-P. Soininen, Linking GENESYS application architecture modelling with platform performance simulation, in *Forum on Specification and Design Languages 2009 (FDL2009)*, Sophia Antipolis, France, pp. 22–24, September 2009, ECSI.

5

Self-Organising Maps: The Hybrid SOM–NG Algorithm

Mario J. Crespo, Iván Machón, Hilario López and Jose Luis Calvo

CONTENTS

5.1 Introduction	120
5.2 *k*-Means	120
5.2.1 Implementation of the Algorithm	121
5.2.2 Iterations	121
5.3 Neural Gas	122
5.3.1 How Was the Algorithm Developed?	123
5.3.2 Sequential Algorithm	123
5.3.3 Neural Gas for Local Modelling	124
5.3.4 Batch Algorithm	125
5.3.5 Variations of Neural Gas	126
5.4 Kohonen's Self-Organising Map	127
5.4.1 Lattice	127
5.4.2 Learning Rule	127
5.4.3 Component Planes and the Unified Distance Matrix	128
5.4.4 Other Versions and Uses for SOM	130
5.5 Hybrid SOM–NG	131
5.5.1 Hybrid Neighbourhood Function	131
5.5.2 Updating Rule	133
5.5.3 Topology Preservation Constant γ and Its Effect	134
5.5.4 Generating Continuous Maps	136
5.5.5 Supervised Version for Local Linear Modelling	137
5.5.6 Using the Supervised Algorithm for Estimation Tasks	141
5.5.7 Using the Supervised Algorithm for Knowledge Extraction	141
5.6 Application to Real-World Data	143
5.6.1 Classification of Iris Data Set	143
5.6.2 Estimation of Oil Consumption for Cars	144
5.6.3 Searching for Local Relevancies	145
5.7 Conclusions	147
References	150

5.1 Introduction

Self-organising maps are competitive learning neural networks that use a set of prototype vectors, also called codewords, neurons or units, to represent in some way the data used to train it. The whole set of prototype vectors is called a codebook as it groups all the codewords.

In competitive learning, the prototypes compete to be the most representative one for each data sample; this prototype is called the 'best matching unit' or 'bmu'. The learning process is centred in the bmu and depending on the cooperative philosophy, or the lack of it, other prototypes are influenced to a lower degree than the bmu.

Self-organising maps are usually unsupervised algorithms, as they do not get an expected output value for each data sample to be reproduced. Otherwise, there are supervised algorithms based on the original self-organising maps that include some kind of supervised learning.

k-Means is a clustering algorithm that can be considered to be one of the first self-organising maps because of its learning method. In this algorithm, the prototypes, called cluster centroids, have no cooperation with each other.

Neural gas (NG) is an algorithm similar to k-means, but it has a cooperative learning rule based on a distance ranking in the multidimensional input space. This algorithm produces the optimum quantisation results as it has no other restrictions to place prototypes.

A different approach is adopted in Kohonen's self-organising map (SOM), which trains how to use a neighbourhood function based on a grid, called output space. That grid tries to fit the shape of the data samples but by preserving the order between prototypes, so adjacent neurons in the output map should be adjacent in the multidimensional input space too.

This chapter will explain the hybrid SOM–NG algorithm and its applications. Classic algorithms will be introduced and explained to allow an understanding of the different features. Then the hybrid algorithm will be presented along with a step-by-step explanation of how a self-organising map algorithm can be created. After explaining the algorithm, some examples will be provided and explained in a how-to way for a deeper understanding of the uses.

5.2 k-Means

This algorithm is a clustering algorithm published by James MacQueen in 1967 [1]. It tries to find k partitions of multidimensional data by relocating the centroid of the located clusters until it reaches convergence. It is a pure competitive algorithm and it works on a 'winner takes all' philosophy.

The main purpose of this algorithm is to minimise the sum of squared distances to the cluster centroid. The cost function for this algorithm is

$$E = \sum_{k=1}^{K} \sum_{x_i \in Q_k} \|x_i - c_k\|^2 \qquad (5.1)$$

where K is the number of clusters, Q_k is the influence region defined by the centroid c_k and x_i are data samples.

Self-Organizing Maps 121

5.2.1 Implementation of the Algorithm

Even when it has a deep and complex mathematical background, the algorithm itself can be explained in a quite simple way. In every iteration, input points are grouped according to which centroid is the closest one. After selecting the data samples for each cluster, the centroid is moved to the centre of the group. The process is represented in the flow chart in Figure 5.1.

With this simple procedure, a partition into Voronoi regions can be done. This algorithm only requires two parameters, the number of desired clusters k and the initial position of the centroids. Selecting the number of clusters is not an easy task. It is common practice to make several reruns with different initialisations to keep the best one.

Figure 5.2 shows examples of right and wrong results obtained with k-means with an obvious clustering problem with four well-defined clusters. The only difference between both cases was the initial position of the centroids.

5.2.2 Iterations

Once a starting point is defined, the k-means algorithm is executed. The most common implementation is the 'Lloyd algorithm'. It is based on the Voronoi tessellation, but calculating the graphic solution is not needed to obtain k-means clustering.

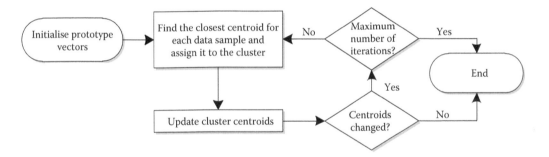

FIGURE 5.1
Flow chart for k-means.

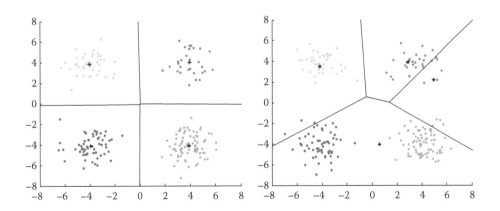

FIGURE 5.2
Examples of k-means.

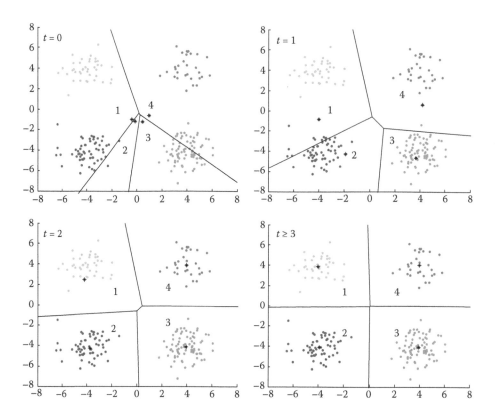

FIGURE 5.3
Iterations using *k*-means.

In Figure 5.3, a step-by-step solution is shown. Initial values ($t = 0$) for centroids were chosen randomly, so the initial solutions are not accurate. In this toy example, the second iteration ($t = 2$) produces an accurate classification, but a stable result is achieved in the third step ($t = 3$). Any further iteration after the third step will produce the same position for the centroids.

With a more detailed view over the figures, it can be seen that before training, centroid '1' covers most of the left clusters and centroid '2' gets only a couple of points of the lower left cluster. This issue is not really a problem as, after the first iteration, centroid '2' covers most of the lower left cluster and lets centroid '1' move towards the upper left cluster.

The illustrative example in Figure 5.3 helps to understand how *k*-means work.

In this example, *k*-means is a good algorithm for achieving a good solution as it has well-separated, compact and hyper-spherical clusters. Examples of bad results using *k*-means are data with concave shape or clusters with small distances if compared to their sizes.

5.3 Neural Gas

Martinetz et al. presented a new algorithm for vector quantisation in 1993 [2]. They called it 'neural gas' as the prototype vectors move with gas-like dynamics (and there is even a

Self-Organizing Maps

section in the article to explain the name). It is a cooperative–competitive algorithm with a 'winner takes most' philosophy. The closest prototype (bmu) is the most affected one for any single data sample, but others are also affected depending on how many prototypes are closer to that data sample.

NG proposes a new way of calculating the influence, instead of using distance measures; for example, Euclidean distance considers distance as a rank. Closer prototypes are more affected, but there is no direct influence of the distance.

5.3.1 How Was the Algorithm Developed?

This algorithm is based on solving a cost function using iterative methods. It is important to be familiar with linear optimisation methods, especially the gradient descent method and Newton's method.

The cost function to optimise is the following:

$$E_{NG} = \frac{1}{2\sum_{k=0}^{N-1} h(k)} \sum_{i=1}^{m} \int_{V} v \cdot P(v) \cdot h(k(v,w)) \cdot \|v - w_i\|^2 \, dP \tag{5.2}$$

where v represents data, with a density distribution $P(v)$ in the input space V. Prototype vectors w_i are a list of m representative points in the input space V. The neighbourhood function h is based on a distance ranking $k(v, w)$ that will be explained later in detail.

Equation 5.2 represents the accumulation of the distance for all points v to all the prototype vectors w_i weighted by the neighbourhood function h. As data are usually given as a set of discrete vectors and not as a density distribution, the integrator should be substituted by the sum of all data samples.

$$E_{NG} = \frac{1}{2\sum_{k=0}^{N-1} h(k)} \sum_{i=1}^{m} \sum_{j=1}^{N} h(k(v_j, w_i)) \cdot \|v - w_i\|^2 \tag{5.3}$$

The cost function described in Equation 5.3 is the sum of all the squared distances from N data samples to m prototype vectors weighted by the neighbourhood function h.

The neighbourhood function is a measure of membership of data sample v_j to the cluster defined by w_i. It should be a decreasing function with values between zero and unity, so an exponential curve is proposed using distance ranking instead of a direct distance measure. Distance ranking $k(v_j, w_i)$ has increasing values from 0 for the closest prototype to $m - 1$ for the furthest one. The mathematical expression for h is

$$h(k(v_j, w_i)) = \exp\left(-\frac{k(v_j, w_i)}{\sigma}\right) \tag{5.4}$$

where σ is the neighbourhood radius that defines the limits of influence. In this neighbourhood function, the effects further than 4σ will be very small, almost negligible.

5.3.2 Sequential Algorithm

The sequential algorithm modifies the prototype vectors using only one data sample, each time fitting the solution in small steps.

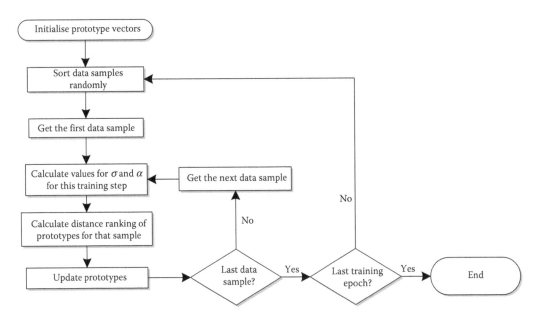

FIGURE 5.4
Flow chart for sequential neural gas.

Using the gradient descent method, the sequential training rule is obtained. Knowing that data samples are constant and supposing that small variations in w_i will not affect distance ranking, the only variable to be optimised is the position of the prototypes. The updating rule is

$$\Delta w_i = -\alpha \frac{dE_{NG}}{dw} = \alpha \cdot h(k(v_j, w_i)) \cdot (v_j - w_i) \qquad (5.5)$$

where α is the step size which weights the influence of the sample for all the prototypes. Equation 5.5 is the final expression to be implemented when programming the algorithm.

The step size α must have values between zero and unity, but the recommended values are 0.5 as the initial value and 0.005 as the final value [2]. The decreasing exponential function $\alpha(t) = \alpha_i(\alpha_f/\alpha_i)^{t/tmax}$ is usual for this purpose.

The neighbourhood radius σ also decreases with time; thus, the influence of further prototypes shrinks with every step. An exponential expression is also recommended: $\sigma(t) = \sigma_i(\sigma_f/\sigma_i)^{t/tmax}$. A good initial value is half of the number of prototypes, that is, $\sigma_i = m/2$, and a very small value is advisable as the final value, for example, $\sigma_f = 0.01$.

Owing to the decreasing influence of data samples, the ones presented earlier tend to have more influence on the final result; to avoid this bias, the data samples should be randomly sorted for each training epoch.

The algorithm works with the flow chart presented in Figure 5.4.

5.3.3 Neural Gas for Local Modelling

The original paper also presented the NG 'Application to Time-Series Prediction', but it is a local modelling tool usable for any kind of data and not only the temporal series.

Self-Organizing Maps

It uses linear models with the following expression:

$$\tilde{y} = y_{i^*} + a_{i^*} \cdot (v - w_{i^*}) \tag{5.6}$$

where $\tilde{y}$ is the estimated value of $f(v)$, y_{i^*} is a reference value for the closest prototype w_{i^*} and a_{i^*} is the estimated gradient in the Voronoi region defined by w_{i^*}.

To obtain y_{i^*} and a_{i^*} for the prototype vectors, a similar process is applied. The target function is

$$E_{NG} = \sum_{i=1}^{m} \sum_{j=1}^{N} h'(k(v_j, w_i)) \cdot (f(v_j) - (y_i + a_i \cdot (v_j - w_i)))^2 \tag{5.7}$$

Equation 5.7 measures the squared estimation error for v_j using the model of w_i weighted by the neighbourhood function. The neighbourhood function is called h' because it can be different from the one used to calculate the prototype position.

Using the gradient descent method again with the reference value y_i, its updating rule is obtained:

$$\Delta y_i = \alpha' \cdot h'(k(v_j, w_i)) \cdot (f(v_j) - y_i) \tag{5.8}$$

Step size α' can also be different from the one used in prototype learning.

Equations 5.5 and 5.8 are very similar. They update their value based on multiplying their step size and neighbourhood function by the difference between the actual value and the prototype. It means that the output value y_i can be understood like a new dimension and it is made different because of its use and not because of its features.

Obtaining the gradient estimator is more tedious, but the concept is the same. Differentiating Equation 5.7 with respect to the gradient vector a_i, the following expression is obtained:

$$\Delta a_i = \alpha' \cdot h'(k(v_j, w_i)) \cdot (f(v_j) - y_i - a_i \cdot (v_j - w_i)) \cdot (v_j - w_i) \tag{5.9}$$

Equation 5.9 learns from the distance from the data point to the prototype vector and the estimation error, also weighted by the neighbourhood function and the step size as usual.

5.3.4 Batch Algorithm

To improve the NG algorithm, Cottrell et al. presented a batch version in 2005 [3]. The batch version of NG takes all the data samples together and updates the prototypes only once per epoch.

Formulation of the batch algorithm starts with the same cost function as the sequential process. In the batch version, Newton's method is used instead of the gradient descent method, and the updating rule is

$$\Delta w_i = -\frac{J(E)}{H(E)} = \frac{\sum_{j=1}^{N} h(k(v_j, w_i)) \cdot (v_j - w_i)}{\sum_{j=1}^{N} h(k(v_j, w_i))} \tag{5.10}$$

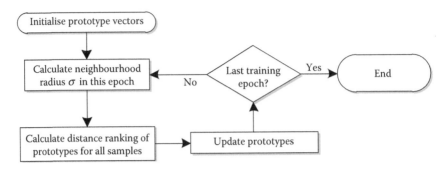

FIGURE 5.5
Flow chart for batch neural gas.

Operating over Equation 5.10, the current value of w_i can be cancelled on both sides. So the final expression for the new value is

$$w_i = \frac{\sum_{j=1}^{N} h(k(v_j, w_i)) \cdot v_j}{\sum_{j=1}^{N} h(k(v_j, w_i))} \tag{5.11}$$

The batch algorithm is simpler than the sequential one as there is no need to sort the samples and there is no step size parameter α. The flow chart is presented in Figure 5.5.

A batch algorithm has faster convergence and is less sensitive to initialisation. The main drawback of this version is the amount of memory needed to make the training in one step. It is needed to calculate the ranking for every data sample and prototype, needing an $N \times m$ matrix. It is also needed to calculate the neighbourhood function for all these combinations of samples and prototypes. There are many strategies for these issues; simple ones chop the data into several chunks or calculate the prototype update one at a time.

5.3.5 Variations of Neural Gas

NG is a capable algorithm and other versions were developed to achieve different kinds of results besides quantisation.

'Growing neural gas' (GNG) is explained in Ref. [4]; this algorithm only requires constant parameters and deciding the amount of prototypes is not needed. GNG starts using only two prototypes and moves them to fit the data distribution, and after a fixed number of iterations, a new prototype can be added if it is needed to achieve the desired quality of the model. The new prototype is added halfway between the prototype with the highest accumulative error and its neighbour with the largest error. The process is repeated until a predefined stopping criterion is fulfilled, such as the number of prototypes or any quality measure. In Ref. [4], edges between adjacent prototypes are also added to create a better representation of data distribution.

A supervised version of NG for classification tasks is proposed in Ref. [5]. It includes a classification term in the cost function to improve the classification accuracy instead of taking into account only quantisation. This term focuses on defining the class borders and lets the NG dynamics spread the prototypes of the same class uniformly among data corresponding to their class.

5.4 Kohonen's Self-Organising Map

A different approach for vector quantisation is Kohonen's SOM, also known as a self-organising feature map or just Kohonen's map. It is based on a neural network that finds the bmu, that is, the closest prototype, from a collection based on a low-dimension map. For practical purposes, there is no need to know the original neural network; modern algorithms just compare the distances to identify the bmu.

The map is usually 2D, but 1D and 3D maps are also known. Higher dimensions are possible, but they do not offer the advantage of dimensionality reduction. One of the most important advantages of the low-dimensional map is the creation of component maps that represent the value of only one of the dimensions in the reference grid. A good reference for this algorithm is Ref. [6] and a deep study is done in Ref. [7].

5.4.1 Lattice

Before a training SOM, a lattice is needed. It defines the topology to be preserved in the multidimensional input space. The lattice is an ordered set of neurons, usually in a hexagonal or rectangular disposition, where the distance between adjacent units is equal to unity (Figure 5.6).

Using the lattice, distances between neurons can be calculated and neighbourhoods can be defined by geometrical rules.

Often, the lattice is referred to as the output space N and each unit is named with its position in the lattice, $N_{i,j}$, where i is the row and j the column.

5.4.2 Learning Rule

The learning rule is based on biological concepts instead of being a mathematical process. There are biological evidences that seem to prove that brain stores information in topological order. From this idea, a topology-preserving rule was proposed.

$$\Delta w_i = \alpha \cdot h(w_i, w_{j*}) \cdot (v_j - w_i) \tag{5.12}$$

where $h(w_i, w_{j*})$ is a neighbourhood function based on the distance in the lattice between w_i and w_{j*}, the bmu for sample v_j. Neighbourhood function is defined as

$$h(w_i, w_{j*}) = \exp\left(-\frac{\|r_i - r_{j*}\|^2}{2\sigma^2}\right) \tag{5.13}$$

(a) 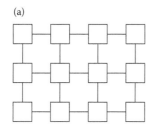 (b)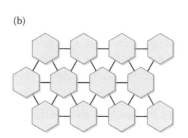

FIGURE 5.6
Rectangular (a) and hexagonal (b) lattices.

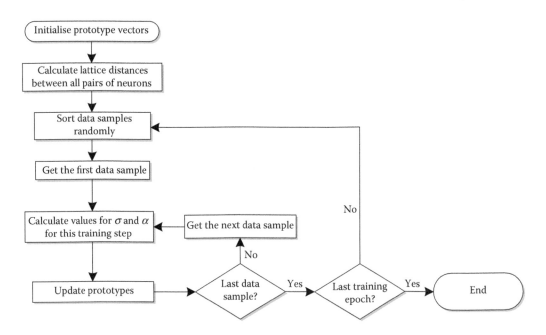

FIGURE 5.7
Flow chart for sequential SOM.

where r_i and r_{j^*} are the coordinates of w_i and w_{j^*} in the grid, respectively, and σ is the neighbourhood radius.

As in NG, only one data sample is processed at a time. This makes the algorithm quite similar, only changing the updating rule to use Equation 5.12. The flow chart is shown in Figure 5.7.

There is also a batch version of SOM. It takes all the training samples in every epoch and gives them the same importance; thus, it makes better models with less training epochs. The updating rule is

$$w_i = \frac{\sum_{j=1}^{N} h(w_i, w_{j^*}) \cdot v_j}{\sum_{j=1}^{N} h(w_i, w_{j^*})} \quad (5.14)$$

and the flow chart is given in Figure 5.8.

SOM is a very fast algorithm as it needs fewer calculations in each epoch. The distance in the output map is calculated only once before the first epoch and it is constant for all the training. In every epoch, the closest prototype is located for each data sample and the neighbourhood functions are calculated using the stored distances.

5.4.3 Component Planes and the Unified Distance Matrix

Component planes are a low-dimensional representation created using the predefined lattice. One map is created for each variable and its value is represented using colour scales. This helps to find patterns and correlations.

Self-Organizing Maps   129

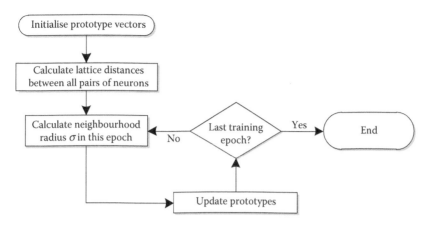

FIGURE 5.8
Flow chart for batch SOM.

Using the very common example of Iris flowers, the maps presented in Figure 5.9 were created.

Figure 5.9 shows the values of the four variables in the data set. The most obvious fact is the correlation between the petal length and width. They have high, medium and low values in the same zones; thus, they have a high positive correlation (they have a Pearson correlation of 0.96). Negative correlation can be detected in the same way, having the same shape but opposite values.

Component planes also help to detect groups; in this example, flowers with big sepals are divided into two groups, one with long, narrow sepals and the other with short, wide ones. Visual inspection from a trained user usually gives some information about the data that can lead to further research.

A special component plane is the representation of the unified distance matrix (U-matrix) [8]. The U-matrix is a matrix with $2m - 1$ rows and $2n - 1$ columns, where m and n are the number of rows and columns, respectively, of the lattice.

Elements in a position $(2i - 1, 2j - 1)$ are associated with the position (i,j) in the lattice. Values between the elements of two positions represent the distance between those elements in the multidimensional input space. Values in the unit positions are a statistic of the surrounding values, usually the average or median, so $u_{i,j}$ is the chosen statistic for all the distances that involve $N_{i,j}$.

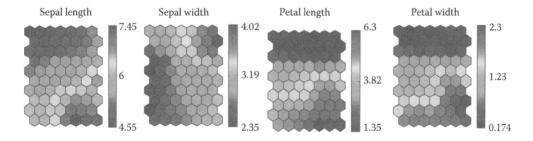

FIGURE 5.9
Iris component planes.

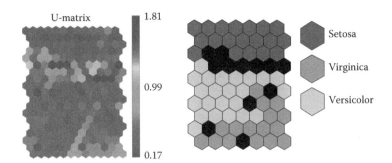

FIGURE 5.10
U-matrix and cluster classification.

A general example of a U-matrix is

$$U = \begin{pmatrix} u_{1,1} & d(N_{1,1},N_{1,2}) & u_{1,2} & \ldots & d(N_{1,n-1},N_{1,n}) & u_{1,n} \\ d(N_{1,1},N_{2,1}) & u_{2,2} & \ldots & \ldots & u_{2,n} & d(N_{1,n},N_{2,n}) \\ \vdots & \vdots & \ldots & \ddots & \vdots & \vdots \\ u_{m,1} & d(N_{m,1},N_{m,2}) & u_{m,2} & \ldots & d(N_{m,n-1},N_{m,n}) & u_{m,n} \end{pmatrix}$$

Using the U-matrix, clusters can be identified by visual inspection.

Figure 5.10 shows the U-matrix and the most repeated class in each Voronoi region. Two very different parts are identified, one corresponding to the setosa class and the other that covers both the virginica and versicolor classes. The setosa class is easily classified because of its small petals as it can be seen in the distance matrix. Also, some interpolating neurons appear in the border and they are not assigned to any class. The virginica and versicolor classes are harder to separate as they do not have well-defined borders.

5.4.4 Other Versions and Uses for SOM

SOM is a widely used algorithm for data processing and is very useful for a number of applications. As a consequence of this use, many different SOM-based techniques were created, some combining SOM with other algorithms and others modifying SOM itself.

Temporal regression is a common requirement in data analysis. A first attempt to use temporal information was the 'temporal Kohonen map' or TKM [9], which included a filter in distance calculation looking for the bmu. An evolution of TKM is the 'recurrent self-organising map' [10] that includes the filter in the bmu search and in prototype adaptation. After these, many variants for time series were created; a review can be found in Ref. [11].

The 'hybrid SOM' is a combination of an SOM as a hidden layer for an multi-layer perceptron (MLP). It has been used for classification [12] and forecasting [13]. In a similar way, an 'SOM of SOMs' or 'SOM^2' is presented in Ref. [14]. SOM^2 is a double layer classifier, it has a 'parent SOM' with a 'child SOM' associated with each prototype. The 'parent SOM' performs a gross classification and then the 'child SOM' does a more accurate one.

All these variations are some of the few interesting ones, but there are many for most of the fields of knowledge, with a special focus on pattern and speech recognition, forecasting and clustering.

5.5 Hybrid SOM–NG

The hybrid SOM–NG algorithm [15] is a combination of SOM and NG, as the name clearly states. The hybridisation is done by combining the neighbourhood functions into a single one that takes into account both the quantisation and the topological orders.

5.5.1 Hybrid Neighbourhood Function

The hybrid neighbourhood function is the following:

$$h_{SOM-NG}(v_j, w_i) = h_{NG}(k(v_j, w_i)) \cdot h_{SOM}(v_j, w_i) \tag{5.15}$$

Both neighbourhood functions are slightly modified to be combined. The NG-based function is

$$h_{NG}(k(v_j, w_i)) = \exp\left(-\frac{k(v_j, w_i)}{\gamma^2 \cdot \sigma}\right) \tag{5.16}$$

where γ is a new parameter included to control the influence of the distance in the multidimensional input space. The other parameters are not modified from Equation 5.4.

The SOM-based function is

$$h_{SOM}(v_j, w_i) = \exp\left(-\frac{s(r_{i*}, r_i)}{\sigma}\right) \tag{5.17}$$

where $s(r_{i*}, r_i)$ is a newly defined topographical ranking. A ranking is needed to measure the SOM and NG neighbourhood radii σ in the same units. It makes no sense to measure SOM using distance in the grid and NG using amounts of prototypes that are closer. So the new topographical ranking is defined as

$$s(r_{i*}, r_i) = \left|\{r_j / d(r_{i*}, r_j) < d(r_{i*}, r_j)\}\right| \tag{5.18}$$

where d is the distance measured in the lattice, r_i is the position of the corresponding neuron to w_i and r_{i*} is the neuron corresponding to the closest prototype to v_j.

Expression (5.18) means that the position in the ranking is equal to the amounts of prototypes that are closer to the winning neuron than the one to be updated. This expression was used previously for the ranking in NG [3], but in the lattice it gets special importance. As the lattice has a fixed distance between the units, there will be more than one unit at the same distance and the value is kept constant during the whole training phase.

The ranking has some special features. The most important one is the range $[0,m)$, with m as the number or prototypes, which is comparable to the ranking in the NG kernel (5.16). A monotonous ranking should not be used because it reduces the robustness of the algorithm in the tiebreaks and units at the same distance should have the same neighbourhood value.

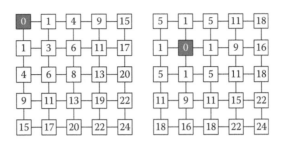

FIGURE 5.11
Examples of distance ranking in a rectangular lattice.

Rankings in a lattice for two different winning neurons are represented in Figure 5.11. The first subfigure (left) is the ranking for a winning neuron in the top left corner. The winning neuron has ranking 0, as usual, because there are no prototypes closer than its corresponding one. The next ones have ranking 1 as only the winning neuron is closer. As was stated above, they have the same distance, so they should have the same ranking.

In the second subfigure, the winning neuron is a different one, so the rankings are different. In this case, there are four units at unitary distance, so they all have ranking 1. The following ones, in the diagonal positions, will have ranking 5 as there are five closer neurons.

The distance matrix is not symmetric; in the examples, $s(r_{i*} = N_{1,1}, r_j = N_{2,2}) = 3$ and $s(r_{i*} = N_{2,2}, r_j = N_{1,1}) = 5$. Even when symmetric measures are preferable, the experimental results are good and the obtained models are similar to the equivalent SOM models.

But the meaning of the ranking is also different. While distance neighbourhood means 'σ neurons away', ranking neighbourhood means 'σ closest neurons'.

In Figure 5.12, both neighbourhood functions are represented. They represent the influence of a data sample whose bmu is the central one in a 25 × 25 map with a distance ranking of 144. In this example, there are 437 neurons in the map with a squared distance lower than 144 and all the neurons are included in the area of influence of the squared distance function. Using the distance ranking, many of the prototypes are outside the influence zone and their modification in those circumstances is negligible.

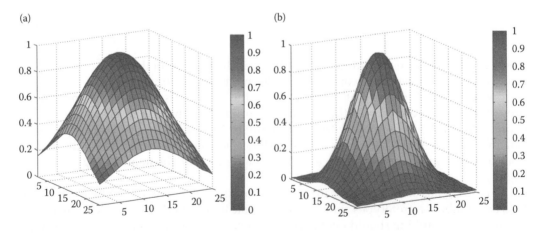

FIGURE 5.12
Neighbourhood function using distance (a) and distance ranking (b).

Self-Organizing Maps

5.5.2 Updating Rule

The updating rule for prototype positions is obtained using the following cost function:

$$E_{SOM-NG} = \sum_{i=1}^{m}\sum_{j=1}^{N} h_{SOM-NG}(v_j, w_i) \cdot \|v_j - w_i\|^2 \qquad (5.19)$$

It is a similar cost function to the one used in NG, but using the hybrid neighbourhood function. The neighbourhood weights the distance between the data sample and the prototype, not only the distance in the input space but also the output lattice. As the neighbourhood functions are multiplied, if one of them has a very low value, the combination will have low value too. Thus, the prototypes that are close to the winning neuron in both spaces will have an important effect.

As the batch versions are faster in convergence and simpler to be implemented, Newton's method will be used. In the mathematical process to obtain the updating rule, it will be supposed that neighbourhood functions are already calculated and without loss of generality only one prototype, named w, will be updated at a time. This can be done because the position of the other prototypes only affects the neighbourhoods.

When training the d-dimensional prototype $w = (w_1, \ldots, w_k, \ldots, w_d)$ with data samples $v = (v_1, \ldots, v_k, \ldots, v_d)$ during training epoch t, the first derivative of Equation 5.19 with respect to a component w_k is

$$\frac{\partial}{\partial w_k} \sum_{j=1}^{N} h_{SOM-NG}(v_j, w) \cdot \|v_j - w\|^2 = \sum_{j=1}^{N} h_{SOM-NG}(v_j, w) \cdot \frac{\partial}{\partial w_k} \|v_j - w\|^2 \qquad (5.20)$$

As neighbourhood functions are known values, only the derivative of the Euclidean distance is needed:

$$\frac{\partial}{\partial w_k}\|v_j - w\|^2 = \frac{\partial}{\partial w_k}((v_{j1} - w_1)^2 + \cdots + (v_{jk} - w_k)^2 + \cdots + (v_{jd} - w_d)^2)$$

$$\frac{\partial}{\partial w_k}\|v_j - w\|^2 = \frac{\partial}{\partial w_k}\left(v_{jk}^2 + w_k^2 - 2 \cdot w_k \cdot v_{jk}\right) = -2 \cdot (v_{jk} - w_k) \qquad (5.21)$$

The second derivative is obvious:

$$\frac{\partial^2}{\partial w_k^2}\|v_j - w\|^2 = \frac{\partial}{\partial w_k}(-2 \cdot (v_{jk} - w_k)) = 2 \qquad (5.22)$$

There is no influence from other components than w_k and v_{jk}, so equations can be written in vector form. Using Newton's method

$$\Delta w = -\frac{J_E(w)}{H_E(w)} = -\frac{-2 \sum_{j=1}^{N} h_{SOM-NG}(v_j, w) \cdot (v_j - w)}{2 \sum_{j=1}^{N} h_{SOM-NG}(v_j, w)} \qquad (5.23)$$

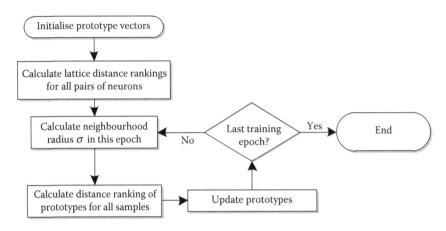

FIGURE 5.13
Flow chart for batch SOM–NG.

A different way of writing Equation 5.23 is

$$2\left(w\big|_{t+1} - w\right)\sum_{j=1}^{N} h_{SOM-NG}(v_j, w) = 2\sum_{j=1}^{N} h_{SOM-NG}(v_j, w) \cdot (v_j - w) \quad (5.24)$$

In Equation 5.24, the current value for w can be cancelled. Renaming $w|_{t+1}$ as w_i, the final expression for updating the prototypes is

$$w_i = \frac{\sum_{j=1}^{N} h_{SOM-NG}(v_j, w_i) \cdot v_j}{\sum_{j=1}^{N} h_{SOM-NG}(v_j, w_i)} \quad (5.25)$$

This process can be applied using the SOM or NG neighbourhood functions to reach their respective batch training Equations 5.11 and 5.14. The flow chart of the algorithm is given in Figure 5.13.

This algorithm has a higher computational cost than the original SOM and NG algorithms as it has to calculate the neighbourhood functions of both of them, but it seems logical as two different objectives are pursued.

5.5.3 Topology Preservation Constant γ and Its Effect

In the NG-based term of the hybrid neighbourhood, a new parameter was included. This parameter γ tunes the effect of the NG part. Low values of γ give great freedom to prototypes and the effect of the lattice is smaller. Otherwise, high values of γ keep the lattice effect and the prototypes move as a group. Intermediate values are the most interesting option for this algorithm, as they have the freedom to adjust to data but in an ordered way. This makes the algorithm a very competitive one as it keeps good behaviour for quantisation and topographical preservation.

Self-Organizing Maps

The quality of the obtained models is measured using statistic indicators. The most important one is the quantisation error; it represents the average distance from data samples to their closest prototype. Mathematically,

$$q_e = \frac{1}{N} \sum_{j=1}^{N} \|v_j - w_{j*}\| \tag{5.26}$$

Topographical error [16] is defined as the proportion of data samples whose two closest prototypes are not neighbours in the lattice. The expression is

$$t_e = \frac{1}{N} \sum_{j=1}^{N} u(v_j) \tag{5.27}$$

where $u(v)$ is 0 if the two closest prototypes are neighbours and 1 otherwise. When using a rectangular lattice, the topographical error can vary depending on the definition of neighbour. The von Neumann neighbourhood defines neighbours as those units with unitary distance; for rectangular lattices, there are four units that fulfil this condition. The Moore neighbourhood defines neighbours as the units within unitary Chebyshev distance; in usual Euclidean distance, it means those with a distance lesser or equal to $\sqrt{2}$, so all the eight units around will be neighbours.

Figure 5.14 shows examples of these neighbourhoods, where neighbours are filled and not-neighbours are left.

Using these quality measures, the effect of parameter γ can be studied. The evolution can be seen only with the measure, but it is also very interesting to make comparisons with the original algorithms.

As algorithms can have small differences between different initialisations, it is advisable to make several runs with different initialisations for both the reference algorithm and the tested algorithm.

To compare with SOM, maps of the same size (10 × 7) were trained using the Iris data set. Using different values of parameter γ will help to understand its effect over the final results (Figure 5.15).

Parameter γ increases the neighbourhood radius in the quantisation. As it grows, the lattice is more important; thus, the quantisation error increases because prototypes do not move as freely as with low values. Experimental results prove that values of γ above 80

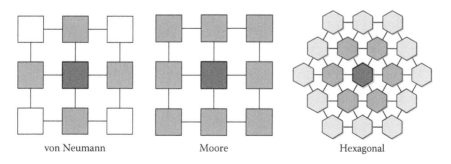

FIGURE 5.14
Examples of neighbourhoods for rectangular and hexagonal lattices.

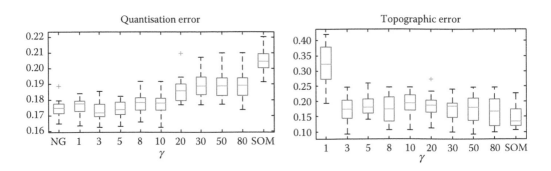

FIGURE 5.15
Quantisation and topographic error for the Iris data set using the SOM–NG algorithm with different values of γ for the same size map.

almost cancel the NG effect and have very similar results to pure SOM. The main advantage of this algorithm is in the intermediate values that keep a competitive result in both quantisation and topographic preservation. In the Iris example, for values of γ between 5 and 20, results are good in both errors.

5.5.4 Generating Continuous Maps

The SOM–NG provides a set of prototypes w_i linked to map units N_i. This model simplifies the Voronoi regions to a single discrete position on the map. This effect may not be desirable if data are going to be projected in temporal series to study evolutions.

The goal of continuous maps is to allow data samples to have different positions in the output map even if they are close to each other. These maps are obtained approximating the function $F(w_i) = N_i$ with a regression technique. A good and very used one is the general regression neural network (GRNN) [17], which approximates a surface without doing an iterative process.

For example, a 400 points punctured sphere will be modelled. For this purpose, a big map (10 × 10) will be used. Topology preservation is important, so the value of γ will be 60, high enough to ensure good topographical preservation and some degree of freedom.

In Figure 5.16, the original data and the obtained model are represented. Every prototype in the model has a position in the 10 × 10 map. Using GRNN with the prototype positions

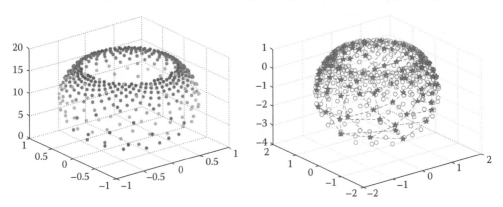

FIGURE 5.16
Punctured sphere data and normalised model.

Self-Organizing Maps

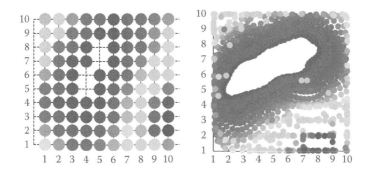

FIGURE 5.17
Discrete and continuous maps.

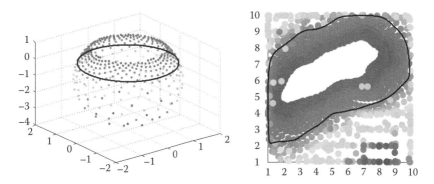

FIGURE 5.18
A circumference plotted in data space and a continuous map.

as inputs and the map position as outputs, a continuous map was created. Projecting a 1200 points sphere, the map was drawn. It represents the surface of the sphere with some discontinuities because the data set does not include a regular surface of the sphere and points in the lower part are more separated than the ones at the top.

The discrete map in Figure 5.17 is a usual component plane, representing the z component for each prototype. Owing to the high value of γ and the strong topological restrictions, some interpolating units appear in the centre of the map. The continuous map is a slightly deformed one, but all the positions in the sphere surface have a place on it. Some of the points in the lower part are located between the lower right and top left units, so the interpolation puts them over other positions in the map to close the sphere.

Figure 5.18 is an example of projection of a continuous path. In the input space is a ring over the sphere surface and in the output map it is an irregular curve as the map does not preserve distances between units; it is proportional to the density of data.

If the data are labelled for different states, the evolution can be followed in the continuous map and not just seeing the state in each moment, probably not detecting the evolution towards an error state until it is too late.

5.5.5 Supervised Version for Local Linear Modelling

After having a good model with well-located prototypes, a good strategy is to use them to create local models. Using local models is a great advantage if data have different working zones. The main drawback is the need to create many models even if they are simpler than a global one [18].

A local linear model uses the following equation:

$$\hat{f}(v) = y_{i^*} + a_{i^*} \cdot (v - w_{i^*}) \tag{5.28}$$

where $\hat{f}(v)$ is the approximation of the target function $f(v)$, y_{i^*} is the reference value learnt for $\hat{f}(w_{i^*})$, w_{i^*} is the closest prototype to data vector v and a_{i^*} is the gradient of the approximated function in the Voronoi region defined by w_{i^*}.

The gradient vector represents the slope of the function in that zone. For d-dimensional functions, the gradient vector is multiplied by the distance vector using the dot product.

The updating rules for y_i and a_i are obtained from a cooperative version of the squared estimation error:

$$E_{SUPSOM-NG} = \sum_{i=1}^{m} \sum_{j=1}^{N} h_{SOM-NG}(v_j, w_i) \cdot \left(f(v_j) - \hat{f}(v_j)\right)^2 \tag{5.29}$$

Again, Newton's method will be used to obtain a batch algorithm. Differentiating Equation 5.29 with respect to y for a single prototype

$$\frac{\partial}{\partial y} \sum_{j=1}^{N} h_{SOM-NG}(v_j, w) \cdot \left(f(v_j) - \hat{f}(v_j)\right)^2 = \sum_{j=1}^{N} h_{SOM-NG}(v_j, w) \cdot \frac{\partial}{\partial y}\left(f(v_j) - \hat{f}(v_j)\right)^2 \tag{5.30}$$

So, the derivative of the squared error is needed:

$$\frac{\partial}{\partial y}\left(f(v_j) - \hat{f}(v_j)\right)^2 = \frac{\partial}{\partial y}(f(v_j) - y - a \cdot (v - w))^2$$

$$= \frac{\partial}{\partial y}(f(v_j) - a \cdot (v - w))^2 + \frac{\partial}{\partial y}y^2 - 2\frac{\partial}{\partial y} \cdot y \cdot (f(v_j) - a \cdot (v - w)) \tag{5.31}$$

Operating in Equation 5.31, the derivative is obtained:

$$\frac{\partial}{\partial y}\left(f(v_j)) - \hat{f}(v_j)\right)^2 = -2(f(v_j) - y - a \cdot (v - w)) = -2\left(f(v_j) - \hat{f}(v_j)\right) \tag{5.32}$$

The second derivative is calculated easily:

$$\frac{\partial^2}{\partial y^2}\left(f(v_j) - \hat{f}(v_j)\right)^2 = 2 \tag{5.33}$$

Self-Organizing Maps

Using derivatives in Equations 5.32 and 5.33, the updating rule is obtained:

$$\Delta y = -\frac{J_E(y)}{H_E(y)} = -\frac{-2\sum_{j=1}^{N} h_{SOM-NG}(v_j, w) \cdot (f(v_j) - y - a \cdot (v_j - w))}{2\sum_{j=1}^{N} h_{SOM-NG}(v_j, w)} \quad (5.34)$$

In this expression, the gradient term of the estimation is symmetrical around the prototype vector w, so the contribution to the sum of one of the halves will cancel the other half [2]. Thus, the term can be cancelled. As was done in the prototype updating rule (5.23), the current value of y can be eliminated on both sides of the equation, leading to the final expression

$$y_i = \frac{\sum_{j=1}^{N} h_{SOM-NG}(v_j, w_i) \cdot f(v_j)}{\sum_{j=1}^{N} h_{SOM-NG}(v_j, w_i)} \quad (5.35)$$

Paying some attention to expression (5.35) will reveal that the output value has the updating rule as prototypes. All the variables have similar features and the difference between the input and output variables is calculated by the user and not by the algorithm. As a result, the reference value will be calculated with the prototypes.

After calculating the reference value, the gradient vectors are needed. The procedure is the same as with the prototypes, but it is a bit more complex. Taking into account only one prototype, the first derivative of Equation 5.29 with respect to a gradient component a_k will be calculated:

$$\frac{\partial}{\partial a_k} \sum_{j=1}^{N} h_{SOM-NG}(v_j, w) \cdot \left(f(v_j) - \hat{f}(v_j)\right)^2 = \sum_{j=1}^{N} h_{SOM-NG}(v_j, w) \cdot \frac{\partial}{\partial a_k}\left(f(v_j) - \hat{f}(v_j)\right)^2 \quad (5.36)$$

The derivative of the squared error is

$$\frac{\partial}{\partial a_k}\left(f(v_j) - \hat{f}(v_j)\right)^2 = \frac{\partial}{\partial a_k}(f(v_j) - y)^2 + \frac{\partial}{\partial a_k}(a_1 \cdot (v_{j1} - w_1) + \cdots$$

$$+ a_k(v_{jk} - w_k) + \cdots + a_d \cdot (v_{jd} - w_d))^2$$

$$-2\frac{\partial}{\partial a_k} \cdot (f(v_j) - y) \cdot (a_1 \cdot (v_{j1} - w_1) + \cdots$$

$$+ a_k \cdot (v_{jk} - w_k) + \cdots + a_d \cdot (v_{jd} - w_d)) \quad (5.37)$$

After a long process, which is not going to be detailed here because of its length, only the terms where a_k appears are retained and derived. The result is

$$\frac{\partial}{\partial a_k}\left(f(v_j) - \hat{f}(v_j)\right)^2 = -2(v_{jk} - w_k)(f(v_j) - y - a \cdot (v_j - w)) \quad (5.38)$$

And the second derivative is

$$\frac{\partial^2}{\partial a_k^2}\left(f(v_j) - \hat{f}(v_j)\right)^2 = -2(v_{jk} - w_k)(v_{jk} - w_k) \tag{5.39}$$

Finally, Newton's method is applied, writing the expression in vector terms:

$$\Delta a_i = -\frac{J_E(a)}{H_E(a)} = \frac{\sum_{j=1}^{N} h_{SOM-NG}(v_j, w_i) \cdot (f(v_j) - y_i - a_i \cdot (v_j - w_i))}{\sum_{j=1}^{N} h_{SOM-NG}(v_j, w_i)(v_j - w_i)(v_j - w_i)} \tag{5.40}$$

Expression (5.40) cannot be simplified, so the gradients are modified with increments instead of obtaining the new value.

Gradients are calculated after the prototype positions and reference values are known. If they are updated at the same time as prototype vectors, there is a risk of learning about data samples that will not be located in the surroundings of the final position of the prototype.

The algorithm is implemented following the flow chart in Figure 5.19.

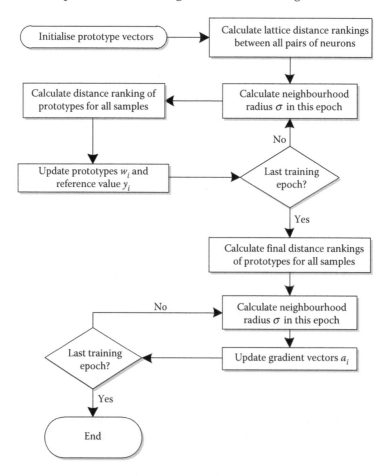

FIGURE 5.19
Flow chart for supervised batch SOM–NG.

Self-Organizing Maps 141

5.5.6 Using the Supervised Algorithm for Estimation Tasks

The supervised algorithm was planned to be a function estimator for multidimensional inputs, but it can be used for other purposes, mainly knowledge extraction.

Estimation quality only cares about the difference between the estimation and the real value, the estimation error. A first example of estimation will be given using a simple trigonometric function: $f(x,y) = \sin(x) \cdot \cos(y)$, defined in the range $[-3, 3] \times [-1.5, 1.5]$.

Obviously, more prototypes and local models will improve the accuracy of the general model. Even when the curve root mean squared error (rmse) versus map size will tend asymptotically to zero, it will reach acceptable results with moderate map sizes (Figure 5.20). Estimations are good even for more complex target functions and the results are similar to the ones obtained using NG with local models. Having a good quantisation helps in making good estimations.

5.5.7 Using the Supervised Algorithm for Knowledge Extraction

The interesting part of keeping topological order is the capability of representing the gradients and its components over component planes, as was done with prototype positions.

Representing the norm of the gradient in a component place will give a map of steadiness. The zones with a low gradient norm are steadier and variations of the input variables will produce small differences on the output value. Otherwise, in zones with a high gradient norm, the output value will have higher differences even with smaller modifications of the input values.

In Figure 5.21, a toy example is represented. It represents a hill with a plain top, where x and y are the input variables and z is the output variable. The plain zones have a gradient

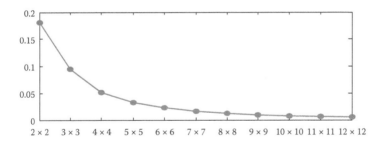

FIGURE 5.20
Root mean squared error for different map sizes.

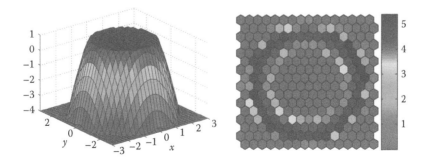

FIGURE 5.21
Data and gradient norm map for a cut hill.

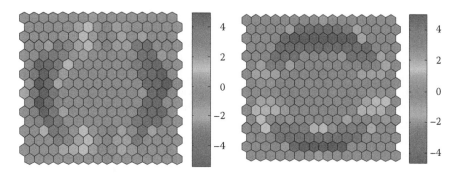

FIGURE 5.22
Gradient component maps for x (left) and y (right).

norm equal to zero, as there are no changes while moving in that region. The slope has a greater norm, as moving to the top of the hill causes a high variation in z.

Knowing how steady a zone is can be interesting but knowing the influence of each variable is also important. Influential variables will have greater values, both positive and negative, in their zones of influence. By representing the components of the gradient vector in a map, local influences can be detected, which is very helpful in knowledge extraction applications.

Gradient components in Figure 5.22 show the influence of each variable in every Voronoi region. In the left half, the gradient component of x is positive because moving to greater x will cause z to grow; otherwise, in the right half, moving towards higher values of x will cause z to decrease, so the gradient component is negative. The same effect occurs with y, but in the perpendicular direction.

Also, a statistical analysis of gradients can help to know which variables are relevant and which are irrelevant. Irrelevant variables will usually have small values in their gradient components. But the researcher should be careful about marking variables as irrelevant when they may have indirect influences, usually being highly related with reference values y_i.

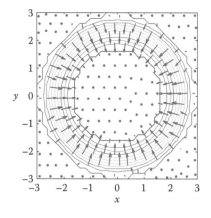

FIGURE 5.23
Gradient map for z.

Using global linear methods, such as Pearson's correlation, will not detect local zones of influence as was done with local models. In a symmetrical data set, like the one in the example, global methods can detect no influence at all when the truth is very different.

A different useful representation is the gradient map, as it is often used in many fields of knowledge. The gradient map is created representing vectors a_i in their prototype positions w_i in a bidimensional map. Owing to the small dimensionality of the representation, variables will often have to be chosen in pairs to be represented.

Gradient vectors in Figure 5.23 point towards the higher value of z. This information is useful for control applications as the optimum direction of growth is known. If the working conditions are $(x, y) = (-2, 0)$ and an increment of z is desired, the map shows that the most interesting direction is increasing x and keeping y constant.

5.6 Application to Real-World Data

The previous sections had some toy examples to support explanations and to allow an understanding of the concepts, but they do not demonstrate the usefulness of the proposed algorithms. The uses of the proposed algorithms will be demonstrated by showing their application in some real-world applications. All data sets are obtained from public sources and can be reproduced by any researcher.

5.6.1 Classification of Iris Data Set

Quantisation features have already been demonstrated in the previous sections. The effect of γ is already known for quantisation error and topographical preservation and now a different use will be presented using the Iris data set.

The unsupervised version of the hybrid algorithm can be used as a classification tool when classes are grouped in the input space. Labelling the prototypes and assuming that all the prototypes in their Voronoi region belong to the same class is a simple and fast way to make a classification. It is not as reliable as a supervised classification algorithm but it is a good complementary use.

Using the Iris data set,[*] presented in the previous sections, the classification accuracy will be studied. As with any other feature, the amount of prototypes is important for accuracy. The data set has 150 samples, so the map should have a small amount of prototypes.

Training with map sizes 5×4, 8×4 and 8×6, the following average accuracies were obtained after five runs of each one. The accuracy is measured as the percentage of samples classified in the right category.

In Figure 5.24, the difference between different map sizes is evident. The more prototypes that are used, the more accurate the model will be. In smaller maps, the effect of γ is also more significant, as the biggest map has a close to plain result in 97% and the smallest one has a wider range of results.

As was stated, the algorithm is not a supervised classification algorithm, but if the results are compared with the algorithms created for that purpose, they are good enough. In Ref. [19], many algorithms were employed to classify the Iris data set and most of them had apparent accuracies of about 98%.

---

[*] UCI Machine Learning Repository, Iris data set (http://archive.ics.uci.edu/ml/datasets/Iris).

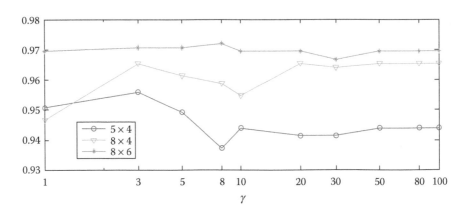

FIGURE 5.24
Classification accuracy versus γ for different map sizes.

5.6.2 Estimation of Oil Consumption for Cars

The supervised version of the algorithm is a good tool for estimation tasks. With the support of well-located prototypes, local models are more accurate than others created with less accurate quantisation methods. The accuracy of the algorithm is demonstrated with an example using a public data set from a UCI repository called 'Auto MPG'.[*] Using these data, an estimation of the fuel consumption is done for different car models. The data set is a collection of 392 cars (after removing those with missing values) with different features. Car origin and manufacturing year are not taken into account as they are not numerical variables and their relationship with consumption is not direct. The following input variables are used: cylinders, displacement, horsepower, weight and acceleration.

The output variable is the fuel consumption measured in miles per gallon. It is important to realise that low values mean high consumption and vice versa.

Data were normalised to zero mean and unity standard deviation to avoid interferences because of the magnitudes, and a 16 × 8 hexagonal map was trained with $\gamma = 10$. Using normalised data samples, it was a quantisation error of 0.193 and a topographic error of 0.207.

The estimation has a root mean squared error of 1.26, a negligible mean error ($\sim 2 \times 10^{-17}$) and a mean of absolute deviation of 0.623 (Figure 5.25).

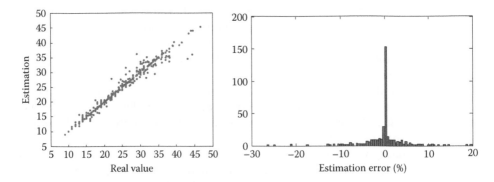

FIGURE 5.25
Estimation versus real values and histogram of the estimation error.

---

[*] UCI Machine Learning Repository, Auto MPG data set (http://archive.ics.uci.edu/ml/datasets/Auto + MPG).

Self-Organizing Maps

The model offers accurate estimations. Close to one half of the samples (47.2%) have an error less than 0.5% and about 84% have an error less than 5%. The accuracy of the algorithm will be compared with a batch version of the local modelling NG with the same amount of prototypes and an MLP with the same number of hidden neurons in one hidden layer.

	SOM–NG	MLP	NG
128 neurons	1.085 ± 0.150	4.399 ± 1.045	1.105 ± 0.137
96 neurons	1.579 ± 0.105	3.805 ± 0.501	1.503 ± 0.082
64 neurons	2.119 ± 0.061	3.637 ± 0.358	2.079 ± 0.082

These results were obtained by running each algorithm 20 times. The SOM–NG algorithm was trained using maps sized 16×8, 12×8 and 8×8, all of them with $\gamma = 10$. For NG runs, the prototypes were located in the same initial positions than in those for SOM–NG. For MLP, the Levenberg–Marquard algorithm was used.

Local linear models have more accurate results and the algorithms are more robust, as their standard deviation is significantly lower than the ones obtained using MLPs.

5.6.3 Searching for Local Relevancies

Using local models makes it easier to detect zones of influence for a variable. Global models try to fit the output in a single function that should work for all circumstances. But in real-world applications, local influences are usual, when certain variables are influential only under certain circumstances. Using energy related data obtained from the word development indicators in The World Bank,[*] a study of the influence of different indicators over CO_2 emissions will be done.

A fast selection of variables, most of them related to electricity generation and consumption, is done and the following ones are selected: combustible renewables and waste (% of total energy), electric power consumption (kWh per capita), electricity production from coal sources (% of total), electricity production from hydroelectric sources (% of total), electricity production from natural gas sources (% of total), electricity production from nuclear sources (% of total), electricity production from oil sources (% of total), electricity production from renewable sources, excluding hydroelectric (% of total), and gross domestic product (GDP) per capita (constant 2000 US$).

Other interesting variables were not selected because of their amount of missing values, but these are enough for our purpose. All the variables are relative because quantity-related influences are not desirable. In the same conditions, a bigger territory will have higher consumptions or productions. As the data set includes data from different zones and years, units per capita and constant currency units are used.

Component planes in Figure 5.26 show that CO_2 emissions are correlated with the electric power consumption and GDP; that influence should be taken into account as it modifies the reference value y and is a great influence. Other component planes show zones of generation from different sources. Each source has a zone of primary influence that covers most of the electricity production. Using only the electricity generation variables, a U-matrix will be calculated to label the different zones (Figure 5.27).

---

[*] Downloaded from http://data.worldbank.org/data-catalog/world-development-indicators.

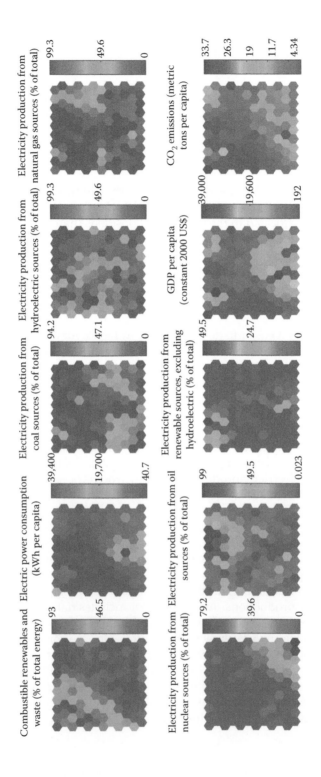

FIGURE 5.26
Component planes for CO_2 emissions data.

Self-Organizing Maps   147

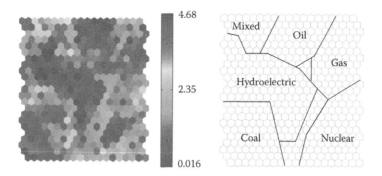

FIGURE 5.27
U-matrix and labelled zones for CO_2 emissions data.

Zones labelled as 'oil', 'gas', 'hydroelectric', 'coal' and 'nuclear' are the zones where one of the sources is predominant. The zone labelled as 'mixed' and the unlabelled ones are not clearly dominated by one of them. Gradient component planes are shown in Figure 5.28.

A fast visual inspection of Figure 5.28 may imply that the gradient component planes give very little or no information at all. This conclusion is incorrect because a more detailed observation can reveal that only one or two prototypes have close to the maximum or minimum value of the colour scale. So, they can be considered outliers or at least not relevant for local analysis.

In addition to their high influence over the reference values for each prototype, the electric power consumption and the GDP variables also have positive gradients with values greater than most of the other variables.

Figure 5.29 represents the values and gradients for nuclear sources. Most of the prototypes have zero value because nuclear plants are not as common as other power plants. Gradients outside the 'nuclear' zone also have close to zero gradients, which fits the general idea. If there are no nuclear plants, they will not modify the CO_2 emissions. The lower left corner also has gradient values close to zero. In those zones, there is a vast majority of nuclear power and small variations that are not as important as in zones with intermediate values. This effect is very similar to the top of the mountain in Figures 5.21 and 5.22.

5.7 Conclusions

In this chapter, the hybrid SOM–NG is described. It combines two well-known algorithms to produce a competitive result that can be adjusted to obtain the best trade-off for each application.

Using the topology preservation constant γ tunes the behaviour of the algorithm, controlling the freedom of the prototypes to move without the restriction of the adjacent prototypes in the grid. Using values of γ in the order of units will keep a low topographical restriction, so quantisation is better and closer to the optimum NG. Otherwise, high values of γ will be more restrictive and have results closer to SOM. But these extreme configurations are not the best application for the hybrid algorithm, as similar results

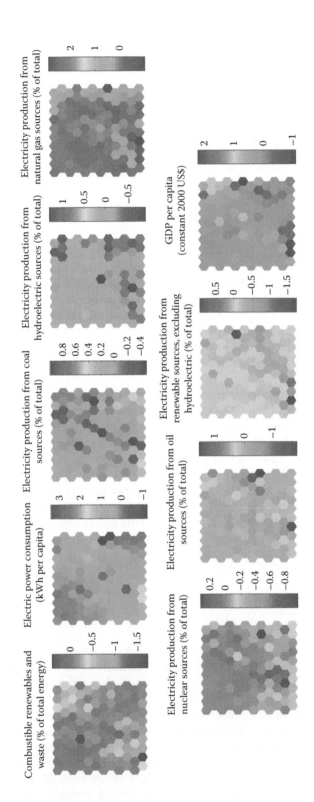

FIGURE 5.28
Gradient component planes for CO_2 emissions data.

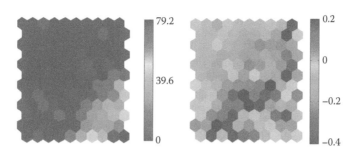

FIGURE 5.29
Component and gradient components for nuclear electricity sources.

can also be obtained using classic NG or SOM. The most interesting results are obtained using values of γ of a few tens, keeping good results in both quantisation and topological preservation.

The hybrid SOM–NG algorithm is good for quantisation, but it can also be used for many other applications, alone or combined with other techniques. The unsupervised version is good for clustering applications as the prototypes keep an ordered representation, which fits the data and can be used for clustering using agglomerative clustering techniques or distance measures like the U-matrix. Clustering prototypes instead of data samples quickens the process as there is no need to check every simple data sample.

The topology preservation of the prototypes over a low-dimensional grid allows component plane representation. Component planes are a representation of the value of a single variable in colour scale using the grid used for topology preservation during the training phase. Component planes are a very useful data mining tool and they allow a trained user to search for non-linear and local correlations.

The topological preservation is also helpful to identify current status in online applications. After acquiring a data for all the expected situations, a status map can be created. Identifying the bmu in a labelled map keeps track of the evolution and helps to predict evolutions. Using an interpolating technique, like GRNN, continuous maps can be created to allow for more accurate tracking.

The supervised version is a good estimator for multidimensional functions. It calculates local models that are useful if the analysed data have different zones of influence or different working states. These models are more accurate than trying to find a general solution that fits the complete data set.

Furthermore, the local linear models are a data mining tool that allows the exploration and visualisation correlations between variables and their gradients using the gradient component planes in combination with the component planes. Gradient maps can be created as well, representing the gradients in the corresponding prototype position. These gradient maps will focus on the zones with more data samples, which are supposed to be the most interesting zones for analysis.

The main drawback of the algorithm is the computational cost. Using a batch algorithm requires working with lots of data at the same time, but the data can be divided into smaller chunks to avoid keeping lots of intermediate results in memory at the same time. However, the amount of operations is still greater than the original algorithms as both neighbourhood functions need to be calculated for each sample–prototype pair.

References

1. J. MacQueen, Some methods for classification and analysis of multivariate observations, in *Proceedings of 5th Berkeley Symposium on Mathematical Statistics and Probability*, Berkeley, 1967.
2. T. M. Martinetz, S. G. Berkovich and K. J. Schulten, "Neural-gas" network for vector quantization and its application to time-series prediction, *IEEE Transactions on Neural Networks*, 4(4), 558–569, 1993.
3. M. Cottrell, B. Hammer, A. Hasenfuss and T. Villman, Batch neural gas, in *International Workshop on Self-Organizing Maps (WSOM 2005)*, Paris, France, 2005.
4. B. Fritzke, A growing neural gas network learns topologies, *Advances in Neural Information Processing Systems*, 7(7), 625–632, 1995.
5. B. Hammer, M. Strickert and T. Villmann, Supervised neural gas with general similarity measure, *Neural Processing Letters*, 21, 21–44, 2005.
6. T. Kohonen, The self-organizing map, *Proceedings of the IEEE*, 78(9), 1464–1480, 1990.
7. T. Kohonen, *Self-Organizing Maps*, Springer, Berlin, Germany, 2001.
8. A. Ultsch and H. P. Siemon, Kohonen's self organizing feature maps for exploratory data analysis, in *Proceedings of International Neural Networks Conference (INNC)*, Paris, 1990.
9. G. J. Chappell and J. G. Taylor, The temporal Kohonen map, *Neural Networks*, 6(3), 441–445, 1993.
10. M. Varsta, J. Heikkonen and J. Millan, Context learning with the self-organizing map, in *Proceedings of Workshop on Self-Organizing Maps*, Helsinki, Finland, 1997.
11. M. Salhi, N. Arous and N. Ellouze, Principal temporal extensions of SOM: Overview, *International Journal of Signal Processing*, 2, 61–84, 2009.
12. F. Nabhani and T. Shaw, Performance analysis and optimisation of shape recognition and classification, *Robotics and Computer-Integrated Manufacturing*, 18(3), 177–185, 2002.
13. M. O. Afolabi and O. Olude, Predicting stock prices using a hybrid Kohonen self organizing map (SOM), in *40th Annual Hawaii International Conference on System Sciences*, Waikoloa, HI, 2007.
14. T. Furukawa, SOM of SOMs, *Neural Networks*, 22(4), 463–478, 2009.
15. I. Machón González, H. López García and J. L. Calvo Rolle, A hybrid batch SOM-NG algorithm, in *International Joint Conference on Neural Networks*, 2010.
16. K. Kiviluoto, Topology preservation in self-organizing maps, in *IEEE International Conference on Neural Networks*, 1996.
17. D. F. Specht, A general regression neural network, *IEEE Transactions on Neural Networks*, 2(6), 568–576, 1991.
18. M. J. Crespo Ramos, I. Machón González, H. López García and J. L. Calvo Rolle, Supervised hybrid SOM-NG algorithm, in *The Fifth International Conference on Advanced Engineering Computing and Applications in Sciences*, Lisbon, 2011.
19. S. M. Weiss and I. Kapouleas, An empirical comparison of pattern recognition, neural nets, and machine learning classification methods, in *Proceedings of the 11th International Joint Conference on Artificial Intelligence*, Detroit, MI, 1989.

6

A Semi-Supervised and Active Learning Method for Alternatives Ranking Functions

Faïza Dammak, Hager Kammoun and Abdelmajid Ben Hamadou

CONTENTS

6.1 Introduction 151
6.2 Learning to Rank 152
    6.2.1 Ranking of Alternatives 153
    6.2.2 RankBoost Algorithm 153
    6.2.3 Semi-Supervised Ranking 153
6.3 Proposition for a Semi-Supervised Method 154
    6.3.1 Adaptation of the RankBoost Algorithm to a Semi-Supervised Ranking of Alternatives 154
    6.3.2 Adaptation of the Algorithm of Selection of Ranking Features 157
6.4 Adaptation of the RankBoost Algorithm to an Active Ranking of Alternatives 159
6.5 Experiments 160
6.6 Conclusion 163
References 163

6.1 Introduction

Learning to rank is a relatively new research area that has emerged rapidly in the past decade. It plays a critical role in information retrieval (IR). Learning to rank is to learn a ranking function by assigning a weight to each document feature, then using this obtained ranking function to estimate relevance scores for each document, and finally ranking these documents based on the estimated relevance scores [1,2]. This process has recently gained much attention in learning due to its large applications in real problems such as IR. In learning to rank, the performance of a ranking model is strongly affected by the number of labelled examples in the training set; therefore, labelling large examples may require expensive human resources and be time consuming, especially for ranking problems. This presents a great need for semi-supervised learning approaches [3] in which a model is constructed with a small number of labelled instances and a large number of unlabelled instances. Semi-supervised learning is a well-known strategy to label unlabelled data using certain techniques and thus increase the amount of labelled training data [4].

Ranking is the central problem for many IR applications. It aims to induce an ordering or preference relation over a predefined set of labelled instances. For example, this is a case of document retrieval (DR), where the goal is to rank documents from a collection based on their relevancy to a user's query. This type of problem is known under the name of ranking for alternatives [1]. The ranking of instances is another type of ranking that comes from IR such as routing information [5].

Since obtaining labelled examples for training data is very expensive and time consuming, it is preferable to integrate unlabelled data in the training base.

Most semi-supervised ranking algorithms are graph-based transductive techniques [6]. These techniques cannot easily extend to new test points outside the labelled and unlabelled training data. Induction has recently received increasing attention.

For an effective use of the semi-supervised learning on large collections of data, Ref. [5] presents a boosting-based algorithm for learning a bipartite ranking function (BRF) for instances. This is an extended version of the RankBoost algorithm [7] that optimises an exponential upper bound of a learning criterion that combines the misordering loss for both parts of the training set. We propose an adaptation of the supervised RankBoost algorithm on partially labelled data of alternatives that can be applied to some applications such as web search. Our algorithm, based on a pairwise approach [8], takes query–document pairs as instances in learning.

Our contribution is to develop two methods for alternatives ranking functions. The first method is a semi-supervised ranking algorithm for alternatives and the second is an active learning method for alternatives ranking functions. The proposed algorithms have an inductive character since they are able to infer an ordering on new examples that were not used for their training [4]. The unlabelled data will be initially labelled by a transductive method (TM) such as the K-nearest neighbours (KNN).

The rest of the chapter is organised as follows: Section 6.2 provides a brief literature review to the related work; we introduce the principle learning to rank and its interest into the IR. We also detail the problem of ranking of alternatives, the RankBoost algorithm and the principle of semi-supervised learning. In Section 6.3, we present our proposal for a semi-supervised method. We propose, in Section 6.4, an active learning method of ranking functions of alternatives. The collections used and experimental results are detailed in Section 6.5. Finally, Section 6.6 concludes the chapter and gives directions for future work.

6.2 Learning to Rank

In learning to rank, a number of queries are provided, and each query is associated with a perfect ranking list of documents; a ranking function assigns a score to each document, and ranks the documents in descending order of the scores [7]. The ranking order represents the relative relevance of documents with respect to the query. In a problem related to learning to rank, an instance is a set of objects and a label is a sorting applied over the instance. Learning to rank aims to construct a ranking model from training data.

Many applications of learning to rank involve a large number of unlabelled examples and a few labelled examples, as expensive human effort is usually required in labelling examples [7].

The issue of effectively exploiting the information in the unlabelled instances to facilitate supervised learning, known as semi-supervised learning, has been extensively studied [2]. We are interested in applying the supervised RankBoost algorithm with this type of learning. Indeed, RankBoost has an inductive character; it is thus able to order a list of examples not seen during the phase of training by inferring an order on this list. In the following sections, we present the principle of the ranking for alternatives, the RankBoost algorithm as well as the principle of a semi-supervised ranking algorithm.

6.2.1 Ranking of Alternatives

Learning to rank is a newly popular topic in machine learning. When it is applied to DR, it can be described as the following problem: assume that there is a collection of documents. In retrieval, giving a query, the ranking function assigns a score to each pair of query–document and ranks the documents in descending order of these scores. The ranking order represents the relevance of documents according to the query. The relevance scores can be calculated by a ranking function constructed with machine learning. This type of ranking is known as the ranking of alternatives [1].

6.2.2 RankBoost Algorithm

RankBoost is a supervised learning algorithm of instances designed for ranking problems. It builds a document ranking function by combining a set of ranking features of a set of document pairs [3].

More precisely, RankBoost learns a ranking feature f_t on each round, and maintains a distribution D_t over the ranked pairs. The final ranking function F is a linear combination of these ranking features:

$$F = \sum_{t=1}^{T} \alpha_t f_t \tag{6.1}$$

Each ranking feature f_t is uniquely defined by an input feature $j_t \in \{1 \ldots d\}$ and a threshold θ_t

$$f_t(x) = \begin{cases} 1, & \text{if } \varphi_{jt}(x) > \theta_t \\ 0, & \text{else} \end{cases} \tag{6.2}$$

where $\varphi_{jt}(x)$ is the jth feature characteristic of x.

Assume that, for all example pairs, one knows which example should be ranked above the other one. The learning criterion to be minimised in RankBoost is the number of example pairs whose relative ranking as computed by the final combination is incorrect.

6.2.3 Semi-Supervised Ranking

Semi-supervised ranking has a great interest in machine learning because it can readily use the available unlabelled data to improve supervised learning tasks when the labelled data are scarce or expensive. Semi-supervised ranking also shows potential as a quantitative tool to understand human category learning, where most of the input is self-evidently unlabelled.

The majority of the semi-supervised ranking algorithms are transductive techniques based on a valuated and non-oriented graph. The latter is formed by gradually connecting the nearest points until the graph becomes connected. The nodes consist of the examples labelled and unlabelled of the training base and the weights reflect the similarity between the neighbouring examples. This graph is built with a method, such as KNN, which allows the labels of the unlabelled examples to be found by exploiting the graph directly by propagating; for example, the labels of the data labelled with their unlabelled neighbours. It thus affects a score for each instance, 1 for the positive instances and 0 for the others. The scores are then propagated through the graph until the convergence. At the end, the scores obtained make it possible to induce an order on the whole of the unlabelled instances [4]. We chose this method in our context to label the unlabelled data in the training set.

In our proposal, these data will be used with the labelled data as inputs which have the advantage of both the inductive and the transductive approaches. We thus propose a semi-supervised algorithm that is able to infer an ordering on new pairs of query–alternatives that were not used for its training. We detail this proposal in the following section.

6.3 Proposition for a Semi-Supervised Method

In training, a set of queries $X = \{x_1, x_2, \ldots, x_m\}$ and a set of alternatives Y is given. Each query $x_i \in X$ is associated with a list of retrieved alternatives of variable size m_i, $y_i = (y_i^1, \ldots, y_i^{m_i})$, with $y_i^k \in \mathrm{IR}$. y_i^k represents the degree of relevance of the alternative k from x_i. A feature vector $\varphi_j(x_i, k)$ is created from each query–document pair (x_i, k) [5].

The ranking function f_t allows a score to be associated for this vector. We thus propose a labelled learning base $S = \{(x_i, y_i)\}_{i=1}^{m}$ and an unlabelled learning base formed with all parts of queries unlabelled $S_U = \{(x_i')\}_{i=m+1}^{m+n}$.

We propose a semi-supervised learning method of ranking functions of alternatives. The principal motivation that led us to the development of this proposal was to find an effective ranking function. It is necessary to have a base of learning that often requires the manual labelling of the alternatives on several examples. The goal is to find the best number of labels to reduce the number of labelled data. For an effective use of the semi-supervised learning on large collections, we adapt a modification of the supervised ranking RankBoost algorithm. We present the model suggested and describe its functionalities as well as the choices of implementation.

Once the ranking function is learnt, it can be used to order any set. It breaks down as follows in two steps: calculating scores for each jurisdiction and order sets based on the score. Noting n the number of instances, the first step requires $O(n \cdot d)$ and calculates the second step that is made with a sorting algorithm effective as quicksort [9], which has an average complexity of $O(n \cdot \log(n))$. Thus, the total complexity is

$$\text{Inference complexity} = O(d \cdot n + n \cdot \log(n))$$

In the following part, we detail the operation of the RankBoost algorithm applied to our context.

6.3.1 Adaptation of the RankBoost Algorithm to a Semi-Supervised Ranking of Alternatives

The adaptation of RankBoost is given in Algorithm 1: we dispose of a labelled training set $S = \{(x_1, y_1), \ldots, (x_m, y_m)\}$, where each example x_i is associated with a vector of relevance judgement $y_i = (y_i^1, \ldots, y_i^{m_i})$, where $y_i^k \in \mathrm{IR}$. m_i denotes the number of alternatives for x_i.

$S' = \{(x_i', y_i'); i \in \{m+1, \ldots, m+n\}\}$ is the second labelled subset obtained from the unlabelled set S_U by using the nearest-neighbours (NN) algorithm.

A Semi-Supervised and Active Learning Method for Alternatives Ranking Functions

At each iteration, the algorithm maintains a distribution λ_t (resp. λ_t') on the examples of the learning base S (resp. S'), a distribution v_t^i (resp. $v_t^{i\prime}$) on the alternatives associated with the example x_i (resp. x_i') and a distribution D_t^i (resp. $D_t^{i\prime}$) over the pairs (query, alternative), represented by a distribution on couples (k, l) (resp. (k', l')) such as $y_i^k \in Y_+$ (resp. $y_i^{k\prime} \in Y_+'$) and $y_i^l \in Y_-$ (resp. $y_i^{l\prime} \in Y_-'$) for each example x_i (resp. x_i').

$\forall i \in \{1, \ldots, m\}, \forall (k, l) \in \{1, \ldots, m_i\}^2$ such as $y_i^k \in Y_+, y_i^l \in Y_-$,

$$D_t^i(k, l) = \lambda_t^i v_t^i(k) v_t^i(l) \tag{6.3}$$

$\forall i \in \{m+1, \ldots, m+n\}, \forall (k', l') \in \{1, \ldots, m_i'\}^2$ such as $y_i^{k\prime} \in Y_+', y_i^{l\prime} \in Y_-'$:

$$D_t^{i\prime}(k', l') = \lambda_t^{i\prime} v_t^{i\prime}(k') v_t^{i\prime}(l') \tag{6.4}$$

These distributions are updated due to the scoring function f_t, selected from the semi-supervised learning of the ranking features algorithm (Algorithm 2) that returns the resulting value of the threshold θ_{res} associated with each characteristic and the possible values that can be associated with f_t, such as

$$f_t(x_i, k) = \begin{cases} 1 & \text{if } \varphi_j(x_i, k) > \theta_{res} \\ 0 & \text{if } \varphi_j(x_i, k) \leq \theta_{res} \end{cases} \tag{6.5}$$

where x_i is the query of index i and k is the index of the alternative associated with x_i.

For each example, the weight α_t is defined by [3]

$$\alpha_t = \frac{1}{2} \ln\left(\frac{1 + r_t}{1 - r_t}\right) \tag{6.6}$$

where

$$r_t = \sum_{k,l} D_t^i(k, l)(f_t(x_i, k) - f_t(x_i, l)) + \beta \sum_{k', l'} D_t^{i\prime}(k', l')(f_t(x_i', k') - f_t(x_i', l')). \tag{6.7}$$

β is a discount factor. We fall back on the situation of supervised learning when this factor is zero.

Algorithm 1: RankBoost Algorithm Adapted to a Ranking of Alternatives: RankBoostSSA

Entry: A labelled learning set $S = \{(x_i, y_i); i \in \{1, \ldots, m\}\}$
A labelled learning set $S' = \{(x_i', y_i'); i \in \{m+1, \ldots, m+n\}\}$ obtained by the KNN method.

Initialisation:

$$\forall i \in \{1, \ldots, m\}, \lambda_1^i = \frac{1}{m}, v_1^i(k) = \begin{cases} \dfrac{1}{p_i} & \text{if } y_i^k \in Y_+ \\ \dfrac{1}{n_i} & \text{if } y_i^k \in Y_- \end{cases}$$

$$\forall i \in \{m+1, \ldots, m+n\}, \lambda_1^{i'} = \frac{1}{n}, v_1^{i'}(k) = \begin{cases} \dfrac{1}{p_i'} & \text{if } y_i^{k'} \in Y_+' \\ \dfrac{1}{n_i'} & \text{if } y_i^{k'} \in Y_-' \end{cases} \qquad v_1^i(k) = \begin{cases} \dfrac{1}{p_i} & \text{if } y_i^k \in Y_+ \\ \dfrac{1}{n_i} & \text{if } y_i^k \in Y_- \end{cases}$$

For $t := 1, \ldots, T$ do

– Select the ranking feature f_t from D_t and D_t'
– Calculate α_t using formula (6.6)

$-\forall i \in \{1, \ldots, m\}, \forall (k,l) \in \{1, \ldots, m_i\}^2$ such as $y_i^k \in Y_+, y_i^l \in Y_-$, update $D_{t+1}^i(k,l)$:

$$D_{t+1}^i(k,l) = \lambda_{t+1}^i v_{t+1}^i(k) v_{t+1}^i(l)$$

$-\forall i \in \{m+1, \ldots, m+n\}, \forall (k',l') \in \{1, \ldots, m_i'\}^2$ such as $y_i^{k'} \in Y_+', y_i^{l'} \in Y_-'$, update $D_{t+1}^{i'}(k',l')$:

$$D_{t+1}^{i'}(k',l') = \lambda_{t+1}^{i'} v_{t+1}^{i'}(k') v_{t+1}^{i'}(l')$$

$-\forall i \in \{1, \ldots, m\}, \lambda_{t+1}^i = \dfrac{\lambda_t^i Z_t^{1i}}{Z_t},$

$$v_{t+1}^i(k) = \begin{cases} \dfrac{v_t^i(k) \exp(-\alpha_t f_t(x_i,k))}{Z_t^{1i}} & \text{if } y_i^k \in Y_+ \\ \dfrac{v_t^i(k) \exp(\alpha_t f_t(x_i,k))}{Z_t^{-1i}} & \text{if } y_i^k \in Y_- \end{cases}$$

where Z_t^{1i}, Z_t^{-1i} and Z_t are defined by

$$Z_t^{1i} = \sum_{k: y_i^k \in Y_+} v_t^i(k) \exp(-\alpha_t f_t(x_i,k)),$$

$$Z_t^{-1i} = \sum_{l: y_i^l \in Y_-} v_t^i(l) \exp(\alpha_t f_t(x_i,l)), \quad Z_t = \sum_{i=1}^m \lambda_t^i Z_t^{-1i} Z_t^{1i}$$

$-\forall i \in \{m+1, \ldots, m+n\}, \lambda_{t+1}^{i'} = \dfrac{\lambda_t^{i'} Z_t^{-1i'} Z_t^{1i'}}{Z_t'},$

$$v_{t+1}^{i'}(k') = \begin{cases} \dfrac{v_t^{i'}(k') \exp(-\alpha_t f_t(x_i',k'))}{Z_t^{1i'}} & \text{if } y_i^{k'} \in Y_+' \\ \dfrac{v_t^{i'}(k') \exp(\alpha_t f_t(x_i',k'))}{Z_t^{-1i'}} & \text{if } y_i^{k'} \in Y_-' \end{cases}$$

… A Semi-Supervised and Active Learning Method for Alternatives Ranking Functions

where $Z_t^{1i'}$, $Z_t^{-1i'}$ and Z_t' are defined by

$$Z_t^{1i'} = \sum_{k':y_i^{k'} \in Y_+'} v_t^{i'}(k') \exp(-\alpha_t f_t(x_i', k'))$$

$$Z_t^{-1i'} = \sum_{l':y_i^{l'} \in Y_-'} v_t^{i'}(l') \exp(\alpha_t f_t(x_i', l'))$$

$$Z_t' = \sum_{i=1+m}^{m+n} \lambda_t^{i'} Z_t^{-1i'} Z_t^{1i'}$$

End

Output: The final ranking function $F = \sum_{t=1}^{T} \alpha_t f_t$

In each iteration t, α_t is selected to minimise the normalisation factors Z_t and Z_t'.

Our goal in this algorithm is to find a function F, which minimises the average numbers of irrelevant alternatives that scored better than the relevant ones in S and S' separately. We call this quantity the average ranking loss for alternatives, $Rloss(F, S \cup S')$ defined as

$$Rloss(F, S \cup S') = \frac{1}{m} \sum_{i=1}^{m} \frac{1}{n_i p_i} \sum_{k:y_i^k \in Y_+} \sum_{l:y_i^k \in Y_-} [[f(x_i, k) - f(x_i, l) \leq 0]]$$

$$+ \frac{\beta}{n} \sum_{i=1}^{m} \frac{1}{n_i' p_i'} \sum_{k':y_i^k \in Y_+'} \sum_{l':y_i^k \in Y_-'} [[f(x_i', k') - f(x_i', l') \leq 0]] \quad (6.8)$$

where p_i (resp. n_i) is the number of relevant alternatives (resp. not relevant); for example, x_i in S and p'_i (resp. n'_i) are the number of relevant alternatives (resp. not relevant); for example, x_i in S'.

[[P]] is defined to be 1 if predicate P is true and 0 otherwise.

6.3.2 Adaptation of the Algorithm of Selection of Ranking Features

The algorithm of selection of ranking features or functions (Algorithm 2) makes it possible to find, with a linear complexity in a number of alternatives, a function f_t that minimises r_t in a particular case where the functions f_t are in {0,1} and created by thresholded characteristics associated with the examples.

Let us suppose that each query x_i (resp. x_i') has a set of characteristics provided by functions φ_j, $j = 1 \ldots d$. For each j, $\varphi_j(x_i, k)$ (resp. $\varphi_j(x_i', k')$) is a real value. Thus, it is a question of using a thresholding of the characteristic φ_j to create binary values. The set of the basic functions to combine east creates by defining *a priori* a set of thresholds $\{\theta_q\}_{q=1}^{Q}$ with $\theta_1 > \ldots > \theta_q$. Generally, these thresholds depend on the characteristic considered.

Algorithm 2: Algorithm of Selection of Ranking Features

Entry:
$\forall i \in \{1, \ldots, m\}$, $(k, l) \in \{1, \ldots, m_i\}$ such as $y_i^k \in Y_+$ and $y_i^l \in Y_-$:

A distribution $D_t^i(k,l) = \lambda_t^i \, v_t^i(k) \, v_t^i(l)$ on the training set S.

$\forall i \in \{m+1, \ldots, m+n\}, \forall (k', l') \in \{1, \ldots, m_i'\}^2$ such as $y_i^{k'} \in Y_+', y_i^{l'} \in Y_-'$:

A distribution $D_{t+1}^{i}{}'(k',l') = \lambda_{t+1}^{i}{}' v_{t+1}^{i}{}'(k') v_{t+1}^{i}{}'(l')$ on the training subset S'.

Set of characteristics $\{\varphi_j(x_i, k)\}_{j=1}^{d}$

For each φ_j, a set of thresholds $\{\theta_q\}_{q=1}^{Q}$ such as $\theta_1 > \ldots > \theta_q$

Initialisation:

$\forall i \in \{1, \ldots, m\}, (k, l) \in \{1, \ldots, m_i\},$

$$\pi(x_i, k) = y_i^k \, \lambda_1^i \, v_1^i(k) \sum_{l:y_i^l \neq y_i^k} v_1^i(l)$$

$\forall i \in \{m+1, \ldots, m+n\}, (k', l') \in \{1, \ldots, m_i'\},$

$$\pi'(x_i', k') = y_i^{k'} \, \lambda_1^{i}{}' \, v_1^{i}{}'(k') \sum_{l': y_i^{l'} \neq y_i^{k'}} v_1^i(l')$$

$r^* \leftarrow 0$

For $j := 1, \ldots, d$ **do**

    $-\,L \leftarrow 0$

    For $q := 1, \ldots, Q$ **do**

$$L \leftarrow L + \sum_{i=1}^{m} \sum_{k:\varphi_j(x_i,k)} \pi(x_i, k)$$

$$+ \sum_{i=m+1}^{m+n} \sum_{k':\varphi_j(x_i',k')} \pi'(x_i', k')$$

        If $|L| > |r^*|$ **then**

            $r^* \leftarrow L$

            $j^* \leftarrow j$

            $\theta^* \leftarrow \theta_q$

            $k^* \leftarrow k$

        end

    end

end

Output: $(\varphi_j^*, \theta^*, k^*)$

At each iteration, the RankBoostSSA algorithm learns a score function inefficient, updates the weights v, v', λ and λ' and calculates the terms D_t and D_t' to estimate α_t. An exhaustive search of the optimal basis function [10] can only be done in $O(K \cdot d(n + n'))$ operations, where n is the number of alternatives labelled by the TM. This step is the most expensive at each iteration. We deduce the total complexity of the algorithm:

$$\text{Complexity [RankBoostSSA]} = O(dK(n + n')) + \text{complex[TM]}$$

We used the k-NN to the labelling. The algorithm must calculate the distance between each example unlabelled and each example labelled ($O(d \cdot n \cdot m)$) and must maintain for each example labelled the k-NN ($O(k \cdot \log k)$). The total complexity is dominated by a step-exhaustive search and the calculation of scalar products when k is small, that is

$$\text{Complexity [knn-RankBoostSSA]} = O(d \cdot K \cdot K \cdot n + d \cdot n \cdot m)$$

Thus, complexity is linear with respect to all model parameters.

6.4 Adaptation of the RankBoost Algorithm to an Active Ranking of Alternatives

We propose in this section an active learning method of ranking functions of alternatives. The principal motivation, which led us to the development of this proposal, was to find an effective ranking function that often requires the manual labelling of the alternatives on several examples. The goal is to find the best entered to label to reduce the maximum number of labelled data. For an effective use of active learning on large collections, we adapt a modification of the supervised ranking RankBoost to the ranking of the alternatives algorithm. We present this model suggested and describe its functionalities as well as the choices of implementation.

We thus propose an algorithm of active learning for the ranking of alternatives by integrating the RankBoost algorithm, called RankBoost_Active.

The approach, which we have advocated in this section, is based on a selective sampling strategy for functions of ranking alternatives using unlabelled data.

We find in the literature two variants of active ranking. The first is to select an entry and label all the associated alternatives. It is suitable, for example, in automatic summarisation. The second variation seeks to select only one pair of entry–alternative. The user indicates whether the alternative is relevant or not with respect to this input. This approach is particularly well suited to applications such as IR, where an entry represents a query and the alternative documents must be ordered according to this entry. The labelling is to order a document from the query.

The first declension was handled by Truong [10] for a proposed active learning method for ranking alternatives. In our proposed model, we choose to use the second declension. We give a set of inputs X and a set of alternatives A. We assume that each instance x is associated with a subset of known alternatives $A_x \subset A$. We consider a training labelled set $S_L = \{x_i, y_i\}$ with x_i, an input, and y_i, a set of labels associated with A_x.

Furthermore, we consider another great set of inputs unlabelled S_U. The objective is to select one or more pairs of entry–alternative and label the alternatives related to best improve the performance of the ranking system.

We can summarise this approach by the following steps:

1. Select a pair of entry–alternative.
2. Ask the user to label the alternative on the function.
3. Learn a new scoring function.

Our selection strategy is based on a set of models called query-by-committee [11,12]. At each iteration, the models are assumed to be consistent with the training set labelled S_U. We consider that the pair entry-alternative is the most informative one for which all models disagree the most on the order induced on the set of alternatives A_x.

To consider the models of the committee, we will construct a random scoring function f^{sa} based on these models.

By definition, this random score function is defined on the committee of models $\{f_1^{sa}, \ldots, f_K^{sa}\}$, obtained by cross-validation on K partitions with size $1/K$. Each model f_K^{sa} is learnt on the labelled base S_L by removing the kth partition for this model. The function f^{sa} is then obtained by randomly choosing a model in the set $\{f_1^{sa}, \ldots, f_K^{sa}\}$ by following a uniform distribution.

Algorithm 3: Active Learning Algorithm for Alternatives

Entry: A labelled learning set $S = \{(x_i, y_i); i \in \{1, \ldots, m\}\}$
– An unlabelled learning set $S' = \{(x_i'); i \in \{m+1, \ldots, m+n\}\}$
– RankBoost algorithm adapted to ranking of alternatives (Algorithm 1)
– k: number of partitions of S
– Nb: number of labelled examples required
– Learn models of the committee and obtain f_K^{sa}
– nbIter $\leftarrow 0$

While nbIter $\leq$ Nb **do**
   – Learn a ranking function with RankBoost on S
   – Select the most useful pair of entry–alternative from S'
   – Ask the expert to label the alternatives based on this pair
   – Remove from S' and add in S with the labels on alternatives
   – Withdraw and add them in S with the labels on the alternatives

End

6.5 Experiments

We conducted experiments to test the performances of the proposed algorithms using the benchmark LETOR (learning to rank). This benchmark dataset was released in the SIGIR 2007 Workshop on Learning to Rank for Information Retrieval (LR4IR 2007). Since then, it has been widely used in many learning to rank papers [8,13]. LETOR was built based on two widely used data collections in IR: the OHSUMED collection used in the information filtering task of TREC 2000 and the '.gov' collection used in the topic of distillation tasks of TREC2003 and 2004. Accordingly, there are three sub-datasets in LETOR, namely OHSUMED, TD2003 and TD2004.

We used the MQ2008-semi (Million Query track) dataset in LETOR4.0 [1] in our experiments because it contains both labelled and unlabelled data. There are about 2000 queries in this dataset. On average, each query is associated with about 40 labelled documents and about 1000 unlabelled documents.

MQ2008-semi is conducted on the .GOV2 corpus using the TREC 2008, which is crawled from websites in the .gov domain. There are 25 million documents contained in the .GOV2 corpus, including hypertext markup language (HTML) documents, plus the extracted text of PDF, Word and postscript files [1].

Each subset of the collection MQ2008-semi is partitioned into five parts, denoted as S1, S2, S3, S4 and S5 to conduct fivefold cross-validation. The results reported in this section are the average results over multiple folds. For each fold, three parts are used for training, one part for validation and the remaining part for testing. The training set is used to learn the ranking model. The validation set is used to tune the parameters of the ranking model, such as the number of iterations in RankBoost. The test set is used to report the ranking performance of the model.

To compare the performance of the two algorithms proposed (semi-supervised algorithm and active algorithm), we evaluate our experimental results using a set of standard ranking measures such as mean average precision (MAP), precision at N and normalised discounted cumulative gain (NDCG).

$$P@n = \frac{\#relevant\ docs\ in\ top\ n\ results}{n} \qquad (6.9)$$

$$MAP = \frac{\sum_{n=1}^{N}(P@n * \text{rel}(n))}{\#total\ relevant\ docs\ for\ this\ query} \qquad (6.10)$$

$$NDCG(n) = Z_n \sum_{j=1}^{n} \frac{2^{r(j)} - 1}{\log(1 + j)} \qquad (6.11)$$

The value of the discount factor that provided the best ranking performance for these training sizes is $\beta = 1$ [7]. Therefore, we use this value in our experiments.

Tables 6.1 and 6.2 show the results of Algorithm 1 on testing the set generated by an assessment tool associated with the benchmark LETOR [1]. RankBoost [7] and RankSVM [14–16] were selected as baselines in these experiments.

These results illustrate how the unlabelled data affect the performance of ranking in the proposed algorithm. We notice a slight improvement in using the criterion $P@n$ (resp. NDCG) for $n = 1, 2, 3, 7$ and $n = 10$ (resp. for $n = 1, 2, 3, 7$ and $n = 10$). The results also show

TABLE 6.1

Performance of RankBoostSSA on Testing Set: $P@n$ and MAP Measures on the MQ2008-Semi Collection

Algorithms	P@1	P@2	P@3	P@4	P@5	P@6	P@7	P@8	P@10	MAP
RankBoost	0.4579	0.4114	0.3916	0.3642	0.3403	0.3210	0.3021	0.2846	0.2487	0.4775
RankSVM	0.4273	0.4068	0.3903	0.3695	0.3474	0.3265	0.3021	0.2822	0.2491	0.4695
RankBoostSSA	0.4586	0.4132	0.3931	0.3644	0.3417	0.3231	0.3028	0.2831	0.2522	0.4794

TABLE 6.2

Performance of RankBoostSSA on Testing Set: $P@n$ and MAP Measures on the MQ2008-Semi Collection

Algorithms	NDCG @1	NDCG @2	NDCG @3	NDCG @4	NDCG @5	NDCG @6	NDCG @7	NDCG @8	NDCG @9	NDCG @10
RankBoost	0.4632	0.4504	0.4554	0.4543	0.4494	0.4439	0.4412	0.4361	0.4326	0.4302
RankSVM	0.3626	0.3984	0.4285	0.4508	0.4695	0.4851	0.4905	0.4564	0.2239	0.2279
RankBoostSSA	0.4651	0.4517	0.4534	0.4463	0.4381	0.4643	0.4414	0.4638	0.4371	0.4346

that our proposed algorithm has an average precision (MAP) better than that found by RankBoost and RankSVM. These results prove the interest of integrating unlabelled data in ranking functions with semi-supervised learning.

The advantage of this model over other models is that it can advantageously exploit unlabelled data. The experiments that we described above clearly showed the contribution of unlabelled data in learning. Technically, the algorithm manages to exploit unlabelled data and reduce labelling errors. Also, note that RankBoost algorithms and our proposed extension are implicitly selecting features. This property may represent an advantage of our method compared to the supervised method, which makes it robust and applicable to a dynamic environment [17].

Tables 6.3 and 6.4 show the results of $P@n$ and MAP measures on the testing set of Algorithm 3: RankBoost_Active.

RankBoost [7] is selected as a baseline in this experiment. For the proposed algorithm RankBoost_Active, the parameter T was determined automatically during each experiment. Specifically, when there is no improvement in ranking accuracy in terms of the performance measure, the iteration stops (and T is determined).

These results illustrate how unlabelled data affect the performance of ranking in the proposed algorithm. We notice a slight improvement in using the criterion $P@n$ (resp. NDCG) for $n = 3$, $n = 5$ and $n = 10$ (resp. for $n = 1$, $n = 2$, $n = 3$ and $n = 5$). The results also show that our proposed algorithm has an average precision (MAP) better than that found by RankBoost. These results prove the interest of integrating unlabelled data in ranking functions with active learning.

TABLE 6.3

Performance of RankBoost_Active on Testing Set: $P@n$ and MAP Measures on the MQ2008-Semi Collection

Algorithms	P@1	P@2	P@3	P@5	P@7	P@10	MAP
RankBoost	0.4579	0.4114	0.3916	0.3403	0.3021	0.2487	0.4775
RankBoost_Active	0.4592	0.4127	0.3932	0.3411	0.3023	0.2526	0.4791

TABLE 6.4

Performance of RankBoost_Active on Testing Set: NDCG@n Measures on the MQ2008-Semi Collection

Algorithms	NDCG@1	NDCG@2	NDCG@3	NDCG@5	NDCG@7	NDCG@10
RankBoost	0.3856	0.3993	0.4288	0.4666	0.4898	0.2255
RankBoost_Active	0.3861	0.3997	0.4373	0.4682	0.4842	0.2249

6.6 Conclusion

In this chapter, we propose a semi-supervised learning algorithm called RankBoostSSA and an active learning algorithm called RankBoost_Active for learning ranking functions for alternatives. These algorithms are able to infer an ordering on new pairs of query–alternatives that were not used for its training. On the one hand, the RankBoostSSA algorithm can implicitly select characteristics. This property can be an advantage for this algorithm to the supervised method. Furthermore, unlike the vast majority of semi-supervised methods, RankBoostSSA is linear with respect to all parameters.

On the other, algorithm RankBoost_Active is able to advantageously exploit the unlabelled data and select the most informative examples for ranking learning because it is based on a selective sampling strategy for functions of ranking alternatives. This selection strategy is based on a set of models called query-by-committee that allows it to be generic. We propose in the stage to follow to supplement the experimental part and to integrate other methods of learning.

References

1. T.-Y. Liu, J. Xu, T. Qin, W.-Y. Xiong and H. Li, LETOR: Benchmark dataset for research on learning to rank for information retrieval. *SIGIR*, 3–10, 2007.
2. J. Xu and H. Li, AdaRank: A boosting algorithm for information retrieval. In W. Kraaij de Vries, A. P. Clarke, C. L. A. Fuhr and N. Kando, eds., *SIGIR*, pp. 391–398. ACM, Amsterdam, 2007.
3. X. Zhu, Semi-supervised learning literature survey. Technical Report 1530, Computer Sciences, University of Wisconsin-Madison, 2005.
4. K. Duh and K. Kirchhoff, Learning to rank with partially-labeled data. In Myaeng, S.-H., Oard, D. W., Sebastiani, F., Chua, T.-S. and Leong, M.-K., eds., *SIGIR*, pp. 251–258. ACM, New York, 2008.
5. M.-R Amini, V. Truong and C. Goutte, A boosting algorithm for learning bipartite ranking functions with partially labeled data. *SIGIR*, 99–106, 2008.
6. D. Zhou, J. Weston, A. Gretton, O. Bousquet and B. Schölkopf, *Ranking on Data Manifolds*. NIPS. MIT Press, Cambridge, Vol. 15, 2003.
7. Y. Freund, R. Iyer, R. E Schapire and Y. Singer, An efficient boosting algorithm for combining preferences. *Journal of Machine Learning Research*, 4: 933–969, 2003.
8. F. Xia, T. Liu, J. Wang, W. Zhang and H. Li, Listwise approach to learning to rank: Theory and algorithm. In *ICML '08*, pp. 1192–1199, New York, NY, USA, ACM, 2008.
9. C. A. R. Hoare, Quicksort. *The Computer Journal*, 5(1): 10–16, 1962.
10. T.-V. Truong, Learning functions ranking with little labeled examples, PhD thesis, University of Pierre and Marie Curie—Paris, 2009.
11. A. McCallum and K. Nigam, Employing EM in poolbased active learning for text classification. In Shavlik, J.W., ed., *ICML*, Madison, pp. 350–358, 1998.
12. B. Settles and M. Craven, An analysis of active learning strategies for sequence labeling tasks. In *EMNLP*, ACL press, Honolulu, pp. 1070–1079, 2008.
13. H. Valizadegan, R. Jin, R. Zhang and J. Mao, Learning to rank by optimizing NDCG measure. In *Proceedings of Neural Information Processing Systems (NIPS)*, Vancouver, Canada, 2010.
14. T. Joachims, Optimizing search engines using clickthrough data. In *Proceedings of the 8th ACM Conference on Knowledge Discovery and Data Mining*, Edmonton, Canada, pp. 133–142. ACM, 2002.

15. R. Herbrich, T. Graepel and K. Obermayer, Large margin rank boundaries for ordinal regression. In Smola, A.J., Bartlett, P.L., Schölkopf, B. and Schuurmans, D., eds., *Advances in Large Margin Classifiers*, MIT Press, Cambridge, MA, pp. 115–132, 2000.
16. A. Rakotomamonjy, Optimizing area under ROC curves with SVMs. In *Proceedings of the ECAI-2004 Workshop on ROC Analysis in AI*, Valencia, Spain, 2004.
17. F. Dammak, H. Kammoun and A. Ben Hamadou, An extension of RankBoost for semi-supervised learning of ranking functions. *The 5th International Conference on Mobile Ubiquitous Computing, Systems, Services and Technologies*, pp. 49–54, Ubicomm 2011. Lisbon, Portugal, 2011.

Part II

Distributed Network Security

7

Tackling Intruders in Wireless Mesh Networks

Al-Sakib Khan Pathan, Shapla Khanam, Habibullah Yusuf Saleem and Wafaa Mustafa Abduallah

CONTENTS
7.1 WMN Intrusion Tackling Schemes: Background 168
    7.1.1 WMN Architecture and Related Background 168
    7.1.2 A Different Perspective of Tackling Intrusion in WMNs 168
    7.1.3 Status of IDSs and IPSs for WMNs 169
    7.1.4 Current Status of Achievements and Our Motivation 170
7.2 Tackling an Intruder with a Tricky Trap 172
    7.2.1 Considered Setting: Network Characteristics and Security Model 172
    7.2.2 PaS Model: The Idea Behind It 173
    7.2.3 Initiating Competition by Using Game Theory 174
        7.2.3.1 Mathematical Model 175
        7.2.3.2 Marking the Intruder and Making a Decision 177
7.3 Analysis of Our Approach 179
    7.3.1 Experimental Results 179
    7.3.2 Marking the Intruder Considering Various Cases 182
7.4 Applicability and Future Network Vision 185
7.5 Wireless IDPS: Features and Differences to Know 186
    7.5.1 Wireless IDPS Security Capabilities 187
    7.5.2 Wireless IDPS Limitations 188
7.6 Potential Research Fields and Concluding Remarks 188
Acknowledgement 189
References 189

This chapter presents a different approach to tackling intruders in wireless mesh networks (WMNs). The traditional approach of intruder detection and prevention suggests purging out intruders immediately after their detection. In our work, we take a different approach and only purge an intruder when it is marked as a direct threat to the network's operation. Our intrusion tackling model is termed the *pay-and-stay* (PaS) model and it allows a rogue node to stay in the network only at the expense of doing some traffic-forwarding tasks in the network. Failing to carry out the required tasks of packet forwarding disqualifies the node permanently, and eventually that rogue entity is purged out. Alongside presenting our approach, we briefly discuss other available literature, essential knowledge on wireless networks, intrusion detection and prevention, and status of intrusion-related works for WMNs.

7.1 WMN Intrusion Tackling Schemes: Background

WMNs [1,2] have recently become a very popular front of research. However, as a type of wireless network, they have several weaknesses that are usually associated with any kind of *wireless* technology. Unlike their wired counterparts, owing to the use of wireless communications, secure authentication with access control and various security issues are very crucial in such networks to ensure proper service to the legitimate users while also preventing a variety of attacks. Most security threats are posed by illegitimate entities that enter or intrude into the network perimeter, commonly termed as *intruders*. Sometimes, a legitimate node could also be compromised in some way so that an attacker-intended task for a *security breach* could be performed. In this chapter we also term any such kind of rogue node or entity as an *intruder*. So, the main objective of this work is to identify any kind of intrusion in a WMN and tackle it in a meticulous manner so that a wide range of security attacks may be deterred and the benefiting network. As we go through the chapter, we will explain the concepts and motivations behind our approach of dealing with this issue.

7.1.1 WMN Architecture and Related Background

The mesh architecture of a wireless network concentrates on the emerging market requirements for building networks that are highly scalable and cost effective. However, WMNs lack efficient security guarantees in various protocol layers [3,4]. There are a number of factors that come into consideration. First, all communications go through the shared wireless nodes in WMNs that make the physical security vulnerable. Second, the mesh nodes are often mobile, and move in different directions and often change the topology of the network. Finally, since all communications are transmitted via wireless nodes, any malicious node can provide misleading topology updates that could spread out over the whole network topology [5,6]. All these points make it difficult to ensure a proper level of security. However, it is desireable to achieve at least some kind of agreed-upon standard for a particular application scenario by identifying the rogue entities within the network. That is why we believe detecting the rogue node (intruder, thus an intrusion event) and tackling the intruder skilfully can keep away different kinds of attacks and keep the network healthy for its proper operation.

7.1.2 A Different Perspective of Tackling Intrusion in WMNs

WMNs consist of mesh routers and mesh clients, where mesh routers form the backbone of the network that provides network access to both the mesh and conventional clients. Mesh clients can connect (see Figure 7.1) either to a backbone or among each other. Hence, the mesh client can access the network through the mesh router in a multi-hop manner. Therefore, any malicious node or intruder can attack the network in the forms of black hole attack, grey hole attack, Sybil attack and so on [1,5]. In all these attacks, the routing packets are deliberately misled towards wrong destinations or network entities. Once the malicious node (here, we will call it an *intruder*) has control over the packet after getting it in its trap, the packet could be modified, fabricated, dropped or forwarded (arbitrarily), all of which are considered major obstacles for secure and guaranteed routing in WMNs. Our idea is that in such an attack scenario, we will allow the node to operate but for its actions, it needs to pay at a high scale so that it is deterred from doing further mischief.

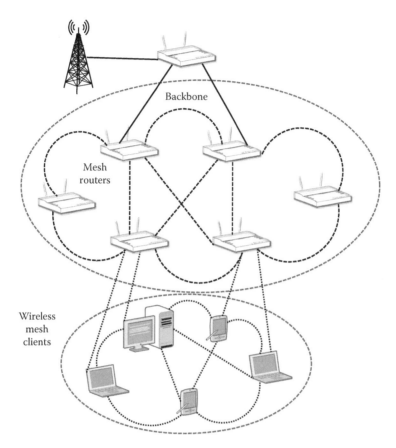

FIGURE 7.1
Hybrid wireless mesh network.

We call our approach the PaS model of intruder tackling as the intruder needs to pay for its stay once it sets itself within the network. We will illustrate how we achieve our goal in the later sections. It should be noted here that we focus on the modelling of intruder/intrusion detection and its efficient tackling; hence, other issues such as physical layer issues, transmission and channel or signal-related issues, core routing issues, cryptographic and key management issues and so on are out of the scope of this work.

7.1.3 Status of IDSs and IPSs for WMNs

Both fields of intrusion detection systems (IDSs) and intrusion prevention systems (IPSs) are insufficiently researched in the case of WMNs. As these types of networks have a mesh backbone that could have devices with a proper amounts of resources and considerably high energy supplies, the usual intrusion detection and prevention mechanisms could often be applied on the basic structure/backbone. However, the problem arises when we want to tackle intrusions in the end-user level (mesh clients or fringe portion). Figure 7.1 shows the typical structure of a WMN, where we show the mesh clients in the network.

The mesh clients could consist of different types of devices ranging from laptops to mobile phones; even a wireless sensor networks (WSNs) could be at the mesh network end as mesh clients. Any effective intrusion tackling mechanism for WSN could be difficult to implement on the sensors because of their lack of proper resources; this matter is treated as a separate research issue, because it is out of the scope of this work. For other types of mesh clients, some kind of intrusion tackling mechanism can be employed. Our work mainly focuses on this area, where a different approach of tackling intrusion could end up being beneficial for the network by allowing an intruder to stay in the network rather than instantly purging it out once it is caught or marked. This matter will be discussed further in the chapter.

7.1.4 Current Status of Achievements and Our Motivation

Until today, the works on intrusion in WMN have been very scarce. Often, the proposed approaches do not offer good solutions to the problem but rather put some kind of scratchy overview. There are a few other works that should also be mentioned because they provide important information and ideas related to the field. Hence, in this section, we present the notable past works that especially focus on this topic and also some other works that are somewhat related to our approach.

RADAR is a reputation-based scheme for detecting anomalous nodes in the WMNs presented in Ref. [7]. The authors use the concept of reputation to characterise and quantify the mesh node's behaviour and status in terms of some performance metrics. The RADAR architecture shows how the reputation of network nodes is maintained. Reputation management is defined as a feedback process that involves the monitoring and tracking of a mesh node's performance and reviewing evaluation reports from the witnesses. A trust network construction algorithm is also presented and the performance is measured by taking into account some critical parameters such as false positive, decision accuracy, response latency and detection overhead. The idea of using reputation is not very different from those that are used in the available literature in other fields, but the way the authors formed their algorithm and architecture for WMNs has been proven to be effective for some scenarios.

Ref. [8] describes an architecture of asymmetric, distributed and cooperative IDSs for WMNs. In this work, the authors mention the notion of the selfish behaviour of the suspected intruder. They use a double-mode mechanism in the detection model for judging the troubling behaviour: (a) the frequency of the node's seizing channel behaviour during the active time of the node and (b) the continuous sampling results of the node's back-off value. Alongside presenting the idea in mathematical form, the authors analysed the whole scheme in terms of throughput ratios and detection delays.

The idea of Ref. [9] is to note down various types of basic information about intrusion detection in WMNs and to propose an IDS architecture. This work, however, is very elementary and the authors also note that they put forward an initial design of a modular architecture for intrusion detection, highlighting how it addresses their identified challenges. Ref. [10] presents an IDS software prototype over a wireless mesh network testbed. The authors implement the idea and the evaluations are presented in a limited range. However, this work is incomplete because there are many unanswered questions, such as what to do with a distributed or large-scale WMN, what to do to ensure real-time analysis and detection of anomalous nodes, and so on.

OpenLIDS is a lightweight IDS for WMNs presented in Ref. [11]. This work shows an analysis of a typical wireless mesh networking device performing common intrusion detection tasks. The authors examine the participating nodes' ability to perform intrusion detection. The experimental study shows that commonly used deep packet inspection approaches are unreliable on typical hardware. So, the authors implement a set of lightweight anomaly detection mechanisms as part of an IDS, called OpenLIDS. They also show that, even with the limited hardware resources of a mesh device, it can detect current malware behaviour in an efficient way.

Ref. [12] presents a very simple model of intrusion detection in WMNs. This work is questionable as the contribution is limited to a vague work-flow diagram with insufficiently done analysis. However, from the objective mentioned in the work, it is understood that the authors targeted designing only a framework without going into any details of the operations. Ref. [13] presents an idea of using a finite-state machine to model intrusion detection in WMNs. By simulation studies, the authors show that under a flooding-combined attacks, the IDS shows a high false alarm rate due to the side effect of flooding on the attacker's neighbours. Hence, the dummy node used in the approach needs more design features to record more information about the monitored node. This work is flawed and a convincing result is yet to be achieved.

In Ref. [14], the authors present a framework for intrusion detection in IEEE 802.11 WMNs. Some intrusion detection agent structures are shown. The concept is mainly shown in the form of some diagrams and where different components work in a cooperative manner. However, the idea appears to have been not thought out very well and somewhat naive. The work presents the primary components (or agents) that should be installed in mesh routers and mesh nodes. The detection of intrusion could be made and an action database could be used for making decisions about any detected intruder. No detailed analysis is presented in the work and it basically touches the surface of the problem.

As is evident from the above-mentioned works on IDSs in WMNs, very few countable papers have been published so far on this topic. Again, none of the above works talked about intrusion tackling or utilising the intruder for the network's benefit before purging it out. Hence, we have come up with the idea of intrusion tackling with the PaS model. While the proposal section outlines the details of the model, there are more related works that inspired or influenced our way of thinking to build up the basic mathematical and theoretical intrusion tackling model.

In Ref. [3], the authors propose an algorithm to specifically defend against security attacks in WMNs where the algorithm makes use of a counter threshold to find the threshold value. This threshold value is compared with the actual number of data packets delivered. If the actual number of data packet is less than the threshold value, then the route is declared to contain malicious node(s), which implies that packet loss is always due to the malicious node(s). Therefore, the path will be excluded from route selection. However, packet loss may occur due to other factors such as mobility and battery power. If we keep excluding the route by assuming that the poor performance routes contain malicious nodes, then we may end up with few or no routes for communications at the end. This method may work on specific settings but is not efficient to encounter security attacks in dynamic topologies of WMNs.

The authors in Ref. [15] advocate using a protocol for carrying authentication for network access (PANA) to authenticate the wireless clients. The PANA model also provides the cryptographic materials necessary to create an encrypted tunnel with the associated remote access router. However, the authentication procedure presented in the paper is tedious and resource consuming. Although the framework talks about protecting the confidentiality of exchanged information and the overall approach is analysed, it has not been tested in a detailed manner that could convince the readers about its efficiency in practical implementation cases.

The authors in Ref. [16] propose a framework of a non-cooperative zero-sum game between genuine and malicious mesh routers and use mathematical tools and models for their approach. This game model solves the problem of grey hole attacks where the malicious node drops a subset of packets that it receives. The game has a source node as the target and a malicious node as the attacker; each competes with the other for limited resources and each node gains depending on its own strategy and that of the other. The attacker benefits from dropping packets and the target gains from forwarding the packets successfully. Our approach adopts a similar game-theoretic model as a part of the total solution. However, the difference is that we circumvent the flaws of this paper's idea by using our own mathematical model and by choosing appropriate parameter values. As an example, Ref. [16] takes 50% of the packet arrival rate to *send buffer* based on which the gains of both nodes vary. Therefore, it may be impractical because, in reality, higher packet arrival rates are expected to minimise packet delays and a large number of nodes should be involved in communications in any WMN.

A novel algorithm named channel-aware detection (CAD) has been adopted in the work presented in Ref. [17]. The authors use two different strategies to detect the grey hole attacks. Their approach detects a potential victim mesh node (i.e. which can be attacked) by hop-by-hop loss observation and traffic overhearing. A comparative performance analysis has been shown to detect and isolate the selective forwarding attackers [25] in multi-hop network scenarios. The probability of misdetection or false alarm is analysed and a design is proposed to limit these to a certain threshold. However, the approach is complicated, focuses on a narrow set of attacks, and is applicable only in some restrictive scenarios. This work basically focuses on the communication and signalling aspects in the physical layer but is related to our work in the sense that some of the ideological concepts helped us in the formulation of our approach, which we will discuss later.

Attention has been devoted to investigating the use of cryptographic techniques to secure the information transmitted through the wireless network. Some other preliminary solutions have been addressed in ad hoc sensors WMNs to prevent different types of malicious attacks [18–20].

Now that we have presented some background knowledge, in the next section, we present our proposed model.

7.2 Tackling an Intruder with a Tricky Trap

It should be noted before going any further that our security model is for *intrusion tackling* or *intrusion handling*, instead of direct *intrusion detection* or *intrusion prevention* in WMNs. The background knowledge noted so far could be useful while explaining our approach to handling the issue. We also clarify some terms in the later sections to explain our position in a better way and to differentiate among the terminologies.

The core concept of our approach is that all the intruders in a network are always harmful for the network. In fact, sometimes, there are ways to obtain benefit out of it or utilise it for the network's welfare or for its own benefit. Keeping this tricky fact in mind, we take a different approach to tackle an intrusion.

7.2.1 Considered Setting: Network Characteristics and Security Model

We assume a hybrid wireless mesh network where different types of devices could form the fringe part or could play the roles of mesh clients. A network model is shown in

Figure 7.1. From the figure, it can be seen that any node in the fringe parts could come and go, that is, they may be mobile, which allows a newcomer or even an *intruder* to try its luck in the network. As we have noted before that even if a node within the legitimate mesh clients acts as an attacker, in our case, we consider it as an intruder; that is, it has lost the legitimacy to stay in the network as a legal participant and is seen as a suspicious entity that has caused an intrusion (illegitimate incursion) within the network perimeter. We assume that standard security components (i.e., cryptographic keys for data confidentiality, security measures, etc.) and other basic intrusion detection mechanisms are present within the network. The basic intrusion detection agents could be installed in any node in the network. Hence, our mechanisms start working after an intrusion is detected or some node is suspected of being an intruder. To capture the whole idea in a single sentence, *We are interested in dealing with the intruder if it is suspected to be such, after it has caused an intrusion rather than purging it out directly from the network.* By 'standard security components' we mean the cryptographic parameters, keys and other security mechanisms that are used in a device that participates in a given network. (The reader is encouraged to go through the basic terms and preliminaries mentioned in Section 7.2 for a recap.)

7.2.2 PaS Model: The Idea Behind It

Once a node in the network is suspected to be an intruder (by any of the standard components installed in the legitimate devices), our model is employed to force the intruder to work for the network. If it works for the network, we see little problem in allowing the node to stay in the network. That is because the routing packets and exchanged data within the network would be protected by other cryptographic measures in place as noted in the previous section. Instead of taking a straight negative decision to defuse it, we give it enough tasks to perform for forwarding any possible network packet to the next hops or to the intended destinations.

If the node is an intruder unwilling to participate in the forwarding process of the packets, we decide finally that the node is not suitable for staying in the network and must be purged out. Otherwise, by putting pressure on it to forward a huge number of packets, we save the network's other resources. Each forwarding takes energy for wireless transmission; hence, if an intruder happily does the network's legitimate nodes' job, we drain its energy or make it pay for its survival/stay in the network. If the node drops the packets randomly or selectively, we catch this with our enforced mechanisms and mark it as a selective forwarder or we charge it for causing *selective forwarding*. To employ this policy, the intrusion tackling agent is installed on each legitimate mesh entity (mesh router or mesh clients).

Figure 7.2 shows an operational diagram of a PaS intrusion tackling model. The intrusion database can be stored in any of the devices with a good amount of storage space or could be partially maintained by each node, that is, each node acts as the intruder tackler for its surrounding nodes. For primary intrusion detection, as noted earlier, any standard scheme could be utilised. Because of the structural dimensions of WMNs, such a strategy is possible, whereas for WSNs or other wireless ad hoc networks, such a strategy may not be used. As shown in the figure, our model gets activated after the IDS does its part; we deal with what to do *after* the intrusion, not *before* the intrusion. The core goal is to maximise or save the utilisation of network resources by putting the burden of packet transmissions on a rogue entity. In case the rogue entity refuses to pay or render the service, we

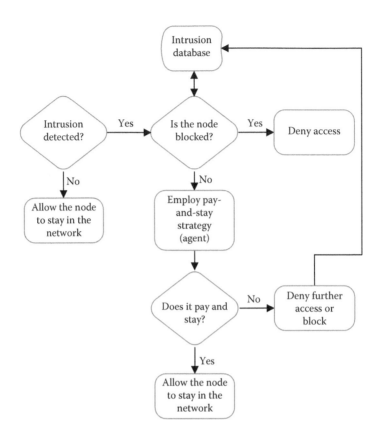

FIGURE 7.2
Operational diagram of our intrusion tackling model.

purge it out from the networks, and thus this is an effort of delicately handling an intruder in a wireless mesh network setting.

The texts below explain how we achieve this PaS strategy for intrusion tackling. There are mainly two phases in our approach. The first phase is a *game theory-based PaS model* and the second phase is *marking the intruder and taking a decision* (that is also a part of the intrusion tackling model). The following sections illustrate our approach in detail.

7.2.3 Initiating Competition by Using Game Theory

Game theory [21] can be defined as a statistical model to analyse the interaction between a group of players who act strategically. Figure 7.3 introduces a usual attack model where there are two players involved, namely, Player_1, which is the source node S, and Player_2, which is the malicious/attacker (in our case, intruder) intermediate node A. Let D be the destination node and N be the finite set of all players. We limit our game to a non-cooperative, incomplete information, and a zero-sum game model [22], where one player wins and the other player loses. Our target is that the intruder should spend more resources in doing wrong with any packet than that are used by the target node to forward packet to the destination. That means the intruder eventually has to pay heavily for its illegal staying within the network. It should be noted that we use the terms *intruder* and *attacker* interchangeably throughout the rest of the chapter.

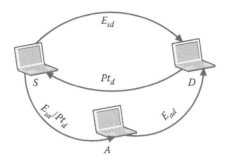

FIGURE 7.3
A diagram showing node S sending packet (directly and via node A) to the destination node D.

7.2.3.1 Mathematical Model

Before presenting the mathematical model of our approach, in Table 7.1, we note the critical notations used in this chapter for ease of reading and to identify of various items at a glance.

Let P_i be the probability to defend the ith node in the network. We assume that v_i is an intermediate node and v_{i-1} and v_{i+1} are the upstream and downstream nodes, respectively. The total probability of defending all N nodes is $\Sigma_{i=1}^{N} P_i$. The energy spent for utility cost is $E_{sd} = \Sigma_{i=1}^{N} P_i$.

The remaining energy is $E_r = 1 - \Sigma_{i=1}^{N} P_i$, where $\Sigma_{i=1}^{N} P_i \leq 1$. Our objective is that the energy that needs to be spent by the intruder to cause trouble in packet flow or forwarding must be more than the energy spent by the victim (which is sending or receiving the packets).

The energy of the sender to send via the attacker could be noted by the equation $E_{sa} = \alpha \Sigma_{i=1}^{N} P_i$, where α is a constant. The success of the attack depends on the value of α. If $\alpha > 1$, the attack succeeds. If $\alpha = 1$, the energy spent by the attacker equals that of the target. When $\alpha = 0$, the attacker cannot attack, and $\alpha < 1$ means that the attacker cannot drop any packet.

TABLE 7.1

Basic Notations and Their Meanings

Notation	Meaning
P_i	Probability to defend the ith node in the network
v_i	Intermediate node
v_{i-1}	Upstream node
v_{i+1}	Downstream node
μ	Packet arrival rate
E_{sd}	Energy spent for utility cost
E_r	Remaining energy
α	A constant
p_a	Probability of transmitting packets via Player_2
p_d	Probability of direct transmission of packets
q_f	Forwarding probability
q_d	Probability of dropping the packet
Pt	Points received

The state of the game is (m, n), where m is the sending buffer of Player_1 and n is the dropping buffer of Player_2. If one packet is present in the sending buffer of m of Player_1, then m will take a value of 1 and n can take a value 0 or d, depending on whether any packet is dropped or not. We also denote μ as the probability that a new packet arrives at the sending buffer of Player_1. There are four possible states of the game: $k_1 = (0,0)$, $k_2 = (0,d)$, $k_3 = (1,0)$, $k_4 = (1,d)$. Therefore, the transition probabilities from one state to another state are calculated as follows:

When $(m = 1)$

$$P_{(m,n)(m+i,n)}(x) = \begin{cases} (1-\mu)(p_d + p_a q_f); & \text{if } i = -1, n = 0 \\ (1-\mu)(p_a q_d); & \text{if } i = -1, n = d \\ \mu(p_d + p_a q_f); & \text{if } i = 0, n = 0 \\ \mu(p_a q_d); & \text{if } i = 0, n = d \end{cases} \quad (7.1)$$

When $(m = 0)$

$$P_{(m,n)(m+i,n)}(x) = \begin{cases} (1-\mu); & \text{if } i = 0, n = 0 \\ (\mu); & \text{if } i = 1, n = d \end{cases} \quad (7.2)$$

where μ is the arrival rate of packets in the sent buffer and x is the joint strategy.

For example, assume that the current state of the system is (1,0). Player_1 (i.e., S) has a packet in its sent buffer. It uses two strategies: transmit the packet directly or transmit via A. If S transmits the packet directly to D, then the states are (0,0) or (1,0) with probability p_d. Otherwise, it transmits packets via Player_2 (i.e., A) with probability p_a. A either drops the packet or forwards it to D. If it drops, then the states become (0,d) or (1,d). If A forwards the packet, then the next states will be (0,0) or (1,0). Note that A is the potential intruder in this case.

The strategy set for Player_1 is $S_1 = \{s_1, s_2\}$, meaning that Player_1 forwards the packet either directly to destination D (s_1) or via A (s_2). Mixed strategies (denoted as x) that correspond to S_1 are $\pi_s(s_1, s_2) = (p_d, p_a)$, where $p_d + p_a = 1$. The strategy set of Player 2 is $A_2 = (a_1, a_2)$. Mixed strategies corresponding to the action of A_2 are $\pi_a(a_1, a_2) = (q_f, q_d)$, where $q_f + q_d = 1$. Here, q_d is the probability of dropping the packet. Hence, $(\pi_s, \pi_a) = (p_d, p_a, q_f, q_d)$.

The destination D gives some points to source S for the transmitted packet. When the source node S sends the packet through the path $S \rightarrow D$, node S receives some points of Pt_d from D. When S transmits packets via A, it receives points of Pt_d from D and it gives A some points, Pt_{sa}. If S does not receive any point from D for the transmitted packet, it means that the packet did not reach D successfully. Each packet transmission from node v_i to node v_{i+1} causes an energy-spending $Ev_i v_{i+1}$. Therefore, depending on the energy spent and points received by the source and attacker, the nodes S and A will remain with the following net utility:

$$U_s = \begin{cases} Pt_d - E_{sd}; & S \text{ transmits directly to } D \\ Pt_d - Pt_{sa} - E_{sa}; & S \text{ transmits to } D \text{ via } A \\ -Pt_{sa} - E_{sa}; & \text{node } A \text{ drops the packet} \end{cases} \quad (7.3)$$

If $(-Pt_{sa} - E_{sa}) < (Pt_d - E_{sd}) < (Pt_d - Pt_{sa} - E_{sa})$, the utility of S will decrease if A drops the packet compared to the utility it receives when a packet reaches D:

$$U_a = \begin{cases} Pt_{sa} - E_{ad}; & A \text{ forwards the packet to } D \\ Pt_{sa} + \beta; & \text{node } A \text{ drops the packet} \end{cases} \quad (7.4)$$

where β is the profit earned by node A. If $(Pt_{sa} - E_{ad}) < (Pt_{sa} + \beta)$, the utility earned from dropping the packet is higher than the utility received from S for transmitting the packet. However, the utility can be calculated from the following equations based on the probability of dropping and forwarding the packets:

$$U_s(x) = \mu(1 - \mu \times p_a q_d)\{p_d(Pt_d - E_{sd}) + p_a(q_f(Pt_d - Pt_{sa} - E_{sa})$$
$$+ q_d(-Pt_{sa} - E_{sa}))\} + \mu^2 \times p_a q_d \{p_d(Pt_d - E_{sa}) \quad (7.5)$$
$$+ p_a(q_f(Pt_d - pt_{sa} - E_{sa})) + p_a(q_d(-Pt_{sa} - E_{sa}))\}$$

and

$$U_a(x) = \mu(1 - \mu \times p_a q_d)\{p_a(q_f(Pt_{sa} - E_{ad}) + q_d(Pt_{sa} + \beta))\}$$
$$+ \mu^2 \times p_a q_d \{p_a(q_f(Pt_{sa} - E_{ad}))\} + \mu^2 \times p_a q_d (p_a q_d (Pt_{sa} + \beta)) \quad (7.6)$$

7.2.3.2 Marking the Intruder and Making a Decision

In this section, we describe a multi-hop acknowledgement (ACK)-based algorithm to detect malicious node(s) doing a selective forwarding attack. Owing to the structure of WMNs, it is possible to use this method. We know that a selective forwarding attack is one of the most dangerous attacks because the packets are dropped randomly, which may contain sensitive data. In this algorithm, multiple nodes need to be selected as ACK points in WMNs. This means that those mesh nodes are responsible for sending an ACK packet after receiving a packet from a source node or the nearest intermediate source nodes. We assume that the WMNs are operating under an ideal channel quality and the majority of the mesh routers are behaving normally. We consider that the packet loss appears only due to malicious activity from an intruder. Moreover, since there may be multiple existing routes from a source mesh node to a destination mesh node and a source node may receive multiple route replies of each of its route requests, we encourage the source node to keep a record of each route for future references. It should be noted here that dealing with physical layer or channel-level matters are out of the scope of this work as we focus on the theoretical framework and mathematical model of the operational concept.

In Figure 7.4, we show the structure of a wireless mesh network where S is the source node and D the destination node. We assume N to be the total number of mesh nodes in the forwarding path. M is the number of malicious nodes among N. Let X be the normal-behaving nodes between each of the two malicious nodes and Y be the number of ACK points in the forwarding path. We consider Z as the percentage of randomly selected check points.

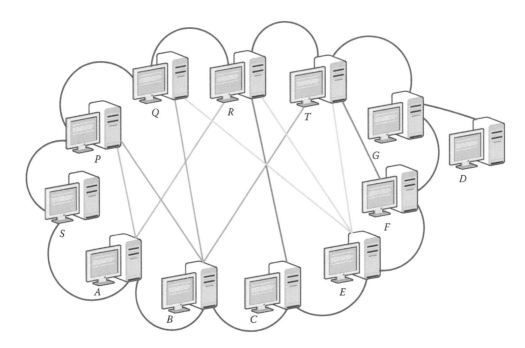

FIGURE 7.4
Multi-hop acknowledgement.

When the source node S sends a route request, it receives several route replies. Let us consider that S chooses the route $SABCEFG \rightarrow D$, where E is the malicious node. We are considering two selected ACK points (i.e., $Y = 2$), namely, B and F. B and F will acknowledge back after they receive the packets from the source mesh nodes. Therefore, the following possibilities may occur if:

Scenario 1: One of the nodes is malicious in the forwarding path.
Scenario 2: One or more nodes are malicious in the forwarding path.
Scenario 3: Both the ACK points B and F are malicious.
Scenario 4: Either B or F is malicious.

This algorithm uses two approaches: hop-by-hop loss observation and traffic overhearing to detect the malicious node on the path of data flow. More specifically, we assume v_i to be an intermediate node and v_{i-1} and v_{i+1} to be the upstream and downstream nodes, respectively. v_{i+1} receives a packet from v_{i-1}; then it updates itself with the packet count history and with the corresponding packet sequence number, it buffers the link layer ACK that it receives for each packet, and then forwards it to v_{i-1} (i.e., downstream node). We denote w_s as the total number of packets that are successfully sent–received by the source S to destination D. $n_{v_i \rightarrow v_{i+1}}$ is the number of packets received successfully by v_{i+1} (i.e., this is the number of successfully received packets from any intermediate node to its downstream node).

Two operations are performed when the mesh router forwards a packet to the downstream node as explained in this paragraph. When each packet is relayed to the downstream traffic, the mesh router or upstream node buffers ACK and overhears the downstream traffic to check whether it (downstream node) forwarded or tampered with the packet. The upstream node observes these two operations and then makes a simple analysis of the scenario.

The downstream node maintains two parameters. They are (a) probability of ACK that we denote as P_{Ack} and (b) probability of no acknowledgement (NACK), P_{NAck}. The probability of ACK (P_{Ack}) is computed as $P_{Ack} = 1 - P_{NAck}$ and the probability of NACK is computed as $P_{NAck} = (n_t + n_d)/n_f$, where n_t is the number of tampered packets, n_d is the number of dropped packets, and n_f is the number of total forwarded packets.

We introduce two packets, PROBE packet and PROBE_ACK, to detect the malicious routers. The PROBE packet is used by source node S with every w_s data packet to the destination node D. When the source node S sends the PROBE packet through the path, each node in the path marks the PROBE packet with the detection parameters and this is termed *packet marking*. A PROBE packet sent to the destination by the source node is also marked by it (i.e., S) with the number of packets that will be transmitted to a particular destination node. When the PROBE packet is passed along the path, each node v_i attaches a mark of its opinion to the downstream node (v_{i+1}). The opinion is calculated by observing the downstream node's behaviour by the transmitter node. The opinion of the downstream node is calculated as follows:

- If ($P_{NAck} > t_m$), it means malicious behaviour
- If ($P_{NAck} < t_m$), it means normal behaviour

where t_m is the monitoring threshold which carries values between 0 and 1. As the PROBE packet is passed through the path, the node also appends the behaviour parameter to the PROBE packet. The behaviour parameter represents the observation of node v_{i+1} about the behaviour of the upstream node v_i. The behaviour of the node is calculated by determining the loss rate of the packets over the link v_i to v_{i+1}. It is calculated by the following formulae:

- If ($L_{v_i \to v_{i+1}} > t_l$), malicious behaviour is detected
- If ($L_{v_i \to v_{i+1}} < t_l$), normal behaviour is detected

where ($L_{v_i \to v_{i+1}} = 1 - (n_{v_i \to v_{i+1}}/n_{v_{i-1} \to v_i})$) is the loss rate of the link that is observed by the node v_{i+1}. t_l is the loss rate threshold that can take any value between 0 and 1. The algorithm will detect the malicious behaviour with higher probability with the lower values of t_l and t_m.

7.3 Analysis of Our Approach

7.3.1 Experimental Results

For the game-theoretic model analysis, we substitute the values for the required energy to transmit packets from S to D either directly or via A and the points earned by source S and A are as follows: $E_{sd} = 0.6$, $E_{sa} = E_{ad} = 0.05$, $Pt_d = 1$ and $Pt_{sa} = 0.3$. We assume that the packet arrival rate μ to send the buffer is quite fast; $\mu = 0.8$ and $\beta = 0.2$.

Using Equations 7.5 and 7.6, we obtained the utility of Player 1 and Player 2. We represented Figures 7.5 through 7.9 of utilities S and A as a function of drop probability using MATLAB® [23]. The packet dropping probability is chosen between 0 and 1. It is observable from Figures 7.5 through 7.9 that the utility of S is decreasing and the utility of A is increasing with the increase of dropping probability.

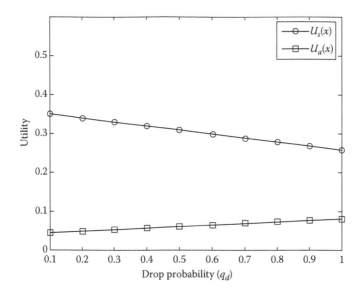

FIGURE 7.5
Increasing the utilities of A and decreasing the utilities of S with respect to different drop probabilities of q_d when $p_d = 0.8$ and $p_a = 0.2$.

Player 2 reaches the maximum utility when source S transfers all the packets via A with the highest dropping probability. It can be seen from Figure 7.9 where $p_a = 1$ and $q_d = 1$ that the maximum utility of $U_a = 0.4$. On the other hand, for $q_d = 1$, Player 1 has its maximum utility $U_s = 0.256$ when the probability of sending packets directly to D increases. The maximum utility of S is shown in Figure 7.5 where $p_d = 0.8$ and $q_d = 0.1$. Figures 7.10 through 7.14

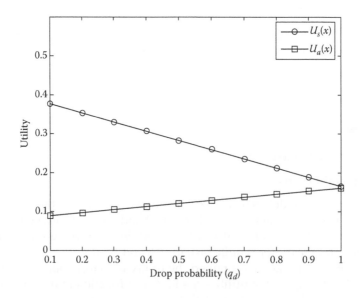

FIGURE 7.6
Increasing the utilities of A and decreasing the utilities of S with respect to different drop probabilities of q_d when $p_d = 0.6$ and $p_a = 0.4$.

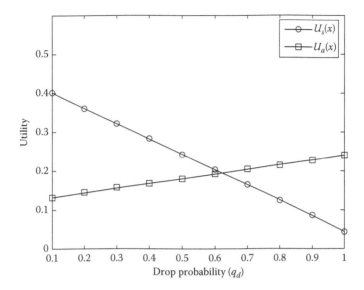

FIGURE 7.7
Increasing the utilities of A and decreasing the utilities of S with respect to different drop probabilities of q_d when $p_d = 0.4$ and $p_a = 0.6$.

represent the utility of S and A as a function of forward probability to A, p_a. The forward probability is chosen between 0 and 1.

The forward probabilities are $q_f = 1$ and $q_f = 0.75$ and the drop probabilities are $q_d = 0$ and $q_d = 0.25$ in Figures 7.10 and 7.11. It is clear that in Figure 7.10 the utilities of S and A are increasing. The maximum utility of S is 0.5 and the maximum utility of A is 0.2. In Figures 7.11 through 7.14, the utility of S is decreasing overall (with a slight bent increase and going

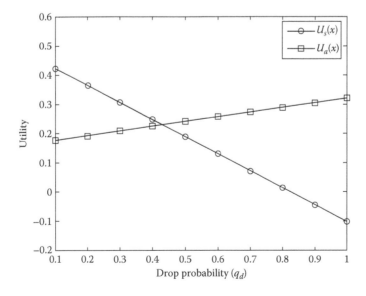

FIGURE 7.8
Increasing the utilities of A and decreasing the utilities of S with respect to different drop probabilities of q_d when $p_d = 0.2$ and $p_a = 0.8$.

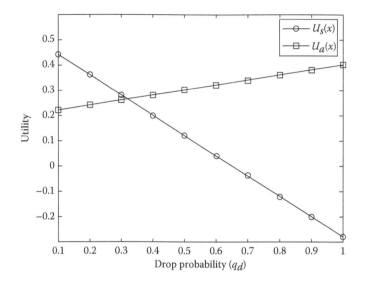

FIGURE 7.9
Increasing the utilities of A and decreasing the utilities of S with respect to different drop probabilities of q_d when $p_d = 0$ and $p_a = 1$.

down in Figure 7.11) and the utility of A is increasing and the maximum utility of A is 0.4; the forward probabilities are $q_f = 0.75$, $q_f = 0.5$, $q_f = 0.25$ and $q_f = 0$ and the drop probabilities are $q_d = 0.25$, $q_d = 0.5$, $q_d = 0.75$ and $q_d = 1$.

7.3.2 Marking the Intruder Considering Various Cases

In the malicious behaviour detection phase, the following possible cases may occur when the upstream and downstream nodes are combined to detect malicious activities:

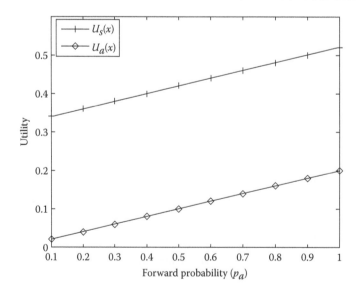

FIGURE 7.10
Increase of utilities S and A as a function of p_b with respect to $q_f = 1$ and $q_d = 0$.

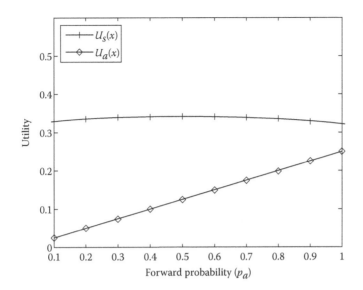

FIGURE 7.11
Increase of utilities S and A as a function of p_b with respect to $q_f = 0.75$ and $q_d = 0.25$.

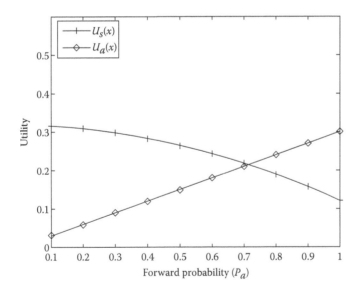

FIGURE 7.12
Increase of utility A and decrease of utility S as a function of p_b with respect to $q_f = 0.5$ and $q_d = 0.5$.

Case 1: If $P_{\text{NAck} v_i \to v_{i-1}} > t_m$ and $L_{v_i \to v_{i+1}} > t_l$. The node v_i either drops or tampers with the packets. The probability of NACK is greater than the monitoring threshold t_m. The node v_{i-1}, the upstream node, will observe node v_i on whether it drops the packets or tampers with it. Node v_{i-1} will increase n_d, which is the number of dropped packets, and also n_t, which is the number of tampered packets. The downstream node, v_{i+1} will observe if loss rate is greater than the threshold t_l, the loss rate threshold. The upstream and downstream will observe if node v_i is misbehaving.

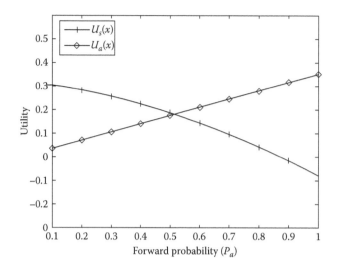

FIGURE 7.13
Increase of utility A and decrease of utility S as a function of p_b with respect to $q_f = 0.25$ and $q_d = 0.75$.

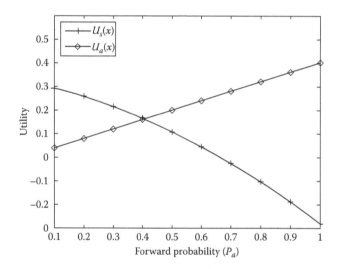

FIGURE 7.14
Increase of utility A and decrease of utility S as a function of p_b with respect to $q_f = 0$ and $q_d = 1$.

Case 2: If $P_{\text{NAck} v_i \to v_{i-1}} < t_m$ and $L_{v_i \to v_{i+1}} > t_l$. In this case, the monitoring threshold is greater than the probability of NACK from node v_{i-1} to v_i. The node v_i is behaving normally. If the observed loss rate of the link from v_i to v_{i+1} is greater than the loss rate threshold, node v_i is misbehaving. According to the upstream node, node v_i is normal but on the other hand, the downstream node can detect if node v_i is misbehaving. To overcome this problem, we need to verify link layer ACKs that are received by the upstream node v_{i-1} for each packet that is forwarded successfully by the node, v_i.

Case 3: If $P_{\text{NAck}\,v_i \to v_{i-1}} > t_m$ and $L_{v_i \to v_{i+1}} < t_l$. In this case, the upstream node v_{i-1} has a greater probability of NACK and is greater than the monitoring threshold t_m. In this case, there is a misbehaving activity at the node v_i. On the other hand, the observed loss rate link from v_i to v_{i+1} is lower than t_l, which is the loss rate threshold. According to the upstream node v_{i-1}, the node v_i is misbehaving and the downstream node will consider node v_i as normal. To overcome this issue, the upstream node can detect the misbehaving node v_i by observing false information in the PROBE packet.

Case 4: If $P_{\text{NAck}\,v_i \to v_{i-1}} < t_m$ and $L_{v_i \to v_{i+1}} < t_l$. The downstream and the upstream nodes do not detect any misbehaving node.

7.4 Applicability and Future Network Vision

Our intrusion tackling model is designed to protect the network from a wide range of security attacks that target the routing mechanisms of the network. The basic IDS mechanisms employed on the devices could notify a suspected intruder and when our model gets activated to handle the case, we ensure the proficiency in dealing with it to use the apparent negative entity for a positive purpose. Only when it is proven to be a serious threat to the network, we retract its permission to participate in the network. The preliminary IDS mechanism acts as the primary defence against intrusion activity and our mechanism acts as the final line of defence against malicious intrusion. The behavioural analysis, to mark the intruder, sets the solid defence strategy to make our model effective in practical scenarios. Owing to the features of WMN with the required amount of resources, this model works fine and proves to be effective, whereas for other wireless networks such as WSN, mobile ad hoc network (MANET) and vehicular ad hoc network (VANET), this model is not directly applicable.

From the higher-level view, however, all the above-mentioned wireless technologies fall under the category of wireless self-organising networks or ad hoc networks. Hence, putting all of them under an umbrella term, we could fit them in various future networking technologies such as pervasive or ubiquitous computing, Internet of things (IoT), future Internet, cloud computing and so on [24]. It is expected that WMN as a wireless network technology will blend well within these emerging technologies and concepts. If it is so, then the basic principle of the PaS model could be applied in various scenarios. Even if the exact model may not be applied, it is possible to think of giving the intruder some waiver to stay in any of the future networks so that when it does more good than evil, the network in an intelligent manner utilises its capacities rather than alienates it without giving it a chance. In fact, a concept such as pervasive computing would allow thousands of computation-enabled devices to work together in a blended environment where it would be extremely difficult to mark any entity as a clear intruder or an unwanted node. This is because all those good and bad entities together would form a pervasive or ubiquitous environment with a complete mixture of human life and various device technologies. Likewise, other future or next-generation computing and network technologies will have various applicable scenarios considering the PaS model's core idea.

7.5 Wireless IDPS: Features and Differences to Know

Intrusion detection and prevention system (IDPS) is basically the combination of detection and prevention mechanisms [26]. Before ending this chapter, we feel that it is necessary for the readers to have some idea about various IDPS technologies commonly used for wireless networking. As our mechanism does not fall under any clear category, we referred to our approach as *intruder tackling* because we let the intruder stay in the network even after finding it out, which is conflicting to any IDS, IPS or IDPS's main objective. The idea behind putting this section here is to clarify these terminologies further so that the readers will be able to make proper distinctions among these mechanisms.

A wireless IDPS monitors wireless network traffic and analyses wireless networking protocols to identify malicious behaviour. However, it cannot identify suspicious activity in the application or higher-layer network protocols (e.g., TCP, UDP) that the wireless network traffic is transferring. It is most commonly deployed within the range of an organisation's wireless network to monitor it, but it can also be deployed to locations where unauthorised wireless networking could be occurring.

Owing to the transmission methods, wireless network attacks differ from those on wired networks. However, the basic components involved in a wireless IDPS are the same as the network-based IDPS: consoles, database servers, management servers and sensors. A wireless IDPS monitors the network by sampling the traffic. There are two frequency bands to monitor (2.4 and 5 GHz), and each band includes many channels. A sensor (here, we mean any kind of sensing mechanism) is used to monitor one channel at a time and it can switch to other channels as needed.

We should mention that most of the wireless LANs (WLANs) use the Institute of Electrical and Electronics Engineers (IEEE) 802.11 family of WLAN standards [24]. IEEE 802.11 WLANs have two main architectural components:

- A station, which is a wireless end-point device (e.g., laptop computer, personal digital assistant).
- An access point, which logically connects stations with an organisation's wired network infrastructure or other network.

Some WLANs also use wireless switches, which act as intermediaries between access points and the wired network. A network based on stations and access points is configured in infrastructure mode; a network that does not use an access point, in which stations connect directly to each other, is configured in an ad hoc mode. Nearly all organisational WLANs use the infrastructure mode. Each access point in a WLAN has a name assigned to it called a service set identifier (SSID). The SSID allows stations to distinguish one WLAN from another.

Wireless sensors have several available forms. A dedicated sensor is usually passive, performing wireless IDPS functions but not passing traffic from the source to the destination. Dedicated sensors may be designed for fixed or mobile deployment, with mobile sensors used primarily for auditing and incident handling purposes (e.g., to locate rogue wireless devices). Sensor software is also available bundled with access points and wireless switches. Some vendors also have host-based wireless IDPS sensor software that can be installed on stations, such as laptops. The sensor software detects station misconfigurations and attacks within the range of the stations. The sensor software may also be able to enforce security policies on the stations, such as limiting access to wireless interfaces.

If an organisation uses WLANs, it most often deploys wireless sensors to monitor the radio-frequency range of the organisation's WLANs, which often include mobile components such as laptops and personal digital assistants. Many organisations also use sensors to monitor areas of their facilities where there should be no WLAN activity, as well as channels and bands that the organisation's WLANs should not use, as a way of detecting rogue devices.

7.5.1 Wireless IDPS Security Capabilities

The main advantages of wireless IDPSs include detection of attacks, misconfigurations and policy violations at the WLAN protocol level, primarily examining IEEE 802.11 protocol communication. The major limitation of a wireless IDPS is that it does not examine communications at higher levels (e.g., IP addresses, application payloads). Some products perform only simple signature-based detection, whereas others use a combination of signature-based, anomaly-based and stateful protocol analysis detection techniques. Most of the types of events commonly detected by wireless IDPS sensors include unauthorised WLANs and WLAN devices and poorly secured WLAN devices (e.g., misconfigured WLAN settings). Additionally, the wireless IDPSs can detect unusual WLAN usage patterns, which could indicate a device compromise or unauthorised use of the WLAN, and the use of wireless network scanners. Other types of attacks such as denial of service (DoS) conditions, including logical attacks (e.g., overloading access points with large numbers of messages) and physical attacks (e.g., emitting electromagnetic energy on the WLAN's frequencies to make the WLAN unusable), can also be detected by wireless IDPSs. Some wireless IDPSs can also detect a WLAN device that attempts to spoof the identity of another device.

Another significant advantage is that most wireless IDPS sensors can identify the physical location of a wireless device by using triangulation—estimating the device's approximate distance from multiple sensors from the strength of the device's signal received by each sensor, then calculating the physical location at which the device would be, the estimated distance from each sensor. Handheld IDPS sensors can also be used to pinpoint a device's location, particularly if fixed sensors do not offer triangulation capabilities or if the device is moving.

Wireless IDPS overcomes the other types of IDPS by providing more accurate prevention; this is largely due to its narrow focus. Anomaly-based detection methods often generate high false positives, especially if the threshold values are not properly maintained. Although many alerts based on benign activities might occur, such as another organisation's WLAN being within the range of the organisation's WLANs, these alerts are not truly false positives because they are accurately detecting an unknown WLAN.

Some tuning and customisation are required for the wireless IDPS technologies to improve their detection accuracy. The main effort required in the wireless IDPS is in specifying which WLANs, access points and stations are authorised, and in entering the policy characteristics into the wireless IDPS software. As wireless IDPSs examine only wireless network protocols and not the higher-level protocols (e.g., applications), generally there is not a large number of alert types, and consequently not many customisations or tunings are available.

Wireless IDPS sensors provide two types of intrusion prevention capabilities:

- Some sensors can terminate connections through the air, typically by sending messages to the end points telling them to dissociate the current session and then refusing to permit a new connection to be established.

- Another prevention method is for a sensor to instruct a switch on the wired network to block network activity involving a particular device on the basis of the device's media access control (MAC) address or switch port. However, this technique is only effective for blocking the device's communications on the wired network, not on the wireless network.

An important consideration when choosing prevention capabilities is the effect that prevention actions can have on sensor monitoring. For example, if a sensor is transmitting signals to terminate connections, it may not be able to perform channel scanning to monitor other communications until it has completed the prevention action. To mitigate this, some sensors have two radios—one for monitoring and detection, and another for performing prevention actions.

7.5.2 Wireless IDPS Limitations

The wireless IDPSs offer great detection capabilities against authorised activities, but there are some significant limitations. The use of evasion techniques is considered as one of the limitations of some wireless IDPS sensors, particularly against sensor channel scanning schemes. One example is performing attacks in very short bursts on channels that are not currently being monitored. An attacker could also launch attacks on two channels at the same time. If the sensor detects the first attack, it cannot detect the second attack unless it scans away from the channel of the first attack.

Wireless IDPS sensors (physical devices) are also vulnerable to attack. The same DoS attacks (both logical and physical) that attempt to disrupt WLANs can also disrupt sensor functions. Additionally, sensors are often particularly vulnerable to physical attacks because they are usually located in hallways, conference rooms and other open areas. Some sensors have antitamper features, which are designed to look like fire alarms that can reduce the possibility of being physically attacked. All sensors are vulnerable to physical attacks such as jamming that disrupts radio-frequency transmissions; there is no defence against such attacks other than to establish a physical perimeter around the facility so that the attackers cannot get close enough to the WLAN to jam it.

We should mention that the wireless IDPSs cannot detect certain types of attacks against wireless networks. An attacker can passively monitor wireless traffic, which is not detectable by wireless IDPSs. If weak security methods are used, for example, wired equivalent privacy (WEP), the attacker can then carry out offline processing of the collected traffic to find the encryption key used to provide security for the wireless traffic. With this key, the attacker can decrypt the traffic that was already collected, as well as any other traffic collected from the same WLAN. As the wireless IDPSs cannot detect certain types of attacks against wireless networks, it cannot fully compensate for the use of insecure wireless networking protocols.

We hope that, from this discussion, it is clear that there are some basic differences between the IDPS technologies of general wireless networking and that of the wireless mesh network.

7.6 Potential Research Fields and Concluding Remarks

This chapter's main focus was to present an intrusion tackling model for WMNs with the idea of utilising the resources of an intruder before taking a final decision of removing it from the network. This approach proves to be useful for WMN and other schemes dealing with intrusion detection or intrusion prevention could be employed side by side. In that case, better protection could be achieved to limit the number of false positives. Also, if applied as the only intrusion tackling module, other security schemes dealing with various types of attacks could work well alongside this mechanism. If the intruders' resources are used for the network and strong cryptographic mechanisms protect the network packets, this model can prove to be one of the best solutions to deal with WMN intrusion. As future works, the idea could be extended to find out a more efficient solution to tackle a huge number of colluding intruders who might make packet drop seemingly a natural event. As none of the previous works dealt with intrusion in this way, this work opens a new frontier to the researchers to work on intrusion tackling rather than direct exclusion by detection or prevention. New models could be developed in this area and numerous ways could be thought of based on the findings presented in this work.

Acknowledgement

This work was supported by Networking and Distributed Computing Laboratory (NDC Lab), KICT, IIUM. Al-Sakib Khan Pathan is the corresponding author.

References

1. Pathan, A.-S.K., *Security of Self-Organizing Networks: MANET, WSN, WMN, VANET*. ISBN: 978-1-4398-1919-7, Auerbach Publications, CRC Press, Taylor & Francis Group, USA, 2010.
2. Bruno, R., Conti, M. and Gregori, E., Mesh networks: Commodity multihop ad hoc networks, *IEEE Communications Magazine*, 43(3), 2005, 123–131.
3. Shila, D.M. and Anjali, T., Defending selective forwarding attacks in WMNs, *IEEE International Conference on Electro/Information Technology 2008 (EIT'08)*, Iowa, USA, 18–20 May 2008, pp. 96–101.
4. Wireless mesh networks and applications in the alarm industry. WhitePaper, AES Corporation, 2007, available at: http://www.aes-intellinet.com/documents/AESINT-WhitePaper.pdf (last accessed 9 January 2012).
5. Akyildiz, I.F. and Wang, X., A survey on wireless mesh networks, *IEEE Communications Magazine*, 43(9), 2005, S23–S30.
6. Deng, H., Li, W. and Agrawal, D.P., Routing security in wireless ad hoc networks, *IEEE Communication Magazine*, 40(10), October 2002, 70–75.
7. Zhang, Z., Nait-Abdesselam, F., Ho, P.-H. and Lin, X., RADAR: A reputation-based scheme for detecting anomalous nodes in wireless mesh networks, *IEEE Wireless Communications and Networking Conference, 2008 (WCNC 2008)*, Las Vegas, NV, USA, 2008, pp. 2621–2626.
8. Li, H., Xu, M. and Li, Y., The research of frame and key technologies for intrusion detection system in IEEE 802.11-based wireless mesh networks, *International Conference on Complex, Intelligent and Software Intensive Systems, 2008 (CISIS 2008)*, Barcelona, 4–7 March 2008, pp. 455–460.
9. Makaroff, D., Smith, P., Race, N.J.P. and Hutchison, D., Intrusion detection systems for community wireless mesh networks, *5th IEEE International Conference on Mobile Ad Hoc and Sensor Systems, 2008 (MASS 2008)*, Atlanta, GA, USA, 2008, pp. 610–616.

10. Wang, X., Wong, J.S. Stanley, F. and Basu, S., Cross-layer based anomaly detection in wireless mesh networks, *9th Annual International Symposium on Applications and the Internet, 2009 (SAINT'09)*, Bellevue, WA, 20–24 July 2009, pp. 9–15.
11. Hugelshofer, F., Smith, P., Hutchison, D. and Race, N.J.P. OpenLIDS: A lightweight intrusion detection system for wireless mesh networks, *Proceedings of the 15th Annual International Conference on Mobile Computing and Networking (MobiCom'09)*, New York, NY, USA, 2009, DOI: 10.1145/1614320.1614355.
12. Yang, Y., Zeng, P., Yang, X. and Huang, Y., Efficient intrusion detection system model in wireless mesh network, *2nd International Conference on Networks Security Wireless Communications and Trusted Computing (NSWCTC 2010)*, 24–25 April 2010, Wuhan, China, pp. 393–396.
13. Wang, Z., Chen, J., Liu, N., Yi, P. and Zou, Y., An intrusion detection system approach for wireless mesh networks based on finite state machine, Draft available at: http://www.cs.ucla.edu/~wangzy/inestablishment/resource/IDS_draft.pdf (last accessed: 24 March 2012).
14. Khan, S., Loo, K.-K. and Din, Z.U., Framework for intrusion detection in IEEE 802.11 wireless mesh networks, *The International Arab Journal of Information Technology*, 7(4), October 2010, 435–440.
15. Cheikhrouhou, O., Laurent-Maknavicius, M. and Chaouchi, H., Security architecture in a multihop mesh network, *5th Conference on Safety and Architectures Networks (SAR 2006)*, Seignosse, Landes, France, June 2006, pp. 1–10.
16. Shila, D.M. and Anjali, T., A game theoretic approach to gray hole attacks in wireless mesh networks, in *Proceedings of the IEEE MILCOM*, San Diego, CA, 16–19 November 2008, pp. 1–7.
17. Shila, D.M., Cheng, Y. and Anjali, T., Channel-aware detection of gray hole attacks in wireless mesh networks, *Proceedings of IEEE Globecom 2009*, 30 November–4 December, Honolulu, HI, USA, 2009, pp. 1–6.
18. Parno, B., Perrig A. and Gligor, V., Distributed detection of node replication attacks in sensor networks, *Proceedings of the 2005 IEEE Symposium on Security and Privacy (S&P'05)*, 8–11 May 2005, pp. 49–63.
19. Sanzgiri, K., Dahill, B., Levine, B.N., Shields, C. and Belding-Royer, E.M., A secure routing protocol for ad hoc networks, *Proceedings of the 10th IEEE International Conference on Network Protocols (ICNP'02)*, 12–15 November 2002, Paris, France, pp. 78–87.
20. Salem, N.B. and Hubaux, J.P., Securing wireless mesh networks, *IEEE Wireless Communication*, 13(2), April 2006, pp. 50–55.
21. Srivastava, V., Neel, J., MacKenzie, A.B., Menon, R., DaSilva, L.A., Hicks, J.E., Reed, J.H. and Gilles, R.P., Using game theory to analyze wireless ad hoc networks, *IEEE Communications Surveys & Tutorials*, 7(4), 2005, 46–56.
22. Javidi, M.M. and Aliahmadipour, L., Game theory approaches for improving intrusion detection in MANETs, *Scientific Research and Essays*, 6(31), December 2011, 6535–6539.
23. MATLAB. URL: http://www.mathworks.com/products/matlab/ [last accessed: 11 May, 2013].
24. Kindy, D.A. and Pathan, A.-S.K., A walk through SQL injection: Vulnerabilities, attacks, and countermeasures in current and future networks, in *Building Next-Generation Converged Networks: Theory and Practice*, ISBN: 9781466507616, CRC Press, Taylor & Francis Group, USA, 2013.
25. Khanam, S., Saleem, H.Y. and Pathan, A.-S.K., An efficient detection model of selective forwarding attacks in wireless mesh networks, *Proceedings of IDCS 2012*, 21–23 November 2012, Wu Yi Shan, Fujian, China, *Lecture Notes in Computer Science (LNCS)*, Vol. 7646, Springer-Verlag 2012, pp. 1–14, 2012.
26. Mohammed, M. and Pathan, A.-S.K., *Automatic Defense against Zero-Day Polymorphic Worms in Communication Networks*. ISBN 9781466557277, CRC Press, Taylor & Francis Group, USA, 2013.

8
Semi-Supervised Learning BitTorrent Traffic Detection

Raymond Siulai Wong, Teng-Sheng Moh and Melody Moh

CONTENTS

8.1 Introduction 191
    8.1.1 BT Protocol 192
    8.1.2 Detecting BT Traffic: Why? 193
8.2 Related Works 194
    8.2.1 Port-Based Detection 194
    8.2.2 DPI-Based Detection 194
    8.2.3 DFI-Based Detection 195
    8.2.4 Combination of DPI and DFI 196
    8.2.5 Other Detection Methods 196
8.3 Proposed System: Intelligent Combination 197
    8.3.1 Offline Training Module 199
    8.3.2 Online Classification Module 200
        8.3.2.1 Databases 200
        8.3.2.2 Online Classification Module Description 200
    8.3.3 Strengths of Intelligent Combination System 201
    8.3.4 Comparison with Other Methods 201
8.4 Performance Evaluation 202
    8.4.1 Experiment Setup 203
        8.4.1.1 Ground Truth Generation 203
        8.4.1.2 Offline Training Module 203
    8.4.2 Online DFI Classifier 204
    8.4.3 Classification Accuracy 204
        8.4.3.1 Training Data and Test Cases 204
        8.4.3.2 Accuracy Measurements and Metrics 205
        8.4.3.3 Accuracy Results 206
    8.4.4 Classification Time 207
8.5 Conclusion 208
References 208

8.1 Introduction

Internet technologies have been rapidly evolving and exponentially expanding for the past few decades. Back in the olden days, to access the Internet, one had to use the analogue phone-line modem. Today, we can easily go online by using broadband access as well as smart mobile devices through cellular or wireless hot spots.

Another technology that has been well developed with the Internet is the peer-to-peer (P2P) systems. Unlike the traditional client–server architecture, P2P networks offer a new distributed model of connecting computers (or hosts). In a P2P network, every node acts as both a client and a server, providing part of the resources needed for the system. Each peer is simply a computer, or a host, connected to the Internet. All the hosts act autonomously to join or leave the network freely, and contribute as part of the network resources in terms of storage (such as files desired by other peers), computing and bandwidth.

Therefore, P2P technology offers an all-new model that allows hosts to efficiently access a large amount of data. Instead of accessing only through the server (data provider), peers now access the desired data with and through many peers while also contributing themselves to provide data and other resources such as processing and bandwidth. There is less competition and more collaboration; as a result, each peer obtains the data more quickly and at less cost.

P2P systems have also brought much challenge to the community. First, P2P traffic has grown at a tremendous rate and reduced the bandwidth available for other applications. The second challenge coming along with the P2P technology is that many people use P2P protocols to download copyrighted files (such as video, games, music, etc.). This is still illegal in many countries, including the United States, the United Kingdom and Japan.

Therefore, detecting and controlling P2P traffic has become an important task for many. For example, Internet service provider (ISP) and enterprise network administrators want to control P2P bandwidth so that they may provide enough bandwidth for other critical applications. P2P traffic detection, however, has become a challenge in recent years due to the development of many intelligent P2P applications. For example, port detection, one of the existing techniques, does not work any longer since many new P2P applications now use user-defined, non-standard ports, or even dynamic ports.

This chapter first offers a literature survey of major classes of P2P traffic detection methods. Next, a new detection method is proposed. *Intelligent combination* is based on deep packet inspection (DPI) and semi-supervised learning is based on deep flow inspection (DFI). By carefully observing the packet pattern in a BitTorrent (BT) flow, we intelligently arrange DPI to be ahead of DFI in the module. This arrangement has greatly speeded up the classification process. From the conducted experiments, the mechanism gives a promising accuracy rate of over 90% while improving the classification time by 15–20% over a double layer, an existing combined DFI/DPI method [1].

The chapter is organised in the following manner. The rest of Section 8.1 discusses the BT protocol and details the needs of detecting BT traffic. Section 8.2 presents the related studies, including the main class of the mechanisms to detect P2P traffic. Section 8.3 describes our new, improved approach, *intelligent combination*. Section 8.4 illustrates the simulation results of the proposed system along with three existing methods. Finally, Section 8.5 concludes the chapter while presenting future directions. This chapter is an extension of two preliminary works [2,3], and is part of the ongoing effort of the authors' research on network protocols [4] and distributed systems [5].

8.1.1 BT Protocol

BT is a P2P file-sharing protocol for distributing a large amount of data (such as video files) over the Internet. It is one of the most popular P2P protocols. The first release of BT was back in April 2001; since then, there have been numerous BT clients developed for different platforms. As of January 2012, BT has 150 million active users, according to BitTorrent, Inc. This implies that the total number of BT users can be estimated at more than a quarter billion per month [6]!

Semi-Supervised Learning BitTorrent Traffic Detection

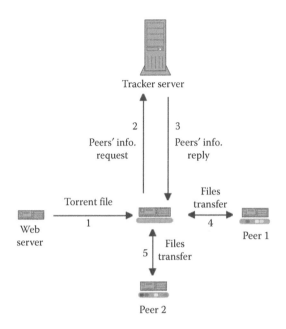

FIGURE 8.1
Steps for a node to join the BT file-sharing network.

The success of BT stemmed largely from its simplicity for users to share large files. Referring to Figure 8.1: To start sharing a file, the person that has the original copy of the file creates another file with an extension of '*.torrent*' and hosts in a web server (step 1). A file is usually divided into segments called *pieces*. Inside this torrent file, it has the pieces information of the sharing files and the URL of the BT *tracker* (see step 2), where a BT tracker is a special kind of *server* that contains the peers' information (i.e., in this case, it has the IP of the computer that has the original file). When another computer downloads the torrent file (usually hosted by a web server), the BT client will contact the tracker (step 2) and add its own information into the *tracker server*. Finally, once the peers' information is obtained from the tracker server (step 3), it starts downloading the pieces of the files and also starts sharing the pieces of the downloaded files (steps 4 and 5).

8.1.2 Detecting BT Traffic: Why?

P2P systems in general, and BT protocol in particular, are used mainly for sharing and downloading large files. According to Klemm et al. [7], BT accounts for approximately 45–78% of all P2P traffic and 27–55% of all Internet traffic as of February 2009. More recently, according to TorrentFreak.com (that posts online news regarding P2P traffic), a new website, Internet Observatory, which offers real-time Internet traffic statistics and illustrates how much bandwidth is consumed by different applications, showed in September 2011 that P2P traffic in Europe accounted for more than 25% of all bandwidth, and about 40% of all the packets sent, and all of them were BT [8]. As such, the bandwidth available for other applications has been largely reduced. Given such enormous traffic flows, many organisations need to control their network flow by detecting and limiting the bandwidth of P2P activities. Furthermore, BT is widely used to transfer copyrighted materials such as movies, video games, music, books and even software. This is unlawful in many countries, including the United States, Japan, Germany and France.

The need for detecting P2P traffic can be viewed from different aspects by various organisations. For an ISP, they may work with the copyright-owning companies to detect illegal downloads. For an enterprise network, the administrators may want to rate-limit the P2P traffic such that it has enough bandwidth for other critical applications. For a local broadband ISP, they may want to limit the cost charged by the upstream ISP. Finally, for regular home users, most of them still use an asynchronous Internet connection service (such as ADSL—asynchronous digital subscriber loop—and cable modem) provided by their ISP. The upstream rate and downstream rate are not equal; in fact, the upstream rate is usually lower than the downstream rate. If the upstream is congested, it will affect the overall Internet experience. This scenario also applies to smart device users connecting through cellular networks that also provide asynchronous connections.

8.2 Related Works

This section surveys the major techniques used to detect P2P traffic. In general, there are several major classes of techniques: (1) port-based, (2) DPI, (3) DFI and (4) some combination of DPI and DFI techniques. Furthermore, there are methods that are outside these four main classes. In this section, we describe the general principles and some major works of these techniques.

8.2.1 Port-Based Detection

The first method is based on transmission control protocol (TCP) and/or user datagram protocol (UDP) ports. The assumption of this method is that different applications use different TCP/UDP ports, and many P2P applications did use fixed ports to communicate among peer computers. One obvious advantage is that it is very easy to implement and almost no computation power is required. Moreover, using traffic identification based on ports not only directly identifies individual P2P applications (e.g., eMule) but also easily eliminates the well-known non-P2P application (such as file transport protocol [FTP], e-mail, etc.).

Unfortunately, to avoid being identified, many newer P2P applications no longer use fixed ports. Instead, they use dynamic techniques utilising user-defined ports, random ports, changing ports and camouflage ports. As a result, this port-based approach has become obsolete.

8.2.2 DPI-Based Detection

The second class is based on DPI. This detection method inspects some particular features of a packet payload to identify P2P traffic. The control messages in a P2P protocol, especially the handshaking messages, usually follow certain features. In other words, they contain some particular patterns in the application layer payload. One example would be the BT protocol. There is always a 'BitTorrentprotocol' string that appears in their handshaking packets.

Liu et al. [9] proposed a DPI algorithm to detect BT traffic. Their algorithm is based on the handshaking message between the BT peers. According to the authors, the BT header of the handshake messages has the following format: < *a character (1 byte)* > < *a string (19 byte)* >. That is, the first byte is a fixed character with value '19', and the string value is 'BitTorrentprotocol'. On the basis of this common header, they use it as signatures for identifying BT traffic [9].

The advantage of this approach is simply the high accuracy and robustness. However, there are several disadvantages. Since the payload needs to be examined, no encryption data can be supported. Also, this method leads to the issue of privacy because of the need to inspect the content of the payload. In addition, high computation power may be required for payload checking in the midst of continuously changing P2P protocols. Finally, sometimes, it is difficult to obtain the characteristics of a protocol from its open-source software.

8.2.3 DFI-Based Detection

The next detection method, DFI, uses the entire flow to detect P2P traffic. As the name implies, the analysis or the classification of P2P traffic is flow based, focusing on the connection-level patterns of P2P applications. Thus, it does not require any payload analysis (as DPI does); therefore, encrypted data packets can be supported. The downside of this approach is the need to extract the connection-level patterns for P2P traffic. Yet, there is no rule of thumb as to which network features should be used.

Le and But [10] used a DFI algorithm to classify P2P traffic. In particular, they focused on the packet length statistics of traffic as the features for their classifier for detecting BT traffic. The four network features used were minimum payload, ratio of small packets, ratio of large packets and small payload standard deviation. Three types of traffic traces were used to train and test their classifier. These include known BT traffic, known FTP traffic and other traffic.

Erman et al. [11] proposed a semi-supervised learning algorithm to classify traffic. Their algorithm involves a two-step approach to train their classifier. The first step is clustering. In this step, the flows having the applications labelled are partitioned with an unsupervised clustering algorithm. The K-means algorithm is used in this step. However, note that this algorithm does not restrict itself to using only the K-means algorithm. The second step is to map clusters to applications and to determine the cluster's label based on the flows label. This mapping is based on the estimation of the probabilities that the labelled flow samples within each of the clusters. It can be estimated by the maximum likelihood estimate, n_{jk}/n_k, where n_{jk} is the number of flows that were assigned to cluster k with label j, and n_k is the total number of (labelled) flows that were assigned to cluster k. As the authors pointed out, the key benefit of the unsupervised learning approach is its ability to identify the hidden patterns [11]. For instance, one can identify new applications by examining flows that form a new clustering.

Chen et al. [12,13] tried to use machine-learning techniques to identify P2P traffic. In one study, they proposed to use a back propagation (BP) neural network algorithm, where they implemented a prototype system that realised both offline learning and online classification [12]. The method has successfully provided an efficient way to measure and classify aggregate P2P traffic.

In another study, Chen et al. [13] presented a support vector machine (SVM) algorithm to detect P2P traffic. The algorithm uses double characteristics, namely, Internet protocol (IP) and IP-port, to identify P2P traffic by means of separating different traffic features. The authors observed from the experimental results that choosing the appropriate traffic features, kernel options, configuration parameters and punish modulus to the SVM algorithm is effective in identifying P2P traffic [13].

Liu et al. [14] focused on analysing the packet length to detect P2P traffic. Their claim is as follows: In P2P applications, there are many small-sized packets as well as many very large-sized packets—close to the maximum transmission unit (MTU) packets. The small-sized packets are often used to transfer messages between the server and the client such as synchronisation and acknowledgement. The large-sized packets are usually used to transfer data (actual sharing content). The authors developed a statistical methodology by analysing packet length and P2P basic characteristics. The method identified P2P flows at the transport layer without relying on the port numbers and packet payload [14].

8.2.4 Combination of DPI and DFI

This class combines the two techniques, DPI and DFI, to increase P2P traffic detection rates. Chen et al. [13] used both DPI and DFI to detect BT packets. They also suggested executing DPI and DFI in parallel to speed up the overall process. The model comprises five parts: traffic-collect module, DPI module, DFI module, coordinated module between DPI and DFI and evaluation module, and is able to enhance the identification accuracy as well as to expand the identifying scope of P2P traffic.

Keralapura et al. [15] used packet delays as a DFI feature along with DPI to increase P2P packet detection rates. The P2P traffic classifier is called a self-learning traffic classifier and it is composed of two stages. The first stage used a time correlation metric algorithm that uses the temporal correlation of flows to different P2P traffic from the rest of the traffic. The second stage used a signature extraction algorithm to identify signatures of several P2P protocols. The work used real network traces from tier-1 ISPs that are located in several different continents and achieved a detection rate above 95% [15].

Wang et al. [1] used both DFI and DPI. They claimed that the combination made both detection algorithms comprise each other and increased the detection rate as a result. Their proposed scheme, named the double layer method, used three steps sequentially: in the first step, traffic detection based on port number is used to filter the common P2P traffic. In the second step, a DFI module is applied to match the characteristics of the data stream against common P2P characteristics. In the third step, those packets that have been identified as P2P will go through a payload characteristic module (aka DPI module); the exact P2P traffic type will then be determined. The resulting detection rate is high. The entire system, however, is complex, and the classification time is long. This algorithm has been chosen to be a part of the performance evaluation study (Section 8.4).

8.2.5 Other Detection Methods

In addition to the above four major classes, several other methods have been proposed to detect P2P traffic. An earlier work by Constantiniu and Mavrommatis [16] used only transport-layer information to identify P2P traffic. Instead of using application-specified information, they made use of fundamental characteristics of P2P traffic, including a large network diameter and the existence of many peers acting both as servers and as clients, to perform port analysis. This is different from many typical P2P traffic detection methods that focused on analysing the host or packets. Instead, the authors proposed to detect P2P traffic by port analysis, as illustrated by the following simple example. There are four nodes in this network: nodes A, B, C and D, initially with no connection (edge). Whenever a connection is made from a node (say i) to another node (say j), node j will be assigned a level that is one step higher than node i. Thus, initially, say node A makes a connection to node B. Node A is assigned level 0, and node B is assigned level 1. After that, B makes a new connection to node C. As a result, node C is assigned level 2. Note that once a level is assigned to a host, it cannot be changed. Therefore, when node D makes a new connection to node A, node A's level cannot be altered. Instead, node D is assigned level −1, which is one level lower to node A. Note that the last connection did not (and should not) change the level value of any nodes in the network.

In this method proposed by Constantiniu and Mavrommatis [16], once the graph is constructed, using the basic characteristics of P2P traffic (i.e., a large network diameter and the existence of many peers acting both as servers and as clients), the following rules are then used to determine if this port is considered as the one used by P2P applications: (1) the number of hosts that act both as servers and as clients ('ClientServers') in the specific port exceeds the ClientServer threshold, (2) the network diameter is at least 2 and (3) the numbers of hosts that are present in the first and last levels of the network exceed the edge-level threshold. (Note that the authors did not specify how to determine these thresholds.) Using the above scheme of port analysis, the method was able to detect both known and unknown P2P traffic.

Zhang et al. [17] proposed a distributed model to detect BT traffic based on peer information. The method records and analyses the peer information obtained from the BT signalling traffic exchanged among peers. The method can accurately identify and control the BT traffic related to these peers even when the traffic is encrypted. As more and more P2P traffic is encrypted, this method is more complete and practical than other methods.

Finally, Basher et al. [18] conducted an extensive analysis of P2P traffic, and did a comparative analysis between web traffic and P2P traffic. In addition, they also did a comparative study between two P2P protocols, namely, BT and Gnutella. Their study used a database that contained 1.12 billion IP packets totalling 639.4 Gb. They also presented flow-level distributional models for P2P traffic. The analysed characteristic metrics are as follows: *flow size*—the total bytes transferred during a TCP flow, *flow inter-arrival time*—the time interval between two consecutive flow arrivals, *duration*—the time between the start and the end of a TCP flow, *flow concurrency*—the maximum number of TCP flows a single host uses concurrently to transfer content, *transfer volume*—the total bytes transferred to and from a host during its activity period, and finally *geography distribution*—the distribution of the shortest distance between the individual hosts and authors' campus along the surface of the earth. The authors expect that this information may be contributed for research in network simulation and emulation experiments [18].

Finally, we proposed a system, *intelligent combination*, which is also on both DPI and DFI methods. While achieving an equally high accurate rate as a double layer [1], based on the BT packet pattern, our method arranges DPI to come before DFI and thereby successfully shortens the classification time. The proposed system is presented in the next section with detailed explanation. The section compares various major algorithms, including DPI [9], DFI [10,14], semi-supervised learning-based DFI [11] and the double layer of combining DFI and DPI [1], and the new proposed intelligent combination is given at the end of Section 8.3.

8.3 Proposed System: Intelligent Combination

The main goal is to classify each packet flow as either BT or non-BT. The proposed system, *intelligent combination*, is depicted in Figure 8.2. The system can be divided into two major

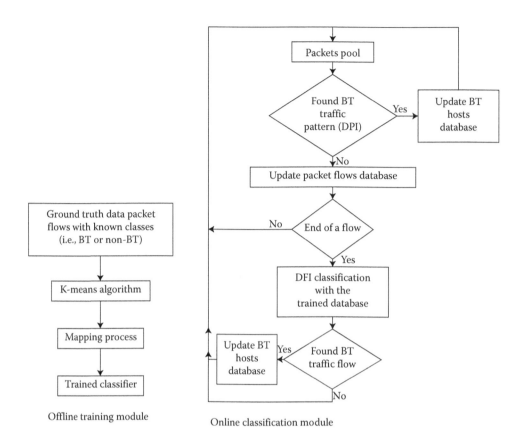

FIGURE 8.2
Proposed intelligent combination system to classify BT packet flows.

parts, namely, the online classification module and the offline training module. Below is an overview of the system; the detailed description is given in the following sections.

1. The *online classification module*: It is an intelligent combination of DPI and DFI.
   a. *DPI*: The module starts with a simple DPI that quickly detects packets with BT character strings, while at the same time recording the host names of these BT packets and updating the BT host name database.
   b. *DFI*: The rest of the packets then go through the DFI portion, which is a classifier based on a semi-supervised learning system [11,19]. It will perform a detailed flow inspection of the features of the unclassified packet flows and classify each as either BT or non-BT.
2. The *offline training module*: It provides for the DFI portion a reliable trained classifier based on the K-means algorithm [11,20] for detecting BT flows.

The arrangement of DPI and DFI, with first a simple DPI followed by a semi-supervised learning DFI classifier, is based on the following rationale: We observe that a BT packet flow typically begins with handshake messages. These messages often include some simple string pattern such as 'BitTorrentprotocol'. If such a pattern is found (matched), obviously a BT flow is to be expected. A simple DPI may be used for such a straightforward pattern matching, and a successful matching can then eliminate the need for a long DFI classification for this flow.

This intelligent combination (arrangement) of DPI coming before DFI not only ensures high classification accuracy, as in other combined DPI/DFI methods (such as the work by Wang et al. [1]), but also greatly improves the overall execution time of the entire classification process (as demonstrated in Section 8.4.5).

In the rest of the section, we first explain the offline training module used for DFI classification. Next, we describe the online classification module. The strength of the intelligent combination system is then discussed. Lastly, a comparison of the proposed scheme with other major methods is presented.

8.3.1 Offline Training Module

The block diagram of the offline training module is given on the left side of Figure 8.2. The module uses the K-means algorithm [11,20], a well-known algorithm in the area of clustering and machine learning, for mapping data objects (in our case, features of packet flows) to classes (BT and non-BT). The module is described below; also refer to Figure 8.3 for an illustration of the classification method using the K-means algorithm.

The offline training module begins with *the ground truth* as the input to the K-means algorithm (step 1, 'Labelled Traffic Flows' of Figure 8.3). The ground truth consists of packet flows with *known classes* (BT and non-BT). We use 10 features to characterise each packet flow, as listed in Table 8.1. These features are extracted from each of the packet flows within the ground truth.

Next, the K-means algorithm divides the packet flows into *clusters* according to the similarities of their 10 features (step 2, 'K-means algorithm' in Figure 8.3). The flows that exhibit closer features will be grouped into the same cluster. At the end of the K-means algorithm, there will be a total of K clusters.

The subsequent step is the *mapping process* (step 3, 'mapping clusters' in Figure 8.3). This process is to map each cluster into a given class (in our case, BT or non-BT). The principle for the mapping decision is simple: given a cluster, if the number of BT packet flows is larger than that of non-BT packet flows, we mark it as a BT cluster. Similarly, we mark a cluster as non-BT if there are more non-BT packet flows within the cluster.

Finally, as shown in step 4, 'trained classifier' in Figure 8.3, when the mapping process is completed, we will have the *trained classifier* (also known as a *trained database*) mapping each cluster (or more specifically, each set of 10 flow features), into a class. The trained

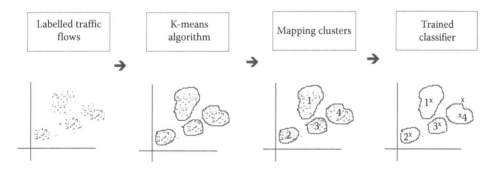

FIGURE 8.3
Overview of the classification method based on the K-means algorithm.

TABLE 8.1

Flow Characteristics Used in the Semi-Supervised Learning of the Offline Training Module

Number	Features
1	Total number of packets
2	Average packet size
3	Total bytes
4	Total header (transport plus network layer) bytes
5	Number of flow initiator to flow responder packets
6	Total flow initiator to flow responder payload bytes
7	Total flow initiator to flow responder header bytes
8	Number of flow responder to flow initiator packets
9	Total flow responder to flow initiator payload bytes
10	Total flow responder to flow initiator header bytes

classifier is then used by DFI of the online classifier module to detect if a packet flow is BT or non-BT.

8.3.2 Online Classification Module

The main goal of the online classification module is to discover all the *BT hosts*. In the following section, we first describe the databases needed, and then the detailed steps of the module.

8.3.2.1 Databases

There are two databases needed. The first is the *BT host database* for all the classified BT hosts. This database is updated whenever a packet or a flow is being classified as BT. The second is the *packet flow database*, which is for keeping track of all the packets that we have seen during the classification process. It is continuously updated as packets are received, so that at the end of a packet flow, it has the complete 10-feature information of this packet flow. This information is then fed into DFI for classification if needed.

8.3.2.2 Online Classification Module Description

The block diagram of the online classification module is given on the right side of Figure 8.2. Initially, a packet flow of unknown class is input to the simple *DPI module*. It is based on the match of a simple string pattern (such as 'BitTorrentprotocol'). This is used to determine if the encountered packet is of BT class. If it is a BT packet, the *database of BT hosts* will be updated immediately.

If no BT string pattern is matched, then the corresponding packet flow information (i.e., total number of packets in flow, average packet size, etc.; features in Table 8.1) will be extracted and input to the corresponding entry of *packet flow database*. If that packet is at the *end of the flow*, we have the entry with completed flow information updated in the packet flow database.

This new packet flow information will then be applied to the *DFI classification with the trained database*. Relying on the trained database and based on the 10 flow features, the DFI classifier will then determine if the new input flow is BT. Recall that the trained database (also called a trained classifier) is obtained from the offline training module as described in Section 8.3.1. Once the classification decision is made, if *the flow is found to be BT*, the BT host database will also be updated accordingly.

Finally, note that the above design is based on its main goal—to discover all the BT hosts. This goal is chosen since BT hosts are often the most important information requested by an enterprise network or an ISP. Alternatively, we can set the goal of discovering all the *BT flows*. Note that our classification system (mainly the online classification module) can be easily modified for this alternative, or other similar purposes.

8.3.3 Strengths of Intelligent Combination System

In this section, we note some major advantages in this approach:

1. *Accuracy*: By employing both DPI and DFI modules, the accuracy of the classification is greatly increased, as demonstrated by other earlier works [1,13,15]. If one module fails to detect, the other can have a chance to look at the packet flow.
2. *Security*: Similar to most existing DIP modules, our simple DPI does not handle encrypted packets. Yet, the subsequent DFI module will be able to handle them since it does not depend upon reading the packet payload (it only needs its features such as payload size).
3. *Efficiency*: Unlike other combined DPI/DFI methods (such as the double layer system by Wang et al. [1]), our system first uses simple DPI to quickly determine if a packet is a BT. DFI is used only if the BT packet pattern cannot be found. The major advantage of this arrangement is that we do not need to wait until the end of the flow to determine the flow type if the BT pattern can be matched in the DPI stage. In other words, we can quickly identify a flow as BT without waiting for the entire flow to complete. This advantage is also clearly demonstrated in the performance evaluation section (Section 8.4.5) when the classification time of the proposed system is shorter than both the double layer and the DFI.
4. *Further speedup*: Since we store the BT hosts in our database, we could further speed up the classification process by first checking if a host is BT. (If there is a match, then both DPI and DFI may be skipped.) This will also avoid repeating the tasks that have been performed. For instance, if the DFI module has already classified a host as a P2P host, then DPI can safely skip packets from that host.

8.3.4 Comparison with Other Methods

In this section, we compare four major approaches with our proposed system, as summarised in Table 8.2. The first one, a DPI method, detects patterns of handshaking messages among BT peers [9]. Since a string ('BitTorrentprotocol') can be found within the handshaking packets, this pattern is used to determine whether a BT client is currently running in the network. The approach is simple and is able to provide acceptable accuracy results. However, the major drawback of this approach is that only non-encrypted packets are supported.

The second major approach, DFI, is based on the packet length statistics of the packet/flow length [10,14]. Since this method is based upon the packet length, encrypted packets

TABLE 8.2

Comparison of Major Approaches for Detecting BT Packets

Methods	Schemes	Implementation Strengths	Implementation Limitation
BT header lookup method [9]	Simple DPI	Simple	
Packet/flow length statistics method [10,14]	DFI	Avoid complex per-packet DPI overhead	Difficult to determine suitable thresholds
Learning algorithm [11]	DFI	Avoid complex per-packet DPI overhead	Initial offline training required
Double layer [1]	(1) Port based, (2) DFI and then (3) DPI	Possible (though unlikely) quick classification through port based	(1) Difficult to determine suitable thresholds for DFI and (2) very complex
Intelligent combination (proposed)	(1) Simple DPI, then (2) DFI	(1) Use simple DPI, (2) possible quick classification through simple DPI and (3) simpler than double layer [1]	Initial offline training required

can also be supported. The philosophy behind this method is that BT and non-BT packets have different packet length distributions. Therefore, only thresholds are needed to determine if the packet is a BT packet. One of the major problems of this approach is that there are some non-BT packets that also have the same length characteristics of BT packets. Hence, it is often difficult to determine the suitable threshold values.

The third approach is based on a learning algorithm upon DFI [11]. It uses classification algorithms to train a database from the packet/flow characteristics (e.g., average packet size, total number of packets, etc.). One problem with this approach is that prior training is required.

The fourth approach, double layer, utilises both DFI and DPI to identify BT packets [1]. DFI is first used to classify whether the packets are P2P, and then DPI is followed to determine the exact type (such as BT) of P2P packets. The major disadvantage of this approach is that it may be difficult to implement under a live network due to the fact that DFI is per-flow based, whereas DPI is per-packet based. Thus, we will need to wait until the end of a flow to perform DFI classification, whereas the BT pattern can only be seen in the handshaking packets, which appear only at the beginning of a BT flow.

8.4 Performance Evaluation

This section evaluates the performance of the proposed method with three existing methods: DPI [9], semi-supervised learning DFI [11] and double layer of a combination of DFI followed by DPI [1]. First, the initial experiment setup is illustrated, including the generation of ground truth (known-class packet flows). Next, the training of the DFI classifier is described, which is a necessary foundation for all the methods evaluated (except the simple DPI). The performance metrics and the main performance results, classification accuracy and classification time are then presented.

8.4.1 Experiment Setup

The main goal of the initial experiment is to capture BT and non-BT packets. There can be one or more PCs behind a router. Inside the PC, BitComet 1.21 is installed as the BT client. A sample torrent file was downloaded for BT packet-capturing purposes. Note that a torrent file contains information about the tracker server, whereas the tracker server contains the peers' information about shared files.

A clear explanation of generating BT and non-BT files (packet flows) is given below (Section 8.4.1.1). The training of a DFI classifier is described in Section 8.4.2, with its specific data that are given in Section 8.4.3.

8.4.1.1 Ground Truth Generation

The ground truth in this project is the packet flows of known classes (i.e., BT and non-BT). To train a classifier, there are two types of packet flows needed to capture, namely, the BT and the non-BT packet flows.

To capture BT packets, we start a sample torrent file, and the BT client will automatically start downloading/uploading its contents. At the same time, we start a packet-capturing program to obtain these packets. Similarly, to capture non-BT packets, we start the packet-capturing program while creating non-BT network activities; these activities include HTTP, FTP and secure shell (SSH).

8.4.1.2 Offline Training Module

The objective of the offline training module (shown on the left side of Figure 8.2) is to create a trained database for the DFI classification used in the online classification module. The offline training module is divided into three separate subprograms for the ease of development. The first two are C programs and the third one is a MATLAB® program (refer to Figure 8.4).

The two C programs are the network packets capture program and the packet features extraction program. They were both built with Cygwin under Windows environment. This combination provided a user-friendly development environment utilising open-source libraries. Packets capture is made possible by the WinPcap [21] library. WinPcap is an open-source library that allows the user to set his/her network interface card (NIC) to operate in 'promiscuous' mode. Thus, all the packets going through the network will be captured.

The first is the network packets capture program. As the names implies, the function of the packet capture program is to communicate with NIC and to capture packets from the network. The captured packets are then stored in a file with PCAP format. The second program, the packet feature extraction program, is used to extract the network features from the packet file. There are 10 features to be extracted, as shown in Table 8.1.

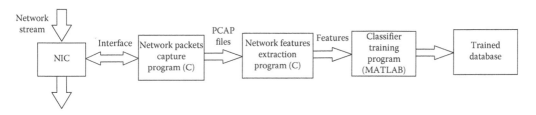

FIGURE 8.4
Offline training flow chart.

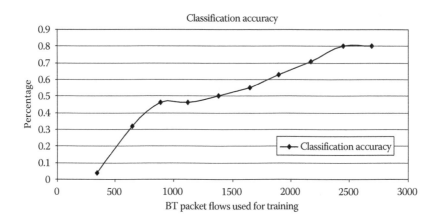

FIGURE 8.5
Classifier accuracy in training the classifier in the DFI module.

Finally, the extracted features will be used to train the classifier. The classifier is based on the K-means clustering algorithm [11,20]. The classifier program is written in MATLAB. The advantage of using MATLAB over C is that MATLAB has a lot of built-in numerical computation routines; the K-means cluster algorithm routine is also supported.

8.4.2 Online DFI Classifier

The DFI module used is based on a semi-supervised learning classifier that would accurately classify BT packet flows. This classifier is first trained with the ground truth generated, as described above. It is then tested against other BT packet flows to observe the accuracy.

We conduct an experiment to estimate the number of packet flows needed to train a reliable classifier for the DFI module, using $K = 400$ in the K-means algorithm [11]. Figure 8.5 shows the classifier accuracy with an increasing number of BT packet flows being used in training the classifier. As expected, the more BT packets that are used to train the classifier, the better the accuracy. As the number of BT packets increases, however, the classifier will be saturated at some point. After that, even when more packets are supplied, the accuracy will not increase significantly.

From Figure 8.5, one can see that the saturation point occurs around 2500 packet flows. Thus, in the following experiments, the number of BT packet flows used for training the DFI classifier should be at least 2500 to ensure high accuracy.

8.4.3 Classification Accuracy

In this section, we first present the training data and test cases. Next, the measurements and metrics are described. The accuracy results are then illustrated.

8.4.3.1 Training Data and Test Cases

In this experiment, we compare the classification accuracy results. On the basis of the results obtained in training the DFI module classifier (Section 8.4.3), we train the classifier with the ground truth of 8000 TCP packet flows of which approximately 3500 are BT–TCP packet flows.

To obtain a fair comparison of the four methods, we design two statistically different test cases for the experiments. In test case 1, there are 60% BT packet flows and 40% non-BT packet flows. (Specifically, it contains 686 BT and 454 non-BT packet flows.) In test case 2, there are about 70% BT and 30% non-BT packet flows (specifically, 408 BT and 167 non-BT packet flows).

8.4.3.2 Accuracy Measurements and Metrics

For just a comparison of the four methods, it is important to take note of both the *positive* and the *negative* results. Therefore, we collect the following four measurements [20].

8.4.3.2.1 Four Measurements
- *True positive (TP)*: The number of BT packet flows that have been correctly detected
- *False negative (FN)*: The number of BT packet flows that have *not* been detected
- *False positive (FP)*: The number of non-BT packet flows that have been falsely identified as BT packet flows
- *True negative (TN)*: The number of non-BT packet flows that have been correctly identified as non-BT

8.4.3.2.2 Performance Metrics
On the basis of the above collected data, the following metrics are calculated [20], listed below, as Equations 8.1 through 8.4, respectively:

$$TPR \text{ (true-positive rate)} = TP/(TP + FN) \tag{8.1}$$

$$TNR \text{ (true-negative rate)} = TN/(TN + FP) \tag{8.2}$$

$$FPR \text{ (false-positive rate)} = FP/(FP + TN) \tag{8.3}$$

$$FNR \text{ (false-negative rate)} = FN/(TP + FN) \tag{8.4}$$

In the above equations, the TPR or *sensitivity*, expressed in Equation 8.1, is defined as the fraction of positive examples (BT packet flows) correctly classified. Similarly, the TNR or *specificity*, expressed in Equation 8.2, is the fraction of negative examples (non-BT packet flows) classified correctly. Furthermore, the FPR, expressed in Equation 8.3, is the portion of negative examples (non-BT packet flows) falsely classified as BT packet flows. Finally, the FNR, expressed in Equation 8.4, is the portion of positive examples (BT packet flows) falsely classified as non-BT.

In addition, two widely used metrics [20] are also employed:

$$\text{Precision, } p = TP/(TP + FP) \tag{8.5}$$

$$\text{Recall, } r = TP/(TP + FN) \tag{8.6}$$

Precision (p), expressed in Equation 8.5, determines, out of the total packet flows that have been classified as BT, the fraction that is actually BT (true or correct classification).

Obviously, the higher the precision, the smaller the number of FP errors committed by the classifier. On the other hand, Recall (r), given in Equation 8.6, measures, among the entire BT packet flows, the fraction that has been *correctly* classified as BT packet flows. A large recall implies that a system with very few BT packet flows has been falsely classified as non-BT. Simply put, the larger these two metrics are, the more accurate the system is.

8.4.3.3 Accuracy Results

Tables 8.3 and 8.4 show the classification results of the four algorithms, DPI [9], semi-supervised learning DFI [11], double layer [1] and the proposed intelligent combination method, for the two test cases, respectively.

First, comparing Tables 8.3 and 8.4, it is clear that they are similar and consistent. Thus, we can trust that the results presented are reliable. Below, we discuss the results of each of the four methods.

Simple DPI [9] has a very high TNR (100%); that is, it has achieved 100% accuracy to detect non-BT traffic flows (including HTTP, FTP and SSH). It is because the DPI method searches for the BT pattern string ('BitTorrentprotocol') explicitly inside the packets, which obviously cannot be found in non-BT flows. On the other hand, unfortunately, it has a very low TPR (31% and 35% in test cases 1 and 2, respectively); that is, it has a very low rate of successfully identifying BT packet flows. It is because the BT pattern string happens mainly in the handshaking messages and it may not appear during the BT data transfer. Owing to its simple, straightforward method, it has a very high precision rate (100%—it does not confuse non-BT as BT), but unfortunately a very low recall rate (31% and 35% in the two test cases); this shows that it does confuse many BT as non-BT. This is very likely due to the fact that it only detects the BT pattern string that occurs mainly in the handshakes but not in data transfer.

DFI [11], on the other hand, has a reasonably high TPR (78% and 73%)—it is able to correctly detect a good portion of BT traffic, and a higher TNR (86% and 87%)—it is also able to correctly detect quite a lot of non-BT traffic. While its precision rate is high (89% and 93%), it does not confuse non-BT as BT; however, it has a slightly lower recall rate (78% and

TABLE 8.3

Classification Accuracy Results (Test Case 1)

Scheme	TPR	TNR	FPR	FNR	Precision p	Recall r
DPI [9]	0.31	1.00	0.00	0.69	1.00	0.31
DFI [11]	0.78	0.86	0.14	0.22	0.89	0.78
Double layer [1]	0.87	0.86	0.14	0.13	0.90	0.87
Intelligent combination	0.87	0.86	0.14	0.13	0.90	0.87

TABLE 8.4

Classification Accuracy Results (Test Case 2)

Scheme	TPR	TNR	FPR	FNR	Precision p	Recall r
DPI [9]	0.35	1.00	0.00	0.65	1.00	0.35
DFI [11]	0.73	0.87	0.13	0.27	0.93	0.73
Double layer [1]	0.85	0.87	0.13	0.15	0.94	0.85
Intelligent combination	0.85	0.87	0.13	0.15	0.94	0.85

73%), and it does sometimes confuse BT as non-BT. This may due to the fact that some BT flows fail to exhibit typical characteristics.

Our proposed intelligent combination method and the double layer method [1] have both exhibited equally high accurate results. Their main difference is in the classification time (to be discussed in the next section). Therefore, we will discuss these two methods together here. They have a very high TPR (87% and 85% in the two test cases) and TNR (86% and 87%). These have resulted in high values of both precision and recall rates. Note that their recall rates (90% and 94%) are the highest among the four methods, which implies that both methods do not easily confuse BT with non-BT packet flows (as DPI [9] and DFI [11] do). This is an important strength of the two methods.

8.4.4 Classification Time

Figure 8.6 shows the packet classification time for various classification methods. Note that the proposed intelligent combination method has the second shortest execution time (only longer than the simple DPI [9]); this is one of the most important performance metrics, discussed in detail below.

We first note that the simple DPI [9] has the fastest classification time due to the fact that its classification is purely based on a string comparison (whereas the other three methods are more computationally intensive). DFI [11], on the other hand, requires the longest execution time. It is a purely flow-based method; each classification theoretically needs to wait till the entire packet flow is completely received and analysed.

The double layer method [1] has the second longest execution time. It is faster than DFI [11] since, before DFI, it applies a simple port-based classification. This step can quickly filter out some BT flows that used popular P2P ports, and therefore speeds up the average execution time.

The proposed intelligent combination method is faster than both the double layer [1] and DFI methods [11]. In particular, it is 15–20% faster than the double layer method. This is because it first uses the simple DPI [9] (that has the fastest execution time), and if there is a match, it records its BT host names into the BT host database systems, which not only shortens the current-flow classification (a quick BT-flow detection), but also helps speed up the future classification time.

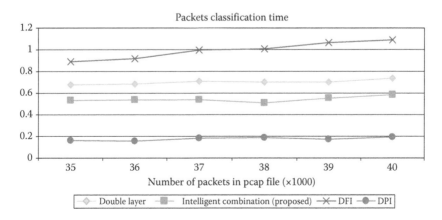

FIGURE 8.6
Packet classification time.

8.5 Conclusion

P2P technology has offered a completely new model on the Internet, changing from the traditional client-and-server model. It allows peers (hosts) to efficiently download large files by acting both as a server and as a client. It has quickly dominated the Internet traffic and has threatened the bandwidth and service available to other applications, especially those that are mission-critical. Furthermore, many use the P2P protocols to illegally download copyrighted files. BT is the most popular P2P protocol available on the Internet. Therefore, it is important to develop efficient, accurate methods to identify and control BT traffic.

By recognising the packet pattern of BT flows, this chapter proposed a new BT traffic identification method that is both accurate and fast. Simply put, combining both DPI and DFI makes it accurate; arranging DPI before DFI makes it fast. The simulation results further support its superior performance. They have shown that, by combining the simple DPI with the DFI that is based on semi-supervised learning, the BT detection rate is as high as that of the comparable method by Wang et al. [1]. In addition, applying an intelligent combination of these two techniques further increases the execution speed, 15–20% faster than that of one of the comparable methods [1]. The principle of putting a simple DPI ahead of a more complex DFI may be applied to other systems that aim to quickly and effectively detect single applications.

For future works, the semi-supervised learning method may be improved and refined. More sophisticated machine learning-based classification engines may be considered, these include Bayes engine and the SVM with either linear or non-linear approaches [19,20]. Furthermore, parameters used in the semi-supervised learning method may be fine-tuned for performance enhancement. The ultimate goal is to apply the BT traffic detection algorithm onto live-network flows.

References

1. C. Wang, T. Li and H. Chen, P2P traffic identification based on double layer characteristics, *International Conference on Information Technology and Computer Science, 2009. ITCS*, 2009, pp. 593–596.
2. R. S. Wong, T.-S. Moh and M. Moh, Poster paper: Efficient semi-supervised learning BitTorrent traffic detection—An extended summary, *Proceedings of the 13th International Conference on Distributed Computing and Networking (ICDCN), Lecture Notes in Computer Science (LNCS),* published by Springer, held in Hong Kong, January 2012.
3. R. S. Wong, T.-S. Moh and M. Moh, Efficient semi-supervised learning BitTorrent traffic detection with deep packet and deep flow inspections, *Proceedings of the 6th International Conference on Information Systems, Technology, and Management (ICISTM),* held in Grenoble, France, March 2012. *Information Systems, Technology and Management.* Series in *Communications in Computer Science.* Vol. 285. pp. 152–163. Published by Springer, Berlin..
4. L. Yang and M. Moh, Dual trust secure protocol for cluster-based wireless sensor networks, *Proceedings of the IEEE 45th Asilomar Conference on Signals, Systems and Computers,* held in Pacific Grove, CA, November 2011.
5. R. Alvarez-Horine and M. Moh, Experimental evaluation of linux TCP for adaptive video streaming over the cloud, *Proceedings of the 1st International Workshop on Management and Security Technologies for Cloud Computing,* co-located with IEEE Globecom, Anaheim, CA, December 2012.

6. BitTorrent and μTorrent software surpass 150 million user milestone. Bittorrent.com. 2012–01–09. http://www.bittorrent.com/intl/es/company/about/ces_2012_150m_users. Retrieved 14 July 2012.
7. A. Klemm, C. Lindemann, M. K. Vernon and O. P. Waldhorst, Characterizing the query behavior in peer-to-peer file sharing systems. *IMC'04: Proceedings of the 4th ACM SIGCOMM Conference on Internet Measurement*, ACM Press, Taormina, Italy, 2004, pp. 55–67.
8. Internet observatory brings real-time P2P traffic statistics, TorrentFreak, September 28, 2011. http://torrentfreak.com/internet-observatory-brings-real-time-p2p-traffic-statistics-110928/. Retrieved 14 July 2012.
9. B. Liu, Z. Li and Z. Li, Measurements of BitTorrent system based on Netfilter, *International Conference on Computational Intelligence and Security*, 2006, pp. 1470–1474.
10. T. Le and J. But, BitTorrent traffic classification, CAIA Technical Report 091022A, Hawthorn, Australia, 22 October 2009, http://caia.swin.edu.au/reports/091022A/CAIA-TR-091022A.pdf.
11. J. Erman, A. Mahanti, M. Arlitt, I. Cohen and C. Williamson, Offline/realtime traffic classification using semi-supervised learning, *IFIP Performance*, October 2007.
12. H. Chen, Z. Hu, Z. Ye and W. Liu, Research of P2P traffic identification based on neural network, *International Symposium on Computer Network and Multimedia Technology, (CNMT)*, Wuhan, China, 2009, pp. 1–4.
13. H. Chen, Z. Hu, Z. Ye and W. Liu, A new model for P2P traffic identification based on DPI and DFI, *International Conference on Information Engineering and Computer Science, ICIECS*, Wuhan, China, 2009, pp. 1–3.
14. F. Liu, Z. Li and J. Yu, Applications identification based on the statistics analysis of packet length, *International Symposium on Information Engineering and Electronic Commerce*, 2009, IEEC'09, pp. 160–163.
15. R. Keralapura, A. Nucci and C. Chuah, Self-learning peer-to-peer traffic classifier, *Proceedings of the 18th International Conference on Computer Communications and Networks (ICCCN)*, San Francisco, USA, 2009, pp. 1–83.
16. R. Constantinou and P. Mavrommatis, Identifying known and unknown peer-to-peer traffic, *Proceedings of the 5th IEEE International Symposium on Network Computing and Applications*, 24–26 July 2006, pp. 93–102.
17. R. Zhang, Y. Du and Y. Zhang, A BT traffic identification method based on peer-cache, *4th International Conference on Internet Computing for Science and Engineering (ICICSE)*, Harbin, China, 2009, pp. 320–323.
18. N. Basher, A. Mahanti, A. Mahanti, C. Williamson and M. Arlitt, A comparative analysis of web and peer-to-peer traffic, *Proceedings of the 17th International Conference on World Wide Web*, Beijing, China, 21–25 April 2008.
19. O. Chapelle, B. Scholkopf and A. Zien, eds., *Semi-Supervised Learning*. MIT Press, Cambridge, USA, 2006.
20. P.-N. Tan, M. Steinbach and V. Kumar, *Introduction to Data Mining*. Pearson, Addison-Wesley, USA, 2006.
21. WinPcap—The industry-standard windows packet capture library, http://www.winpcap.org/.

9

Developing a Content Distribution System over a Secure Peer-to-Peer Middleware

Ana Reyna, Maria-Victoria Belmonte and Manuel Díaz

CONTENTS

9.1 Introduction 212
9.2 Overview of the Content Distribution System 213
    9.2.1 Protocol Overview 214
    9.2.2 Content Distribution 214
        9.2.2.1 Task Assignment 214
        9.2.2.2 Checking Mechanism 215
    9.2.3 Incentive Mechanism 215
        9.2.3.1 Coalition Payment Division 215
        9.2.3.2 Responsiveness Bonus Computation 216
    9.2.4 Simulation Results 217
    9.2.5 Related Work 219
9.3 Overview of the Middleware 220
    9.3.1 SMEPP Abstract Model 221
        9.3.1.1 Peers and Groups 221
        9.3.1.2 Services and Contracts 222
        9.3.1.3 Service Discovery 222
        9.3.1.4 API 222
    9.3.2 Using SMEPP 222
        9.3.2.1 Managing Peers and Groups 224
        9.3.2.2 Managing Services 224
    9.3.3 Security in SMEPP 227
        9.3.3.1 Security Primitives 227
        9.3.3.2 Security Implementation: Software 227
        9.3.3.3 Security Implementation: Hardware 228
    9.3.4 Related Work 228
9.4 Design and Implementation 229
    9.4.1 Step 1: Peer and Service Code 229
    9.4.2 Step 2: Content Discovery Service 230
    9.4.3 Step 3: Responsiveness Bonus Management Service 231
    9.4.4 Step 4: All together: Implementation 233
        9.4.4.1 Peer Code 233
        9.4.4.2 Service Contracts and Code 234

9.5	Deployment and Validation	234
9.6	Discussion and Challenges	234
9.7	Conclusion	238
References		238

9.1 Introduction

Peer-to-peer (P2P) content distribution systems have had a lot of success as P2P paradigm increases download speed significantly. P2P paradigm abandons central servers to give way to a network model where all nodes play the role of server and client simultaneously. P2P protocols enable content distribution in a cost-effective way, as they do not require a centralised provider to handle all the demands, and therefore, the bandwidth is considerably increased because these protocols can use the bandwidth of all the clients to download instead of using only the bandwidth of the server. In addition, this alleviates the load of the server, which has a positive impact on systems performance. An excess of clients in a traditional client server paradigm meant problems of scalability; in the P2P paradigm, it means greater capacity. Indeed, many popular TV streaming systems are based on P2P technology precisely because of these benefits (such as PPLive [1] and TVAnts [2]).

However, the development of this kind of application must consider, in addition to the complexity in distributed software development, that the performance and availability of P2P content distribution systems rely on the voluntary participation of their users, which is highly variable and unpredictable. Hence, any proposed solution must consider user behaviour and provide custom mechanisms to make the system feasible. In this sense, empirical studies have shown that a large proportion of participants share little or no files. This phenomenon, known as 'free-riding', has been observed and measured in P2P systems [3,4] and is still an open issue in content distribution systems. In Ref. [5], we presented a new coalition formation scheme based on game theory concepts, called COINS (coalitions and incentives), which formally proves how coalitions improve P2P systems performance, encouraging participants to contribute resources, receiving in return a better quality of service. Empirical results obtained through simulations illustrated how our approach encourages collaborative behaviour, preventing the free-riding problem, and improves the overall performance of the system. This mechanism was a theoretical proposal, whose features were demonstrated only through simulations. In this chapter, we show the real implementation of this mechanism, offering a step-by-step guide to the development of content distribution systems, continuing on from the ideas presented in Ref. [6].

As said, P2P content distribution systems are complex systems that require a considerable amount of work in their development. Beyond proper distributed software difficulties, concerns about user behaviour must be considered. Our proposal requires a mechanism for look-up, a mechanism to store, update and query the reputation of clients, in addition to the mechanism to distribute the content on the network. This makes the implementation of our proposal an attractive and interesting exercise to delve into problems and possible solutions in implementing secure distributed systems using a P2P middleware. To facilitate the development of distributed systems, new tools and methodologies capable of abstracting all the underlying complexity should be used. A middleware can simplify and reduce the development time of the design, implementation and configuration of applications, thus allowing developers to focus on the requirements of their applications. In Ref. [7], we presented SMEPP (secure middleware for embedded peer-to-peer systems), a new middleware especially, but not exclusively, designed for embedded peer-to-peer (EP2P) systems. This middleware was designed to overcome the main problems of existing domain-specific middleware proposals. The middleware is secure, generic and highly customisable, allowing it to be adapted to different devices, from PDAs and new-generation mobile phones to embedded sensor actuator systems, and domains, from critical systems to consumer entertainment or communication. Choosing an embedded middleware has the additional advantage of allowing the application to run on mobile devices as well as on servers, which, given the popularity of mobile applications nowadays, is a significant advantage.

In this chapter, we take advantage of SMEPP middleware to implement our coalitions and incentives mechanism for distributed content distribution. On the one hand, we prove the suitability and advantages of using a P2P middleware, and on the other, we demonstrate that our mechanism can be developed as a real distributed application, showing step by step how to implement our proposal on the middleware. The chapter is structured as follows: First, the content distribution system proposed for the implementation is explained in Section 9.2. Next, the middleware chosen for the implementation is presented in Section 9.3, highlighting its features and introducing how to use it. Then, in Section 9.4, the design and implementation of the system is explained following a tutorial style. In Section 9.5, deployment constraints and validation tests are introduced. A short discussion on the development of a P2P application is presented in Section 9.6, and finally the conclusions drawn are presented in Section 9.7.

9.2 Overview of the Content Distribution System

In content distribution systems, we distinguish three different interacting subsystems: content discovery, incentive mechanism and content distribution. The content discovery mechanism is responsible for finding the content in a distributed network, whereas the incentive mechanism is intended to achieve fairness and avoid abuses, and the content distribution determines the way that the task of downloading a file is performed between various peers. The central idea of our proposal [5], called COINS, is sharing the task of downloading a file between a set of peers making up a coalition. In addition, it promotes cooperation and discourages free riders through an incentive mechanism, which is based on the game theory concept of *core* and its fairness is formally proved. The main contribution is in the incentive mechanism, where providers are rewarded for their participation in the coalition, entailing a better quality of service when they download. The coalition, from the point of view of the downloader, is beneficial because it reduces the download time if compared with downloading from a single peer, and also from the point of view of the uploader (provider) since its burden is alleviated because the total task is divided between the members of the coalition. In this way, each peer that participates in a coalition is lending 'bandwidth' to other coalition peers, in exchange for profit or *utility*. The reward that a provider obtains by performing a task inside a coalition is calculated using the game theory concept of *core*. The *core* ensures that each coalition participant receives a fair utility in return for the bandwidth that it supplies. In our model, these utilities are used to compute the *responsiveness bonus* (Rb), which represents the overall contribution of the peer to the system. Therefore, this value will determine the quality of service of each peer. The higher the Rb, the better the quality of service; this is the key to encouragement.

To better understand the content distribution system proposal, three sections will be presented. First, in Section 9.2.1, the protocol overview explains the interactions between the different peers in the system. Next, two key points will be explained, the content distribution and the incentive mechanism, in Sections 9.2.2 and 9.2.3, respectively. Finally, a summary of the simulation results is presented in Section 9.2.4, and some related work is presented in Section 9.2.5.

9.2.1 Protocol Overview

Each peer can simultaneously play three different roles: downloader, participant or manager (the manager is the peer responsible for finding the participants, forming the coalition and checking its performance). In short, the download process starts when a peer decides to download a file. To download this file, the downloader has to find file providers in the network (content discovery). Once the providers are found, a coalition manager is elected. The manager's selection does not imply centralisation, because any potential participant can become the manager with equal probability. Next, the manager sends offers to the potential candidates (the rest of the providers). Each provider calculates the bandwidth it intends to offer, according to its load, and sends the answer to the manager. Once the manager has received all these answers (or a timeout is reached), it has to divide the task between the potential coalition members (those participants who answered). This process, called *Task Assignment*, establishes the coalition itself, and after this, the download itself starts. When a participant receives the task, it starts sending the corresponding data. During the download, the downloader periodically sends acknowledgement information to the manager, who runs a *checking mechanism* to guarantee the quality of service in the coalition, adapting to network traffic and helping to avoid any attacks from malicious peers (such as free-riding). After these checks, the Rb of all the members of the coalition is updated using the utilities obtained after the *coalition payment division* (incentive mechanism).

9.2.2 Content Distribution

The content distribution has to define the way the coalition distributes the content; in our proposal, this has two key points: the task assignment, which determines the task or the part of the file that each participant has to send to the downloader, and the checking mechanism, which guarantees that the coalition is working according to the performed assignment.

9.2.2.1 Task Assignment

Given a collection of providers, the task assignment has to determine the task that each provider will be responsible for; this is the input bandwidth that each participant will provide to the coalition. If there are few participants (under a threshold), no selection has to be made; otherwise, only some providers will be chosen for the coalition.

To do this, and to determine the input bandwidth of a participant, the *progressive filling algorithm* is used. This algorithm provides the *max–min fairness* [8]. A bandwidth allocation is max–min fair if and only if an increase of the input bandwidth of a peer x within its domain of feasible allocation is at the cost of decreasing some other input bandwidth of a peer y. Hence, it gives the peer with the smallest bidding value the largest feasible bandwidth.

9.2.2.2 Checking Mechanism

The checking mechanism makes the system less vulnerable to peer failures, churns and network congestion problems, while it ensures the quality of service of the coalition. The mechanism works as follows: During the download of a file, the downloader sends acknowledgement information to the manager with a predefined frequency. The manager calculates the difference between the bytes sent and the ones that should have been sent (according to the task assigned to each participant). If this difference exceeds a predefined threshold, the coalition is reconfigured to provide better quality of service. Moreover, the manager also checks that the downloader Rb is high enough to keep downloading. The central idea is that if the coalition is not working as expected or the downloader is abusing the system, it is cancelled.

Since the updates of Rb values are calculated by the manager and are based on the acknowledgement sent by the downloader, the downloader could avoid the penalty if it sends a fake acknowledgement. But the checking mechanism performed by the manager will stop the coalition if the acknowledgement is too small, and therefore the downloader will not be penalised, but neither will they receive the file.

9.2.3 Incentive Mechanism

The incentive mechanism is responsible for rewarding participants and discouraging selfish behaviours. This mechanism is based on game theory concepts and relies on two formulas: the coalition payment division, which determines the reward obtained as a function of the task performed, and the responsiveness bonus computation, which determines the responsiveness value of a peer, Rb, that represents its overall contribution to the system (can be seen as a reputation value).

9.2.3.1 Coalition Payment Division

The hallmark of our mechanism is that the coalition payment division ensures fairness, thanks to the game theory concept of *core*. This means that peers will not be negatively affected if they have a lower capacity. The details of this are explained in the following paragraph.

Let us call coalitional value $V(S)$, to the total utility or profit of a coalition S. For each peer in the coalition, $P_i \in S$, we must distribute $V(S)$ between the peers, and assign an amount or utility (x_i) to each peer $P_i \in S$. The problem is to distribute $V(S)$ in a stable and fair way so that the coalition peers have no reason to abandon it.

First, we must calculate $V(S)$. The profit obtained by S is calculated as the difference between the time required for the download with just one uploading participant (only P_0, the manager) minus the time it takes with the coalition S (all the participants, including the manager). Then, the coalitional value is given by the following equation:

$$V(S) = t_0 \frac{\sum_1^n b_i^{in}}{\sum_0^n b_i^{in}} \quad \text{where } t_0 = \frac{\text{File size}}{b_0^{in}} \quad (9.1)$$

where t_0 is the time that it would take P_0 to upload the whole file (P_0 being the only uploader or provider), b_0^{in} the upload bandwidth of P_0 and b_i^{in} the upload bandwidth of the remaining participants of S.

Second, we use the *core* to distribute V(S) between the coalition members. A utility distribution belongs to the *core* if there is no other coalition that can improve on the utilities of all of its members. The stable utility division (x_i) to every peer $Pi \in S$ is given, then, by the following equation (details in Ref. [5]), where b_0^{out} is the download bandwidth of P_0.

$$x_i = \begin{cases} t_0 \dfrac{\left(\sum_1^n b_i^{in}\right)^2}{\left(\sum_0^n b_i^{in}\right)^2} & \text{if } i = 0 \\ t_0 \dfrac{b_0^{out} b_i^{in}}{\left(\sum_0^n b_i^{in}\right)^2} & \text{if } i \neq 0 \end{cases} \qquad (9.2)$$

9.2.3.2 Responsiveness Bonus Computation

As has been said, peers with higher utility will get a better quality of service. In our approach, the utility accumulated by each peer (Rb_i) is proportional to the resources it supplies, and it is calculated as a heuristic function of x_i. The value of Rb_i will be reduced when P_i acts as a downloading peer, and incremented when it is a provider or uploading peer. The heuristic uses the x_i values obtained by P_i by means of U_{pi} (upload points) and D_{pi} (download points). U_{pi} and D_{pi} accumulate the utility obtained by each coalition formation process in which P_i participates.

Let us call F_{si} the number of files shared (the total size in bytes) by a peer P_i. The Rb_i value of the peer is calculated using the following equation:

$$Rb_i = \begin{cases} 1 & \text{if } (U_{pi} - D_{pi}) \geq 0 \\ 0 & \text{if } (U_{pi} - D_{pi}) < 0 \wedge U_{pi} = 0 \wedge F_{si} = 0 \\ 1 & \text{if } (U_{pi} - D_{pi}) < 0 \wedge U_{pi} = 0 \wedge F_{si} > 0 \\ \dfrac{U_{pi} \cdot \gamma}{D_{pi}} & \text{if } (U_{pi} - D_{pi}) < 0 \wedge U_{pi} > 0 \end{cases} \qquad (9.3)$$

The Rb_i values are between zero and one. The interpretation of this formula is that if the peer uploads more than it downloads, it gets the maximum value; this is also true when it is not uploading but sharing. If it is neither uploading nor sharing, its Rb_i is set to zero. In any other case, it is calculated as the ratio of the upload point to the download point (the γ parameter allows us to regulate the relation to increase/decrease the penalty/reward).

Therefore, we use this value to decrease the download bandwidth $Rb_i \cdot b_b^{out}$ (using it as a multiplier of the download bandwidth of the peer P_b, when it wants to download a file). Initially, the Rb_i of the peers is 1; a higher responsiveness bonus (Rb_i closer to 1) will mean that P_i will be able to use most of its bandwidth capacity. Otherwise, an Rb_i closer to 0 will reduce its bandwidth capacity (in fact, it could even avoid creating the coalition for the download when it is 0). Thus, our incentive mechanism penalises the selfish behaviour of the peers, and provides incentives for collaborative behaviour.

9.2.4 Simulation Results

In Ref. [5], we presented full simulation results. These experiments confirmed the benefits of using our mechanism. On the one hand, download times are improved, and on the other, free riders are stopped; this leads to an improvement of the system's effectiveness.

Our own simulator was used to run the experiments. It was configured to simulate a P2P network of 1000 peers during 2000 units of simulated time (steps). All peers had the same bandwidth capabilities. The collection of files shared in the network was defined with different sizes (from 10,000 KB to 90,000 KB), and a random number of copies (between 5 and 500) of these were delivered through the network at the start of the simulation. Each peer had a random number of initially stored files, and the objective of the simulation was that each peer downloaded the files that were not initially stored. Our simulations considered three types of users (or behaviours): free riders (FR), collaborative (C) and adaptive (A). The first do not share at all, the second share as much as possible, and the last only share if they want to download. Depending on the behaviour of each peer (that is randomly assigned in each simulation), the downloads will be addressed in different ways.

To analyse the impact of the different behaviours on the system, the experiments were run with two different populations. The first one was run without adaptive users: 50% FR, 50% C and 0% A, called Population 1. The second was run with adaptive users: 40% FR, 30% C and 30% A, called Population 2. In addition, to analyse the impact of the use of coalitions, simulations were run with and without incentive policies: no coalitions (NC), where no incentive mechanism was considered; and coalitions (C), which implemented our proposal. Also, an Emule–like credit system was simulated. After repeating the simulation experiments 100 times, we took the average to give the results. Two main metrics were considered: downloaded bytes and average download time.

In Figure 9.1, the total bytes downloaded for the different scenarios are shown. At first glance, COINS is downloading less in both scenarios when compared with NC. In both populations, indeed, the total amount of bytes downloaded was reduced, by 50% in Population 1, and by 39% in Population 2, with respect to NC. With respect to Emule, COINS performs better in Population 1 but worse in Population 2. But it is necessary to undertake a deeper analysis and see the bytes downloaded for each kind of user behaviour considered in simulations, since COINS is designed to stop free riders; hence, it makes sense that these kinds of users are responsible for the greater part of this reduction. In Figure 9.2, Population 1 and Population 2 downloaded bytes are shown for each behaviour. In Population 1, when coalitions were used (COINS), the total amount of bytes downloaded was reduced to 50% with respect to NC, but the figure demonstrates that 84% of

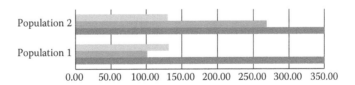

	Population 1	Population 2
■ Emule	131.63	130.09
■ COINS	101.35	268.72
■ NC	385.85	571.95

FIGURE 9.1
Total downloaded bytes (Mb).

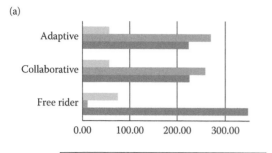

	Free rider	Collaborative	Adaptive
■ Emule	74.36	55.74	55.84
■ COINS	10.58	258.14	269.67
■ NC	346.65	225.30	223.85

	Free rider	Collaborative
■ Emule	65.41	66.22
■ COINS	4.05	97.29
■ NC	207.36	178.49

FIGURE 9.2
Downloaded bytes (Mb) in Population 1 (a) and Population 2 (b).

this reduction was due to the free riders' detection. This showed how the algorithm prevents free riders from abusing the system, thereby avoiding the overhead of the system resources. In Population 2, when adaptive users were introduced, the benefit of using coalitions was higher (than in Population 1). The total amount of bytes was reduced by 39% with respect to NC, where 83% was due to the free riders' detection. Similar results justify the reduction in total downloaded bytes if compared with Emule. In addition, comparing adaptive users in coalitions in both populations, the total amount of downloaded bytes were increased by 20% using Population 2, proving that adaptive users benefit the system. Note that in Population 2 there were fewer free riders and collaborative users; therefore, there were less shared files in the network. This justifies the smaller amount of total bytes downloaded with respect to Population 1.

In addition to the analysis of the downloaded bytes, the average download time offered even better results. In Figure 9.3, the average download time in the different scenarios is shown for both populations. Experiments showed that by using coalitions the average download time was smaller. As expected, the benefit of using coalitions is increased as the file size grows. When adaptive users were introduced, the download times were

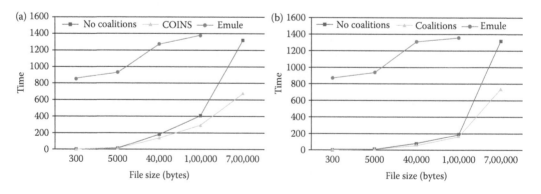

FIGURE 9.3
Average download time (time vs. bytes) for Population 1 (a) and Population 2 (b).

improved in Population 2 when compared with Population 1. These results demonstrated the effectiveness of our incentive mechanism.

9.2.5 Related Work

The incentive mechanisms in P2P networks for content distribution [9] have been classified into different categories: monetary payments, reputation-based or reciprocity-based mechanisms. Our approach belongs to reciprocity-based mechanisms: peers that contribute more get a better quality of service. The other publications also included in this category are Refs. [9–14]. In this, we can distinguish reciprocity-based on mutual-reciprocity (symmetric) or indirect-reciprocity (asymmetric).

In mutual reciprocity, BitTorrent [11] was the first widely used P2P file download protocol to incorporate a fairness enforcement. Its 'tit-for-tat' policy of data sharing works correctly when the peers have a reciprocal interest in a particular file. However, BitTorrent does not preserve information on peer contributions between different download sessions. And it limits the peer download capacity to its upload capacity, and thus the achievable download performance is reduced. In our approach, the download capacity is not reduced to its upload capacity, and in addition, by means of Rb, the information about peer contribution is preserved between different download sessions. Something similar happens in Emule [10] because the mutual reciprocity-based approaches do not fit with the asymmetric nature of a collaborative relationship, since the decision of the peer to upload to another peer is based on the direct exchange of data/services between them. Unlike these approaches, COINS encourages cooperative behaviour by forming coalitions of peers that help each other to download files, and therefore any peer with idle bandwidth can participate in a coalition even if the peer is not interested in the same content.

The indirect reciprocity-based approaches, such as Refs. [12–14] or our approach, consider peers' overall contribution to the network and therefore encourage cooperation. 2Fast [12] is based on creating groups of peers that collaborate in downloading a file. From this point of view, the idea is similar to our proposal; however, the system does not enforce fairness in the collector and helper peers. Neither is it specified how the helper may reclaim its contributed bandwidth in the future. The only incentives mentioned are promises for future collaborations. In addition, they study the system's performance based on the download speed but do not provide results related to free-riding, although they affirm that this phenomenon is prevented.

In Ref. [13], the authors propose a new scheme for service differentiation in decentralised P2P systems. In this, as in COINS, the contribution is based on the number of peer uploads and downloads. However, in our approach, the contribution is computed using the core, which guarantees that the peers in the coalitions receive fair utility in return for the bandwidth that they supply. Furthermore, our results have proved that having coalitions of peers improves the download time.

In Ref. [14], a distributed framework is proposed in which each peer monitors its neighbours, and the free riders are located and dealt with appropriately. To do so, each peer records the number of messages coming from and going towards its neighbours. This framework may overload the run of the system and the free riders can plan different attacks to bypass this policy, such as continuously changing its neighbours and so on. However, and unlike this work, stopping free riders is not the main goal of our approach, but rather it is to increase the effectiveness of the download mechanism while encouraging cooperative behaviour.

9.3 Overview of the Middleware

As presented, we propose the use of SMEPP to develop our content distribution system. This middleware is the main result of the 4-year European project with its same name, coordinated by the University of Málaga [7]. The middleware is service oriented and its architecture is component oriented. Its application programming interface (API) provides an abstraction that significantly leverages the programming burden, as we will demonstrate in the next section. As its name suggests, security is one of the main concerns of the middleware, which has been addressed since its conception, leading to a built-in security middleware.

As mentioned, distributed applications have an inherent complexity due to the problems related to latency, reliability, partitioning, ordering, security and so on. SMEPP aims to alleviate this complexity by offering a secure, generic and highly customisable middleware that allows its adaptation to different devices (PC, laptops, PDAs and motes: Iris, micaZ), operating systems (Windows NT, Windows Vista, Linux, Windows Mobile 6, TinyOS and Android), platforms (JavaSE, JavaCDC, .NET Framework), networking interfaces (802.11, Bluetooth, infrared, cellular, Ethernet, Zigbee, etc.), networking protocols and underlying hardware platforms and domains (from critical systems to consumer entertainment or communication). More detailed information relating to SMEPP implementation can be found in Ref. [15]. SMEPP's main features are as follows:

Heterogeneity: SMEPP solves the heterogeneity of devices, offering three different and interoperable suites, for the three different levels of devices (motes, PDAs, laptops); it also solves the heterogeneity in software, as the model is language and platform independent, SMEPP has been implemented in different languages offering the same API. In addition, the heterogeneity in connectivity is also tackled because peers can be connected through different network interfaces or protocols and SMEPP peers are capable of establishing communications between the different networks.

Configurability: SMEPP is configurable; its component-based architecture enables the combination of different component implementations to fit the middleware implementation to the application or device needs. In addition, SMEPP offers a configuration tool that automates the process of combining the components. These components can be implemented in different ways; this also makes the middleware extensible.

Scalability: Scalability is addressed in different parts of the middleware architecture. For example, SMEPP is based on the notion of group. The creation of groups can help reduce the network traffic and the division of the system into smaller logical groups, which helps to manage the complexity. In addition, the events mechanism provided by SMEPP helps to obtain a loosely coupled system where producers and consumers of events do not require a direct connection. SMEPP adapts to different scenarios by selecting different component implementations.

Efficiency: Related to efficiency, the energy efficiency component of the *nesC* SMEPP implementation saves sensors' energy by keeping the radio off whenever possible, for example, when the sensors do not expect to receive or send data. This component manages the radio by means of user duty cycles, which are implicitly determined by the subscribe messages (these messages contain the subscribe rate and an expiration time), and a management duty cycle, which enables the sensors in the group to exchange control messages (in particular, join and subscribe messages). Each of these duty cycles defines periodic intervals when the radio should be turned on by all the sensors.

As stated, the central idea of this chapter is to use this middleware to implement the presented content distribution system. To use it, first, we need to briefly introduce the SMEPP model, which defines the entities involved and how they relate in P2P environments. It is also important to introduce the common API provided to access the middleware functionality and to explain how the service discovery works in SMEPP, since it will be a key feature used in the implementation; all of this is presented in Section 9.3.1. Then, in Section 9.3.2, we go to a practical exercise where we explain how to use the main primitives of the API and, in addition, describe how to develop SMEPP applications. In Section 9.3.3, the security concerns covered by SMEPP are summarised. We conclude with a short related approach in Section 9.3.4.

9.3.1 SMEPP Abstract Model

As said, the SMEPP model is based on a service-oriented architecture, where services enclose the application functionality. Briefly, SMEPP considers peers, services and sessions as entities. Peers run on nodes; each peer is a process and each node is a device. Services are published (and consumed) within a group (of peers). The interaction is carried out through service invocations, messages or events. Thereafter, this section explores the basic concepts that we will use in our implementation: peers, groups and services. More detailed information can be found in Ref. [16].

9.3.1.1 Peers and Groups

An SMEPP peer is a process running on a connected device, which is using SMEPP middleware, thus invoking the API. A process becomes an SMEPP peer once it has invoked the *newPeer* primitive, and therefore it has received its identifier. SMEPP defines an abstract service model to establish how to discover, publish and interact with (available) services and how peers can communicate with each other (Figure 9.4).

To deal with security, services are offered inside groups. Organising peers into groups allows us to establish the joining mechanism, which validates user credentials, so that we can guarantee a determined level of security in each group and therefore in services.

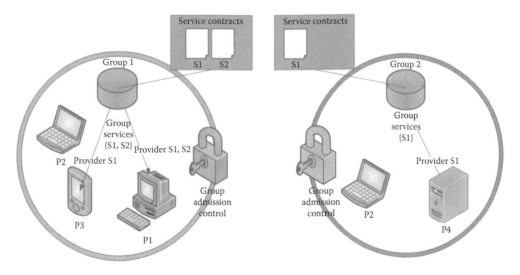

FIGURE 9.4
SMEPP basic concepts.

9.3.1.2 Services and Contracts

Every SMEPP service needs an implementation together with a contract; this is an XML file that contains its description, and it includes the interface (signature) and the semantic description (behaviour), and can optionally include extra-functional properties. This description enables the service invocation and also its orchestration or composition.

The contract provides descriptive information of the service, while the implementation is the executable service (e.g., a Java service) exposed to the middleware through grounding. A service contract describes 'what the service does' (viz., the service signature), 'how it does it' (viz., the service behaviour), and it may include other extra-functional service properties (e.g., QoS). These contracts play an important role in service discovery, as we will show in the following section.

9.3.1.3 Service Discovery

Although service discovery is not essential to understand SMEPP, we include this section because the way it works is key in the proposed implementation of the content distribution system, as will be shown in Section 9.4.

SMEPP uses an overlay network that represents a virtual network of nodes and logical links that is built on top of the existing network with the purpose of implementing an efficient service and group description sharing that is not available in the physical network. The service discovery protocol is built on top of this structured overlay network. The overlay network is based on DHT and implements the Chord protocol [17]. In short, Chord defines a virtual ring and a key space, so that each peer that joins the ring becomes responsible for a specific range of the key space. When a service is published, SMEPP generates a key from the contract (by the hash of its contract file); this key determines which peer in the group is responsible for this service, for storing its contract, since the service will always be offered by the publisher peer. As the overlay network determines the range of keys a peer is responsible for, each key determines where the contract is stored, and since every peer can calculate the key associated with a given service, enables a fast search mechanism.

9.3.1.4 API

The SMEPP model offers a definition of API for developers to interact with the middleware, programming either services or peers. SMEPP API provides a suitable level of abstraction that hides the details concerning all the supporting implementation.

The SMEPP model is language and platform independent and allows for efficient implementations in terms of memory and energy consumption. In addition, support tools for simulation and verification have been deployed, which are useful in the design process of EP2P applications based on the SMEPP abstract model. Table 9.1 summarises SMEPP API primitives grouped by functionality.

9.3.2 Using SMEPP

The key to using SMEPP lies in its API. Currently, SMEPP provides a Java API (for programming code to be run on laptops and on limited-capacity devices such as smart phones) and a *nesC* API (for programming sensor networks). The *nesC* API does not implement service management primitives; thus, sensors do not use services, but just events.

TABLE 9.1

SMEPP API

Peer Management Primitives	
newPeer	Creates a new SMEPP peer and assigns a *peerId* identifier to the peer
getPeerId	Returns the *peerId* of the caller peer (or the *peerId* of the peer providing the caller service)
Group Management Primitives	
createGroup	Creates a new SMEPP group and assigns a *groupId* identifier to the group
joinGroup	Peer enters a group
leaveGroup	Peer exits a group
getGroups	Returns the list of the existing groups
getGroupDescription	Returns the description of a group
getPeers	Returns the list of the peer members of a group
getIncludingGroups	Returns the list of the groups where a peer belongs to
getPublishingGroup	Returns the group in which a service is published
Service Management Primitives	
publish	Publishes a service and assigns a *peerServiceId* identifier to the service
unpublish	Unpublishes a previously published service
getServices	Returns the list of the services matching some requirements
getServiceContract	Returns the contract of a service
startSession	Opens a new session with a (session-full) service
Message Handling Primitives	
invoke	Invokes an operation of a running service
receiveMessage	Receives an invocation message and starts the execution of the invoked operation
reply	Sends an output response message to the invoker of a request-response operation
receiveResponse	Retrieves the result of a request-response operation
Event Handling Primitives	
event	Raises an event
receiveEvent	Receives an event
subscribe	Subscribes to an event
unsubscribe	Cancels a previous event subscription

The *nesC* API is provided by a specific version of the SMEPP middleware for sensor networks, which supports sensor requirements such as small memory, small computing power and battery power.

An SMEPP program specifies either a peer code or a service code. Converse to services, peers can create, search, join or leave groups and (un)publish services; however, peers cannot wait for (nor reply to) incoming requests. Consequently, the SMEPP primitives divide into three groups: primitives to be used only by peers; primitives to be used only by services; and primitives to be used by both. For example, the group management primitives (*getPeers* excluded) and the service management primitives *publish* and *unpublish* can be used only by peers. The message handling primitives (*receiveMessage* and *reply*) are instead available to services only, while the *newPeer* primitive can be called only by applications to become peers.

9.3.2.1 Managing Peers and Groups

First, SMEPP peers are created through the invocation of the *newPeer* primitive.

```
Credentials myCredentials = new Credentials("");
PeerManager peer = PeerManager.newPeer(myCredentials, configFile);
```

SMEPP API offers two basic classes to interact with the SMEPP middleware, namely, *PeerManager* to program SMEPP peers, and *ServiceManager* to program SMEPP services. The result of a *newPeer* invocation is indeed a *PeerManager* object, which provides access to the primitives reserved to peers. The *newPeer* primitive requires two input parameters, namely, credentials, to authenticate the peer and a configuration file, including some information of the peer. For the sake of simplicity, the peer uses empty credentials.

Typically, SMEPP peers will join or create groups to collaborate with other peers and/or to provide services. The peer searches for groups that match a specified group description by invoking the *getGroups* primitive that returns a (possibly empty) array of *groupId* group identifiers.

```
GroupDescription groupDescr1 = new GroupDescription("group1",
                              new SecurityInformation(1),"Gr1");
groupId[] groupIds = peer.getGroups(groupDescr1);
```

The optional *GroupDescription* parameter can be used as a filter to return only the groups that match some requirements. The sample peer is specifically searching for groups with the name 'group1', security information of level 1 and with the description 'Group one'.

The *GroupDescription* type can also be used to create new groups by invoking the *createGroup* primitive.

```
gid = peer.createGroup(groupDescr1);
```

The *createGroup* primitive creates a new SMEPP group featuring the *groupDescr1* group description and assigns to it a *groupId* group identifier. The group description specifies the security level of the group, thus restricting the access to the group and its visibility.

Alternatively, peers can join existing groups. First, the peer can look for groups matching the specified description using the *getGroups* primitive; then the peer joins any of the groups returned by using the *joinGroup* primitive.

```
peer.joinGroup(gid);
```

9.3.2.2 Managing Services

Publish: To publish an SMEPP service, the peer must basically provide its contract and its grounding. The contract provides descriptive information on the service, while the grounding provides access to the executable service (e.g., Java class). The peer then invokes the publish primitive to publish the service inside a group.

```
PeerServiceId psid = peer.publish(gid,
    ContractLoader.loadFromFile("myServiceContract.xml"),
    new SMEPPServiceGrounding(MyService.class),null,null);
```

Contracts: An SMEPP service contract is expressed using XML and its structure is validated by an XML schema file. Briefly, a contract contains the name and the category of the service, signature (types, operations, events) and service type (state-less, session-less or session-full). The publish primitive requires a *ContractType* object, which can be simply obtained by invoking the methods exposed by the *ContracLoader* class.

```
ContractLoader.loadFromFile("myServiceContract.xml");
```

Implementation: The publish primitive also requires the service grounding; this is the handle to access the executable Java service that contains the implementation of the service. The implementation of a simple service that listens to a message *myMessage* that has a string as input should look like the following:

```
public class MyService implements Runnable {
    private final ServiceManager svc;
    public MyService(ServiceManager svc,
                    Serializable constrValue) {
        this.svc = svc;
    }
    @Override
    public void run() {
        try {
            System.out.println("Service started!");
            while (true) {
                ReceivedMessage message = svc.receiveMessage
                  ("myOperation", new Class[]{String.class});
                System.out.print("invocation received: ");
                Serializable[] data = (Serializable[]) message.data;
                System.out.println("received message = "+ data[0]);
            }
        }
        catch (InterruptedException ex) {
            //manages the catched exception
        }
    }
}
```

The service constructor inputs an instance of the *ServiceManager* class, which provides access to the functionalities reserved for SMEPP services. When the service starts, it waits for a *myOperation* invocation message, which has a string as argument or input. Note that the *receiveMessage* primitive waits until an invocation message (including the requested inputs) for the *myOperation* operation is received, and it returns a *ReceivedMessage* object, which contains the identifier of the invoker (viz., a peer or [a session of] a service) and the data inputs of the operation, if any. The *receiveMessage* primitive may also input a timeout to wait for the operation invocation message until either a message is actually received or the timeout expires.

Unpublish: The provider peer keeps its services until they are unpublished using the *unpublish* primitive.

```
peer.unpublish(psid);
```

Service discovery: SMEPP services can be state-less, session-less and session-full. In the following, we discuss how to invoke *state-less* services; additional information for the other service types can be found in Ref. [15].

We can suppose a peer program that creates a new SMEPP peer, searches for groups that match a specified group description and joins one of them. Then, it invokes the *getServices* primitive to search for services that match some given requirements, specified by means of a *QueryType* object.

```
PeerServiceId[] psids = peer.getServices(null,null,template,null,1);
```

The first two input parameters of the *getServices* primitive are a group identifier *groupId* and a peer identifier *peerId*. *groupId* and *peerId* restrict the service discovery to the group *groupId* and to the peer *peerId*, respectively.

The third parameter of the *getServices* primitive is a *QueryType* object that includes a contract template—that is, a partially specified SMEPP contract that describes the functionalities that the desired services have to provide—and provides methods to customise the contract matching. If no *QueryType* object is specified, *getServices* searches for any services. A *QueryType* object can be obtained by invoking the *loadFromFile* method exposed by the *QueryLoader* class.

```
QueryLoader.loadFromFile("myServiceTemplate.xml");
```

The *getServices* primitive returns the *PeerServiceId* identifiers of the matched services. If *getServices* succeeds, the peer invokes the desired operation of the retrieved *psid* service (e.g., *myService*).

```
Short result = peer.invoke(psid,
                    "myOperation",
                    new Serializable[]{"a parameter"});
```

Invoking services: The invoke primitive takes as input the *psid* identifier of the service to be invoked, as well as the name and the list of the parameters of the specific *psid* operation to be invoked.

Since *myOperation* is a one-way operation, it invokes primitive returns when the invoked *psid* service does a corresponding *receiveMessage* of the *myOperation* operation. The invoke primitive may also input a timeout to wait for a corresponding *receiveMessage* until either *psid* actually calls the *receiveMessage* of the *myOperation* operation or the timeout expires. Operations can also be *request-response*; in this kind of operation, the sender waits for an answer from the receiver (*receiveResponse*), and the receiver answers the invocation using the *reply* primitive.

9.3.3 Security in SMEPP

As stated, security is a key concern in SMEPP; in this section, we highlight the main achievements in this area. Traditional security infrastructures cannot be easily adapted to EP2P systems since most of them are based on trustable servers that provide authentication and authorization. The goal of SMEPP in this area is to develop security services for the middleware, taking into account the constrained characteristics of these systems. SMEPP offers a security infrastructure for EP2P systems, which is integrated inside the middleware. In addition, secure routing and cryptographic protocols and primitives are provided inside SMEPP.

9.3.3.1 Security Primitives

SMEPP defines a set of security primitives that consider specific attacks for EP2P systems and devices (attacks on data aggregation, impersonation, node compromise, etc.). These primitives include:

- New non-centralised authentication and authorisation mechanisms based on groups.
- A new ID-based signature scheme. ID-based signature schemes are especially useful for the resource-constraint devices to provide authentication and integrity protection, since they are unable to support the deployment of PKI.
- Network audit and reporting mechanisms.

9.3.3.2 Security Implementation: Software

Security primitives and services have been integrated in the middleware. One of the most important innovations of SMEPP is the STAR secure routing protocol, which provides routing capabilities in mobile environments permitting the adaptation of the security level (depending on the application and/or the capabilities of the participating devices). All the primitives have been designed to minimise energy consumption (an in-depth analysis for each of the primitives is also available in the project website). The list of supported security primitives includes the following:

- Hashing (SHA-1)
- Message authentication code (HMAC using SHA-1)
- Symmetric key encryption and decryption (AES-128, CBC mode)
- Asymmetric key encryption and decryption (hybrid mode using ECC and AES-128 CBC mode)
- Digital signature (Schnorr's scheme)

- Lightweight certificate
- Random number generator

9.3.3.3 Security Implementation: Hardware

Some primitives have been implemented in hardware to minimise the overhead of these primitives in devices with constrained resources. The most relevant hardware developments are

- The ORIGA chip: a low-cost authentication device based on elliptic curve cryptography (Figure 9.5).
- An RFID tag supporting asymmetric cryptography: this RFID prototype is the first worldwide RFID tag supporting asymmetric cryptography.

9.3.4 Related Work

When choosing a middleware to develop a distributed P2P application, you can opt for a general-purpose middleware such as JXTA platform [18] or Microsoft's Peer-to-Peer networking [19]. These platforms are not adapted to mobile devices, which nowadays is an important disadvantage given the relevance that mobile computing is taking. For instance, in Ref. [20], the authors propose the implementation of a content distribution system over JXTA. In recent years, some work has also been done in the field of middlewares for the development of applications in embedded peer-to-peer systems. To the best of our knowledge, none of this work covers all the areas covered by SMEPP, such as security, peer-to-peer communication, services and so on.

CARISMA [21] is a mobile peer-to-peer middleware that exploits the principle of reflection to enhance the construction of context-aware adaptive mobile applications. CARISMA models mobile distributed systems as an economy, where applications compete to have a common service delivered according to their preferred quality-of-service level; in this economy, the middleware plays the role of an auctioneer, collecting bids from applications and selecting the policy that maximises social welfare. Although the economic model for QoS proposed in CARISMA is quite promising, we miss the security issues. SMEPP could integrate this economic model to similarly maintain QoS in the network.

Kavadias et al. [22] present a P2P technology middleware architecture enabling user-centric services deployment on low-cost embedded networked devices. Intelligence in home environment management, composed by the use of wireless communication terminals, requires the presence of mechanisms for services authentication, billing, discovery

FIGURE 9.5
ORIGA chip.

and execution. The authors propose a middleware architecture, seamlessly integrating communication interfaces and control functions in the form of user applications.

Hwang and Aravamudham [23] describe the scalable inter-grid network adaptation layers (signal) middleware architecture that integrates mobile devices with existing grid platforms to conduct P2P operations through proxy-based systems. The approach is envisioned to enhance the computational capabilities of the mobile devices by making the grid services accessible for these devices. Signal addresses issues related to device heterogeneity, low bandwidth, high latency connectivity, extended periods of disconnection and software interoperability. The Signal middleware proposes an interface with open grid service architecture (OGSA) and it employs web services and Globus technologies.

Proem [24] is a mobile P2P middleware that aims to solve the problems of application development for mobile ad hoc networks. The main issues to be tackled in mobile ad hoc environments are the tolerance of the network to peer failures, the changing topology of the network due to the mobility of peers, peer re/connection and disconnection and the decentralisation of the network.

9.4 Design and Implementation

SMEPP is based on peers, groups and services. Hence, the first step in the implementation of our content distribution system is to identify the functionality that will be performed by peers and the functionality of the services, in a way that they successfully cover the functionality required by the proposed system. COINS provides a mechanism to form coalitions and to reward, but it can be implemented in different ways, according to the different P2P architectures (centralised, pure or hybrid), since the *Rb* value can be stored in a central trusted server, calculated using gossip-style algorithms [25] or stored using a hybrid P2P network, for instance. The same goes for content discovery. Hence, the second and third steps will be to explain the implementation of these two subsystems, respectively, content discovery and incentive management, where we opt for a fully distributed or pure P2P approach. Finally, in the fourth and last step, we put it all together to have an overview of complete implementation.

9.4.1 Step 1: Peer and Service Code

In our approach, each peer works simultaneously as a participant, manager or downloader, and therefore this functionality must be included in the peer code. Furthermore, in SMEPP, services are offered inside groups, and therefore before publishing services (that still are not defined), peers will have to join a group. In the design stage, services have to be modelled, and therefore the functionality that will be provided by services has to be separated from the functionality of the peer. A practical exercise is to write in a table the functionality of the application and organise it (Table 9.2).

As shown in Table 9.2, different behaviours will not need different peer codes. We will use a single peer code that will receive as input the SMEPP's peer configuration file1[*]. The peer will have to be able to raise the *new download event*; this is to start a download. This can be easily offered through a simple GUI. Every peer will have to offer the same services:

---

[*] As explained in Section 9.3.

TABLE 9.2

Behaviours and Functionality

	Participant	Manager	Downloader
Peer code	New peer	New peer	New peer
	Join group	Join group	Join group
	Publish services	Publish services	Publish services
			New download event
Service code	Wait coalition offer	Wait management offer	Find file
	Evaluate offer	Evaluate offer	Select manager
	Upload the file part	Find participants	Download the file
		Form a coalition	
		Upload the file part	

at least we need one to download and one to upload (together implement the coalition protocol) and one more to implement the incentive mechanism.

To decide the functionality of the service or services, we need to understand the protocol, that is, the messages that peers send and receive to perform a download. The following diagram shows a simplified version of the messages sent between the different peers in a download. These steps were introduced in Section 9.2.1, and are summarised and simplified in Figure 9.6.

Basically, a service is a thread that waits for the different messages, or operation invocations, runs the corresponding code and probably answers the message or operation. As seen in Section 9.3.2.2, operations can be one-way or receive-response. It is very useful to determine the message that each service will be responsible for, and if the peer or another service is the invoker. This is summarised in Table 9.3 (for the sake of simplicity, operations are the ones shown in the simplified diagram). This information is written in the service contract, which essentially contains the operations and their parameters.

To sum up, we propose to model COINS using three different services:

- *Coins protocol service*: This service enables the download and the manager functionality.
- *Sharing service*: This service encapsulates the functionality of the participants.
- *Incentive mechanism service*: This service is responsible for enabling the update and the query of the *Rb* value of each peer.

9.4.2 Step 2: Content Discovery Service

As has already been mentioned in Section 9.2, the content discovery is responsible for finding the peers that provide a specific file in the network. Different content distribution systems implement this in different ways; for instance, BitTorrent [11] uses the torrent file that the user has to obtain, typically using search engines. Emule [10] stores this information in servers, and allows the user to search among them (hybrid). In COINS, service discovery can also be implemented in different ways; we opt for a full distributed content discovery taking advantage of SMEPP's service discovery.

As stated, SMEPP generates a key for each service that is published, by the hash of its contract. We can define a template contract file, such as Figure 9.7, just by setting a different profile name for each file. This way, if a peer publishes this contract, it means that it is sharing the file, and the content discovery is implicitly performed by the middleware's service discovery. This way, if two peers share the same file, they will publish the same contract,

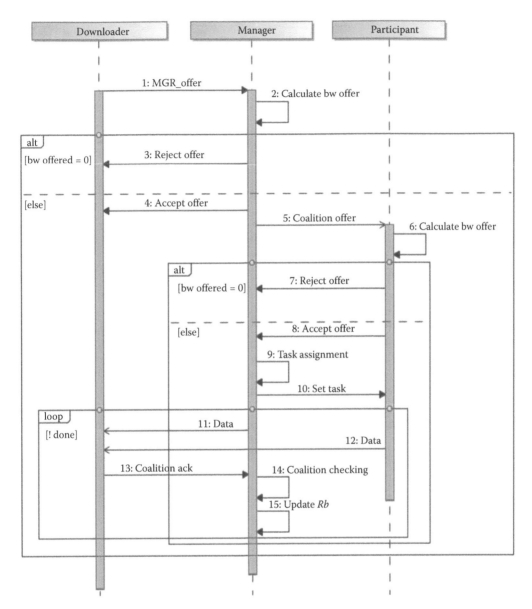

FIGURE 9.6
Simplified sequence diagram.

which will result in the same key and therefore the same responsible peer. Thus, peers that share the same file will publish the same service on the same peer.

9.4.3 Step 3: Responsiveness Bonus Management Service

The incentive mechanism proposed in COINS requires the implementation of a service that is able to update and query the Rb value of any peer. Rb represents the overall contribution of a peer, and therefore it has to be updated each time a peer participates in a coalition (as

TABLE 9.3

Service Operations/Messages

COINS Protocol Service	Type	Invoker
Manager offer (1)	One-way	Downloader—coins service
Offer answer (3/4)	Receive-response	Participant—sharing service
Coalition Acknowledge (13)	One-way	Downloader—coins service
New Download (0)	One-way	Peer code
Sharing Service	**Type**	**Invoker**
Coalition offer (5)	Receive-response	Manager—coins service
Cancel	One-way	Manager—coins service
SetTask (10)	One-way	Manager—coins service
Incentive Mechanism Service	**Type**	**Invoker**
Query Rb	Receive-response	Manager—coins service
Update Rb	One-way	Manager—coins service

a downloader or as a participant or manager; typically, the former will decrease the Rb and the latter will increase it). In addition, every time a manager sets up a coalition for a downloader, it has to query the downloader's Rb, to determine the bandwidth that the coalition will provide. As in the case of content discovery, there are several choices to implement this. We can opt for a centralised storage server, or the pure distributed scenario using distributed consensus algorithms or storing this information among peers.

As anticipated, we opt for the distributed scenario. We take advantage of the structured overlay network in a similar way as for the approach of the implementation of content discovery. The central idea is that every peer has to delegate the task of storing and updating its Rb value to another peer. Using the unique identifier of the peer and a hash function, a peer can find the peer responsible for storing the value of any other peer in the network. The Rb management functionality is encapsulated into a service; this service will be responsible for storing and updating a peer's Rb. When a peer joins the group, it must publish an Rb service with its ID (as was proposed for files in content discovery). To update or query an Rb, a peer just needs to invoke the middleware primitive *getServices*, specifying the ID of the peer that wants to update or query in a contract template.

```
<Contract
   xmlns:xsi='http://www.w3.org/2001/XMLSchema-instance'
   xmlns:xs="http://www.w3.org/2001/XMLSchema"
   xmlns='http://www.smepp.org/schema/Contract'
   xsi:schemaLocation='http://www.smepp.org/schema/Contract
../SMEPPContracts/XSD%20SMEPP%20contracts/Contract.xsd'>
    <Profile name="file_1" category="testing"/>
    <Signature>
        <interface>
        </interface>
    </Signature>
    <Behavior type="session-full"/>
</Contract>
```

FIGURE 9.7
File shared sample contract.

9.4.4 Step 4: All together: Implementation

This closing section pulls together what has been presented in the previous sections or steps. As stated in Section 9.4.1, the SMEPP implementation of COINS requires the peer code and the service code, which are shown in the following sections.

9.4.4.1 Peer Code

Each SMEPP peer starts invoking *newPeer*. Then, to be able to invoke services, it must join the COINS group, and then, it has to publish the services (Figure 9.8).

```
// CREATE NEW SMEPP peer
 String configFile = args[0];
 Credentials myCredentials = new Credentials("");
 PeerManager peer =
   PeerManager.newPeer(myCredentials, configFile);
//Join Group (find and join)
 GroupDescription myP2PGroup =
   new GroupDescription("COINS",
   new SecurityInformation(1), "COINS");

GroupId[] groupIds =
   peer.getGroups(myP2PGroup);
//(...)
 GroupId gid = groupIds[0];
 peer.joinGroup(gid);
//PUBLIHS SERVICES
//Main Service
 PeerServiceId psid =
   peer.publish(gid,
ContractLoader.loadFromFile("CoinsService.xml"),
   new SMEPPServiceGrounding(CoinsService.class),
   null,
   null);
//Rb management service
   psid =
   peer.publish(gid,
   ContractLoader.loadFromFile("RbMgr.xml"),
   new SMEPPServiceGrounding(RbMgr.class),
   null,
   null);
//File Sharing services
  foreach (file f in sharedFiles){
  String fContract =
  GenerateContract(f,"SharingS.xml");

 psid = peer.publish(gid,
   ContractLoader.loadFromFile(fContract),
   new SMEPPServiceGrounding(SharingS.class),
   null,
   null);
   //Invoke local service from GUI to start downloads
  }
 }
```

FIGURE 9.8
Peer code.

9.4.4.2 Service Contracts and Code

In Section 9.4.1, COINS was modelled based on three services, showing the different operations each service should offer. As introduced in Section 9.3.3.2, services need an SMEPP contract and its grounding. According to this, the contract of the COINS service should look like as shown in Figure 9.9.

The service code has to wait for the invocation of any of the messages or operations and performs the corresponding actions, similarly to the example code presented in Section 9.3.3.2 that implements *myService*. In Figure 9.10, the service code for operation *ManagementOffer* is illustrated; this is the first operation received by the manager that starts the coalition formation algorithm.

To sum up, SMEPP simplifies the implementation of this kind of application, as the above code shows not only by abstracting the underlying complexity but also by offering an efficient look-up mechanism for file discovery. Moreover, it tackles the security issues internally, without additional effort.

9.5 Deployment and Validation

The proposed implementation has been tested on a small scale in lab tests. However, the scalability and the performance of the application are given by the middleware, SMEPP, the scalability of which has already been validated and whose performance has already been measured. Since our algorithm does not cause any overhead, larger tests were found to be unnecessary. In addition, SMEPP middleware had already been validated with two applications: an e-health telephony application and a radiation monitoring application for nuclear power plants. Also, a live streaming application was developed to test performance in more stressing scenarios. Detailed results of these tests and the applications can be found in Ref. [26].

Regarding deployment, to run this SMEPP application, you need to instal it on different devices that will form the P2P network. These devices must be equipped with some kind of connectivity, and at least 2.5 Kbytes of RAM and 50 Kbytes of ROM (including middleware and application) in the SMEPP version for sensor networks. In in-lab tests (Intel core 2 DUO 2.93 GHz and 4 GB RAM), the application performed well; the profiler in Figure 9.11 shows that less than 20 MB of heap memory is needed to run the application.

Our SMEPP application just needs this and a light database to save permanent data (in this version, a text file is enough), and the SMEPP middleware, which is included within the application jar file. The application offers a GUI, shown in Figure 9.12, through which the user will be able to look for, find and join SMEPP groups; once the peer joins the COINS group, the interface allows them to share, search for and download files shared within the group.

9.6 Discussion and Challenges

Trends in distributed computing have been shifting from the pure distributed approach to the hybrid and centralised approaches over the last decade. This has also been reflected in file sharing applications, which of late have found a reliable partner in hosting servers,

```xml
<Contract
   xmlns:xsi='http://www.w3.org/2001/XMLSchema-instance'
   xmlns:xs="http://www.w3.org/2001/XMLSchema"
   xmlns='http://www.smepp.org/schema/Contract'
   xsi:schemaLocation='http://www.smepp.org/schema/Contract
../SMEPPContracts/XSD%20SMEPP%20contracts/Contract.xsd'>
    <Profile name="COINS" category="testing"/>
    <Signature>
        <types>
            <JavaTypes>
            <JavaType name="string" class="java.lang.String"/>
            <JavaType name="int" class="java.lang.Integer"/>
            <JavaType name="double" class=" java.lang.Double"/>
            <JavaType                                      name="serviceId"
class="org.smepp.datatypes.smm.StringContainer"/>
                <!-- -->
            </JavaTypes>
        </types>
        <interface>
            <operation name="START_DWN" xsi:type= "one-way">
                <input type="string"/>
            </operation>
            <operation name="MGR_OFFER" xsi:type= "request-response">
                <input type="string"/>
                <output type=" double "/>
                <output type=" double "/>
            </operation>
            <operation name="COALITION_OFFER" xsi:type= "one-way">
                <input type="string"/>
            </operation>
             <operation name="COALITION_OFFER_ANSWER" xsi:type= "one-way">
                <output type=" double "/>
                <output type=" double "/>
            </operation>
            <operation name="TASK" xsi:type= "one-way">
                <input type="int"/>
                 <input type="int"/>
            </operation>
            <operation name="COALITION" xsi:type= "request-response">
                <input type="int"/>
                <output type="int"/>
                <output type="int"/>
            </operation>
            <operation name="DWN_ACK" xsi:type= "one-way">
                <input type="string"/>
            </operation>
            <operation name="CANCEL" xsi:type= "one-way">
            </operation>
            <operation name="RECONFIGURE" xsi:type= "one-way">
            </operation>
            <operation name="DATA" xsi:type= "one-way">
                <input type="string"/>
                <input type="int"/>
            </operation>
        </interface>
    </Signature>
    <Behavior type="state-less"/>
</Contract>
```

FIGURE 9.9
COINS service contract.

```
ReceivedMessage m = svc.receiveMessage("MGR_OFFER",new Class[]{String.class});
sender = m.caller;
String source_list = (String) m.data[0];
ServiceId senderService = (PeerServiceId)
PeerServiceId.fromString((StringContainer)in_data[0]);
if (sender != null){
   System.out.println("[COINS SERVICE] Received MGR_OFFER FROM "+sender);
   Integer bw = CalculateBwOffered();
   svc.reply(m.caller, "MGR_OFFER",new Serializable[]{bw,p.m_id},null);
   System.out.println("[COINS SERVICE] Answer Sent bw "
                     + bw +",id " + p.m_id + " to"+ sender);
   if (bw>0){
     Process p = new Process(sender.getPeerId(),
     ProcessTypeEnum.MANAGER, svc.getPeerId());
     m_managementProcesses = new Process[1];
     m_managementProcesses[0] = p;
     //read source list
     System.out.println("RECEIVED SOURCES: " + source_list);
     String[] source_listSplit = source_list.split("@");
     PeerServiceId[] sources = new PeerServiceId[source_listSplit.length];
     org.smepp.datatypes.smm.StringContainer auxS = new StringContainer();
     for (int i=0; i<source_listSplit.length; i++){
       auxS = new StringContainer();
       if (source_listSplit[i].length()>0){
         auxS.setValue(source_listSplit[i]);
         sources[i] = (PeerServiceId) PeerServiceId.fromString(auxS);
       }
     }
     //SEND OFFERS
     if (sources.length >0)
         String offer = new Offer(senderService, bw);
         for (int i=0; i<sources.length; i++){
           svc.invoke(sources[i],
                      "COALITION_OFFER", new Serializable[]{offer});
         }
     /// ...
```

FIGURE 9.10
Service code example.

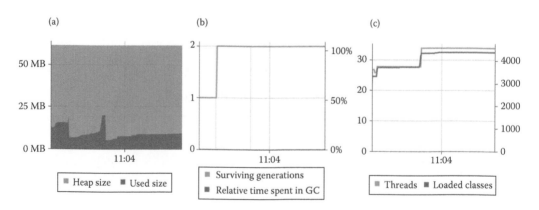

FIGURE 9.11
Profiler. (a) Memory heap (total and used); (b) memory (garbage collector); (c) threads (current and maximum).

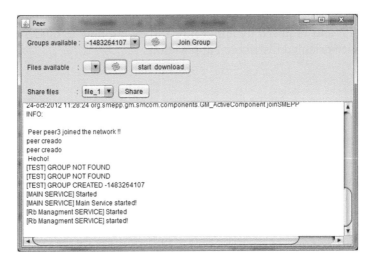

FIGURE 9.12
Application GUI.

leaving aside P2P proposals. But VoIP and mainly video streaming applications have kept P2P alive despite trends towards centralisation. Curiously, the centralisation and the use of servers in the first P2P file sharing application (Napster) caused its downfall, not because of performance, but rather for sharing of copyrighted content. However, this was not a problem for hosting servers, until Megaupload was shut down in January 2012 by the U.S. courts and the FBI after 7 years of existence. Hence, file sharing has returned to the distributed approach. As seen, these shifts are not promoted by researchers, and are mainly guided by the legal issues surrounding them. Here, two trends appear, one that tries to guarantee author rights and one that tries to guarantee user's privacy, and both of them are challenges in P2P.

P2P is still one of the hotter technologies on the Internet, but despite the huge variety of applications it has been proposed for, and indeed used for, none have proliferated or triumphed beyond file sharing applications. One of the reasons for this lack of applications may lie in the difficulties in the development stage of P2P, owing to a lack of suitable frameworks or middlewares capable of adapting to application constraints. General-purpose frameworks, such as JXTA and Microsoft P2P, may be hard to learn and the benefit of the provided flexibility is paid for with a worse performance. Opting for a specific domain solution can also be discouraging since there are a wide range of approaches and proposals that should also be analysed and may also be hard to learn. In this sense, a standardised solution should be adopted to facilitate the choice of which P2P middleware best fits the application's requirements. Even though the community has been clamouring for standardisation in P2P almost from its conception, it is currently still to come.

P2P still has many challenges ahead; we find particularly important those associated with the development, where a trade-off between flexibility, abstraction, adaptability and performance must be found. File sharing remains the main application of P2P, but it also has open issues like those related to legality. In our solution, we use SMEPP middleware to implement a formally proved, fair coalition system that guarantees the system's health. We show how using a specific domain middleware, such as SMEPP, eases the burden on the developer. Nevertheless, SMEPP provides tools, guides and examples all of which are useful for developers who are new to the application development of SMEPP; moreover, this work can be taken as a starting point. It helps in understanding the level of abstraction offered by SMEPP and required by applications, showing how to model the file sharing system as a service-oriented distributed application.

9.7 Conclusion

In this chapter, we presented the implementation of a content distribution system using SMEPP middleware. Our mechanism, COINS, is based on game theory and takes into account the rational and self-interested behaviour of peers. The central idea is that incentives encourage participation; each time a participant contributes in a coalition, they receive a reward. The fairness of the rewards division within a coalition is guaranteed by means of the game theory concept of *core*. These rewards are accumulated in the *Responsiveness bonus*, which represents the overall contribution of the peer to the system, and this is used to increase or decrease the quality of service for the downloads the peer performs. This way, our approach manages to promote cooperation and therefore reduces the free-riding phenomenon. Moreover, simulations showed that download times are improved.

As mentioned, P2P content distribution systems are complex systems that require a considerable amount of work in their development. When addressing the implementation of distributed applications, it has been demonstrated that the use of a middleware simplifies the implementation issues. We proposed the use of SMEPP, a secure middleware for embedded peer-to-peer, which brings two advantages in addition to the ease of the development: on the one hand, the security is built-in, so we can easily develop our incentive mechanism without worrying about security concerns, and on the other, we can use the application not only on PCs or servers but also in mobile devices, which is quite interesting given the popularity and the intensive use of these devices nowadays.

In this chapter, we have proved the suitability and the advantages of using a P2P middleware, and we have demonstrated that our mechanism can be developed as a real distributed application. We have shown how to use the middleware and how to implement our proposal using the middleware. We have presented the design and implementation of the system, in a tutorial style, discussing the main design issues. For instance, we have shown the implementation of the fully distributed content discovery that takes advantage of the middleware's service discovery, which greatly eases the implementation.

References

1. PPLive [Online] http://www.pptv.com/en/index.html. Last accessed 01/07/2012.
2. TVAnts [Online] http://tvants.es/. Last accessed 01/07/2012.
3. Saroiu, S., Gummadi, P.K. and Gribble, S.D. Measurement study of peer-to-peer file sharing system. In: Kienzle, M.G. and Shenoy, P.J. (eds), *Multimedia Computing and Networking*, pp. 156–170. SPIE, San Jose, CA, USA. 2002.
4. Handurukande, S.B., Kermarrec, A.-M., Le Fessant, F., Massoulíe, L. and Patarin, S., Peer sharing behaviour in the edonkey network, and implications for the design of server-less file sharing systems. *SIGOPS Operating Systems Review*, 40, 359–371, 2006.

5. Belmonte, M.V., Díaz, M., Pérez-de-la-Cruz, J.L. and Reyna, A. COINS: COalitions and INcentiveS for effective Peer-to-Peer downloads. *Journal of Network and Computer Applications*, 36, 487–497, January 2013.
6. Belmonte, M.V., Díaz, M. and Reyna, A. Coalitions and incentives for content distribution over a secure peer-to-peer middleware. In *AP2PS 2011, The Third International Conference on Advances in P2P Systems*, pp. 71–78, 2011.
7. Díaz, M., Garrido, D. and Reyna, A. SMEPP and the internet of things. In *Workshop on Future Internet of Things and Services*. CD.ROM, 2009.
8. Bertsekas, D., Gallager, R. and Humblet, P. *M.I. of Technology. Center for Advanced Engineering Study. Data Networks.* New York, USA: Prentice-Hall, 1987.
9. Karakaya, M., Korpeoglu, I. and Ulusoy, O. Free riding in peer-to-peer networks. *Internet Computing, IEEE*, 13, 92–98, 2009.
10. Kulbak, Y., Bickson, D. et al. 2005. The Emule protocol specification. http://www.cs.huji.ac.il/labs/danss/p2p/resources/emule.pdf. Last accessed 01/07/2012.
11. Cohen, B. BitTorrent protocol specification. http://www.bittorrent.org/beps/bep_0003.html. Last accessed 01/07/2012.
12. Garbacki, P., Iosup, A., Epema, D. and van Steen, M. 2fast: Collaborative downloads in p2p networks. In *Peer-to-Peer Computing, IEEE International Conference on*, pp. 23–30. IEEE Computer Society, Los Alamitos, CA, USA. 2006.
13. Karakaya, M., Korpeoglu, I. and Ulusoy, O. Counteracting free riding in peer-to-peer networks. *Computer Networks*, 52, 675–694, 2008.
14. Mekouar, L., Iraqi, Y. and Boutaba, R. Handling free riders in peer-to-peer systems. In *Agents and Peer-to-Peer Computing: 4th International Workshop, AP2PC 2005*, Utrecht, The Netherlands, July, pp. 58–69. Springer-Verlag, New York, USA. 2006.
15. Díaz, M., Garrido, D., Reyna, A. and Troya, J.M. SMEPP: A secure middleware for P2P systems. In: *Horizons in Computer Science*, Thomas S. Clary (ed.), vol. 3. Nova Publisher, London, UK. 2010.
16. Brogi, A., Popescu, R., Gutiérrez, F., López, P. and Pimentel, E. A service-oriented model for embedded peer-to-peer systems. In *Proceedings of the 6th International Workshop on the Foundations of Coordination Languages and Software Architectures*, Lisbon, Portugal, 8 September 2007.
17. Stoica I., Morris, R., Karger, D.R., Kaashoek, M.F. and Balakrishnan, H. Chord: A scalable peer-to-peer lookup service for internet applications. In *SIGCOMM*, pp. 149–160, 2001.
18. JXTA web. http://jxta.kenai.com/. Last accessed 01/07/2012.
19. Microsoft P2P Networking web. http://technet.microsoft.com/en-us/network/bb545868.aspx. Last accessed 01/07/2012.
20. Kim, Y. and Ahn, S. A reliable contents distribution middleware system for peer-to-peer networks. In *International Conference on Information Networking (ICOIN)*, pp. 442–446, IEEE, 2012.
21. Capra, L., Emmerich, W. and C. Mascolo. CARISMA: Context-aware reflective middleware system for mobile applications. *IEEE Transactions on Software Engineering*, 29(10), 929–945, 2003.
22. Kavadias C.D., Rupp S., Tombros S. L. and Vergados D.D. A P2P technology middleware architecture enabling user-centric services deployment on low-cost embedded networked devices. *Computer Communications* 30, 527–537, 2007.
23. Hwang, J. and Aravamudham, P. Middleware services for P2P computing in wireless grid networks. *Internet Computing, IEEE*, 8(4), 40–46, 2004.
24. Kortuem, G. Proem: A middleware platform for mobile peer-to-peer computing, *ACM SIGMOBILE Mobile Computing and Communications Review*, 6(4), 62–64, 2002.
25. Boyd, S., Ghosh, A., Prabhakar, B. and Shah, D. Randomized gossip algorithms. *IEEE Transactions on Information Theory*, 52(6), 2508–2530, 2006.
26. SMEPP Project. http://www.smepp.net/. Last accessed 01/07/2012.

Part III

Applications and Trends in Distributed Enterprises

10

User Activity Recognition through Software Sensors

Stephan Reiff-Marganiec, Kamran Taj Pathan and Yi Hong

CONTENTS
10.1 Introduction 243
10.2 Background 245
    10.2.1 Sensors 246
    10.2.2 Context Models 247
    10.2.3 XML to OWL Mapping 249
10.3 Context Model and Reasoning 250
    10.3.1 Design Considerations 250
    10.3.2 Context Model 251
        10.3.2.1 Entity Ontology (Person, Organisation, Device) 252
        10.3.2.2 Place Ontology 253
        10.3.2.3 Time Ontology 253
        10.3.2.4 Activity Ontology 253
    10.3.3 Inference Procedures for Context Reasoning 254
    10.3.4 Information Retrieval 254
10.4 Context Acquisition 255
10.5 Mapping 256
    10.5.1 Methodology 257
    10.5.2 Direct Match 257
    10.5.3 Synonym Match 258
    10.5.4 Hypernym Match 259
    10.5.5 No Match 260
10.6 Sample Scenario 260
    10.6.1 Queries 261
    10.6.2 Rules 261
    10.6.3 Degree of Confidence 262
10.7 Conclusion and Future Work 262
References 263

10.1 Introduction

Context-aware systems have been investigated for quite a while, and there are many aspects of such systems that are readily used today. Typical systems often provide location-aware services to users (such as targeted advertising for mobile users through SMS (short message service) messages offering discounts at restaurants in the vicinity of the user). However, sensing the activity of the user (i.e., what a user is doing) has been very hard to tackle. Most of the context gathering is achieved through the use of hardware sensors (e.g., tracking devices). The systems based on software sensors only process the user data according to the websites visited or the relevant documents [1–3] or tell the online status [4], which is either updated by the user or automatically through logging in or out. The activities a user undertakes otherwise are not understood. Some of the advertising services (such as Google in its Gmail interface [5]) look at the description and subject of the email and advertise, no matter whether that advertisement is relevant to the users in their given situation and hence possibly missing the intention of advertising.

A typical understanding of a sensor is 'a device that converts a physical phenomenon into an electrical signal. As such, sensors represent part of the interface between the physical world and the world of electrical devices, such as computers' [6]. This view has a very hardware-oriented angle. Hardware sensors typically measure a specific phenomenon such as a GPS (global positioning system) location or the ambient temperature. However, they could be more complex; for example, a camera might track a user's position and, through a connection to models, can identify whether a user is sitting or standing or even where on a screen he is looking. It should be said that for the latter to work, complex setups are needed, and hence a high cost occurs with the deployment of hardware sensors. In addition, it is not really possible to conclude what exactly a user is doing: they might look at their email client, but are they reading private or work emails, and which work project does the email relate to? This is especially relevant in a world with knowledge workers working on many projects simultaneously, answering emails and phone calls relating to each of these rather than just focusing on one task. Understanding the time spent on specific tasks automatically could lead to very fine-grained understanding of the time projects take and hence will allow for very precise invoicing to clients and good predictions for efforts for new projects.

Returning to the view of sensors as devices that capture some physical phenomenon, one could argue that it is a very narrow view. And indeed, a sensor could also be software, as long as it acquires data that capture real-world snapshots [7].

In today's world, there is much exchange of messages between users and the networked applications that they use; in the context of service-oriented computing, this exchange often involves structured messages defined in SOAP (simple object access protocol), which contain data on the service invoked, the operation of those services requested and possibly a payload of attributes sent to the service or received in reply. Using software sensors, we can capture this wealth of data and put it to use in understanding a user's activities as well as in supplementing data gathered from hardware sensors. The challenge is to ensure that the raw data will be converted into structured data (typically populating a context model) where it then becomes knowledge, which can further be reasoned about using some inference mechanism.

For example, if a user is using a calendar service and is sending data for adding an event, the data sent will make explicit that it is an `AddEvent` request to a calendar, and possibly one can even see details such as the title of the event, the time (from–to), the location, invited participants and so on. These data are carried through a SOAP message and is usually interpreted by the receiving service. Our approach is to extract those data and store them in an RDF (resource description framework) triple form so that automatic reasoning can be performed. For example, from event data obtained from a calendar, weather/traffic conditions of the location and the location of a speaker, one could deduce whether the event is likely to start on time.

For this to work effectively, we need to (a) be able to capture the SOAP message and extract data from it and (b) instantiate and update an ontology with new data. This chapter addresses both aspects. We discuss the data extraction and introduce a mapping methodology that can map data from the XML (extensible markup language) payload in SOAP

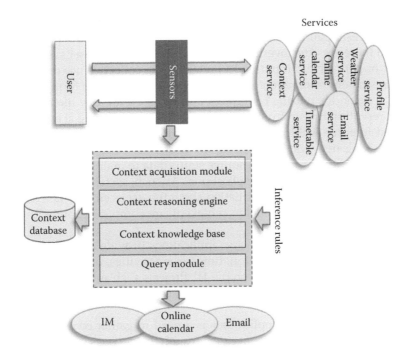

FIGURE 10.1
Architecture for context-aware system using software sensors.

messages to an existing OWL (web ontology language) ontology and update the instances along with the hierarchy it is based on should there be a need. The latter is building on work presented in Ref. [8]. To place this methodology in context, the respective context model detailing the data model to store and process the data and the infrastructure for context-aware systems using software sensors to sense the activity context is also presented. Figure 10.1 shows the overall architecture employed in our work, with this chapter focusing on the sensors and context acquisition aspects shown in the figure.

This chapter is structured as follows: Section 10.2 will discuss the developments of sensors, context models and XML to OWL mappings. We will then present our adopted (and adapted) context model and discuss design considerations and knowledge generation by reasoning in relation to the model in Section 10.3. The following two sections (Sections 10.4 and 10.5) introduce the data extraction and the data mapping methodology, which are then applied to an example in Section 10.6. We conclude with a summary and an outlook on future directions and open questions in Section 10.7.

10.2 Background

There has been a growing interest in context-aware systems, significantly fostered by the way in which people use computing devices these days. In an always-connected, always-on environment with smart phones and other mobile devices making their way into our lives in ways overshadowing the impact of desktop and home computing, there is a desire by manufacturers to provide differentiating features and by users to seek ever new applications and services. In addition to context-aware systems, a number of other terms have been used, and these are often overlapping when it comes to end-user systems. Ubiquitous computing [9] or pervasive computing [10] refers to the collective use of computers available in the physical environment of users, perhaps embedded in a form invisible to users representing a vision for putting computers out into the everyday life and environment, instead of representing the environment in the computers. Looking at embedded computers in cars and Internet-enabled TVs and refrigerators, this is already much of our daily reality now, seeking new ways of making the user's life easier. Sentient computing [11] refers to systems 'using sensors and resource status data to maintain a model of the world, which is shared between users and applications'. As such, systems try to build a model of a part of the world from sensory information about the user's circumstances and environment. This idea is very closely related to context-aware computing, but with an emphasis on the world model [7].

Existing context-aware middleware and applications provide and exploit various types of contextual information about location, time, user activities, user's preferences, profiles of users, devices and networks and so on [12–14]. Projects such as ECOSPACE [15] or inContext [16] provide middleware platforms, taking the context-aware applications into collaborative work environments.

What is common in all these applications is a need to gather context data as the most basic information to drive the whole system structures. Of course, it is not just gathering these data, but also storing them in ways that allow knowledge generation. We will now look at the aspects that are most pertinent to the work presented: sensors, context models and data to knowledge mapping (for XML and OWL).

10.2.1 Sensors

According to Klein, a sensor is a transducer (front-end hardware) that 'converts the energy entering its aperture into lower frequencies from which target and background discrimination information is extracted in the data processor' [17]. Likewise, very physical definition is provided by 'A sensor is a device that converts a physical phenomenon into an electrical signal. As such, sensors represent part of the interface between the physical world and the world of electrical devices, such as computers' [6]. Interestingly, the latter continues to take this full circle, by also considering the way the sensing system can influence the environment: 'The other part of this interface is represented by actuators, which convert electrical signals into physical phenomena' [6].

However, more generically, a sensor is a device (in the widest sense) that can acquire data about the environment. In that wider sense, sensors might be hardware or software, or indeed the combination of both. So, for example, a location tracking device or the computer clock accessed using an operating system function would both be considered sensors (one for location, the other for time) [7].

Using sensors to gather context information, also called context sensing, refers to the process of acquiring context information. Ferscha et al. [18] define two types of context sensing mechanisms:

- *Event-based sensing* monitors the environment for occurrence of events and sends a notification to the applications upon detection of an event.
- *Continuous sensing* explores the real-world information continuously and provides streams of observations.

Both these types have advantages and disadvantages and one might be more suitable than the other in specific scenarios. Considering sensing the context of a taxi as an example, one could say that the location of the taxi is best sensed in a continuous fashion (as it constantly changes if the taxi is moving) while the fact whether the taxi is occupied or not is more event based (a passenger getting in or out). There are a number of interesting questions regarding handling the data amounts occurring in such settings. For complex event processing, Ref. [19] is being employed to deal with streaming data, and the work in Ref. [20] focuses on determining ways to provide sensor data to applications with essentially zero latency.

In general, when considering continuous sensing, there are questions as to how often sensors shall be polled or at what frequency data should be sent—sending all information to all users is not possible due to the vast amount of data generated. What makes this more complex is that context is more or less dynamic. Consider, for example, locations: these are very dynamic for a person but quite stable for a desktop device or printer, and absolutely fixed for rooms. The current approach is to preset different polling rates by the user or by the application at the start-up [21]. These polling rates are based on people's experiences. Brewington and Cybenko [22] developed a formal model for monitoring information sources by using previously observed information lifetimes to estimate change rates. Their model was developed for web pages but is applicable to many information-monitoring problems, including context sensing using software. More recent complementary work formalised a push–pull architecture with a filtering mechanism that specifies when services or sensors should deliver certain types of information [23]. Context-sensing techniques need to be reliable and robust, which is sometimes very hard to achieve: consider active badges [24] as an example. If the badge is not worn by the user but left in the office, then the location determined by the system will be false. This requires systems to be built in such a way that information has lifetimes (old context data are probably less reliable), confidence of correctness (some data are more trusted) and redundant (several sensors should provide insight that jointly allows one to arrive at conclusions).

A study by Clarkson [25] looked at gathering data using sensors in real life with a view to identifying patterns. Data were collected using two cameras, a microphone and an orientation sensor worn by a user for 100 days—all of these are, of course, again hardware sensors. Quantitative analysis techniques were used to extract patterns in daily life showing that a 'person's life is not an ever-expanding list of unique situations' [25], but that there is a regularity in daily life's situations. Such regularity is, of course, what allows for data mining and reasoning techniques to generate knowledge from context data successfully.

As we mentioned earlier, systems based on software sensors are very rare—and often they are about adapting websites or using online status information, rather than the much wider use of and more rigid definition of sensors proposed here. Existing work uses user data based on websites visited or relevant documents [1–3], which are automatically gathered, for example, through shared cookies. In case of using online status [4], the user will either have to update this manually or it is changed through logging in or out. In the Mobisaic web information system [26], context information is extracted from the dynamic URLs, allowing for documenting which web pages change automatically by looking at the change of other variables.

10.2.2 Context Models

A well-defined data model helps with aspects of processing and storing data more effectively, and this also holds for a context-aware system. The data concerned here are, of course, context data, which have a complicated structure and intricate links that are often exploited not only by the structure of the data model but also by superimposed reasoning rules that generate the more interesting knowledge. Additionally, the system is often distributed, so different parts of the system that might have been written independently will access and store data in the same data store. There are several models in existence, often specialised on the main context aspects that the researchers are concerned with. Furthermore, the approaches use different modelling techniques, leading to distinct advantages, making comparison more complex. We will now look at a number of models used to represent, store and exchange context data.

Schilit et al. [27] used key–value pairs to model the context by providing the value of a context entity to an application as an environment variable. Markup scheme models are based on a hierarchical-based data structure, which extends on this simple model by using tags with attributes and content. They are usually based upon XML-type languages such as RDF/S (RDF Schema) and have the advantage of easy tool access, but lack of formality and expressiveness. The unified modelling language (UML) is a language to model the context using UML diagrams. Bauer et al. [28] and Henricksen et al. [29,30] defined a context model for in air traffic management in a graphical way using UML. The resulting object-oriented models provide encapsulation and reusability, where object-level details are encapsulated (hidden) to other components. A first logic-based context modelling approach has been published by McCarthy [31], introducing context entities as abstract mathematical entities, which are helpful in artificial intelligence. Ontologies are commonly used to specify concepts and their interrelations [32] and context models based on ontologies have been first proposed by Otzturk and Aamodt [33] and have since proven to be useful and popular in this domain.

Ref. [33] proposed an ontological model because of a need to normalise and combine knowledge from various domains and thus formalise it and allow one to reason about the knowledge. The context of this context model was not context-aware systems, but rather a psychological study depending on contextual information. The CoBrA system [34] uses a broker-centric agent architecture to provide runtime support for intelligent meeting rooms. It offers a set of concepts based on ontologies to characterise entities such as places, persons and objects. Most of this work focuses on describing people and their locations; it also considers time. However, it does not consider the activity or the motivation (the why).

Gu et al. [35] also followed an ontology-oriented approach by developing a generic context ontology. This model is quite complete in its own way, but it lacks division in the upper and lower ontologies. This decision, trivial as it might sometimes seem, allows one to distinguish generic concepts from more domain-specific ones and thus eases specialisation of ontologies. For example, in contrast to our work, activity context is not considered in much detail in their work. However, activity context is a very rich source of information for advanced adaptation as was shown for collaborative systems in the inContext project [16].

What has crystallised over the last decade is that ontology-based approaches for context modelling are very promising as they offer the needed structure, while providing ease of extensibility and reusability. All this is provided at a level of formality open to reasoning while still providing simplicity offering a good understanding of the models. There are a number of standard languages for defining ontologies, predominantly OWL, which provide ready access to tools and also readily tie in with web services (in fact, research on semantic web services uses OWL at its heart to introduce semantics to what are otherwise just syntactically described services).

10.2.3 XML to OWL Mapping

One need specific to software sensors is the need to convert from data observed in structured forms to other structured forms. Specifically, we require methods that can map data from XML payloads in SOAP messages to OWL ontologies. There is existing work addressing the problem of mapping from XML schemas to OWL ontologies (or their respective RDF descriptions). This work falls into two broad groups: work that creates ontology models based on the XML schemas [36,37] and work that matches an XML instance document into a related pre-existing ontology [38]. This body of work provides a sound understanding of the relation between elements in XML and their ontological counterparts, and we summarise the key relations in Table 10.1 and Figure 10.2.

This existing work maintains the class hierarchy given through the tree structure of the XML documents without maintaining their actual structure. In relation to our work, these issues become somewhat preliminary as they might help to establish the initial ontology models. The second benefit is the conceptual relation of the concepts. However, for our work, we wish to reuse existing context models and are more concerned with mapping the actual XML instance data into the context model. For this we need, and this is a contribution presented later, a way to identify the right model element to assign data to and map the actual data. We also do not wish to extend the ontology model with new concepts, but rather keep that stable, while at the same time being able to insert as much information on XML schemas that are encountered as part of service exchanges. Thus, the problem addressed here is more concerned with matchmaking at a vocabulary rather than structural level, and hence the need to use a lexical database, such as WordNet [39], arises, so that we can check and acquire the results accordingly.

TABLE 10.1

Relationships between XML Elements and OWL Elements

XML	OWL
Represented by trees with labelled nodes	Represented by triples (i.e., subject–predicate–object)
Tree structure	Concept hierarchy
Schema	Model
Instances	Instances
<tag>	Resource/literal (depends if the tag has a value)
Nested tags	Resource with object properties (if the tag has no value)

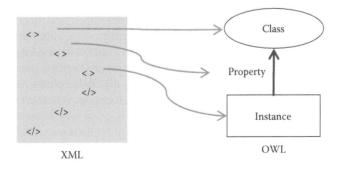

FIGURE 10.2
Mapping XML elements to OWL.

10.3 Context Model and Reasoning

Context modelling specifies the context-related entities and relationship and the constraints between each other. Taking an example of our scenario, a user's activity could be derived from his profile, calendar, timetable and email services. A context model provides the structure and methods to process the context data, which can be saved for later use. Context is quite wide ranging and includes, for example, a user profile information, the user's location or planned activities, but generally it is quite varied [14]. After acquiring the user's context, data need to be stored (at least while it is valid and of use), it will be reasoned upon or mined to extract knowledge to provide information needed by the context-aware systems or its users. A number of context-aware computing applications need information to be exchanged between entities, which might be user or services, and the context model should also support that interoperability.

We presented a generic semantic model for use with software sensors in a service-based, context-aware environment. It provides general upper-level classes, properties and relations between them so that lower ontologies can easily be created for particular domains. Such a generic model of context is beneficial for systems or services that are interacting with each other and the end users, who ultimately try to accomplish their daily tasks more easily in the context-aware environments utilising software sensors. The use of standard representation languages makes sharing, reuse and extension of context models and context data easy. In addition, the model provides the ability to infer new knowledge from gathered context information.

The generic model of context presents the context in a form that provides general-level upper classes, properties and relations such as `Place` and `Resources` capturing the aspects that are always useful when considering context information, independent of domain. These can then readily be extended using lower ontologies for specific domains, maintaining the overall structure and relationships (if one wants, conservatively extending the model). The model will allow the retrieval of knowledge, which will be at higher levels of abstraction than the data gained from the sensors; when considering activities, we like to obtain information on the 'situation'.

10.3.1 Design Considerations

The semantic web provides standards for structural, syntactic, meta-data and ontology languages. We have selected OWL to develop our context model because it is an international standard for information representation with good tool support. OWL allows interoperability between systems and services, provides inference mechanism by providing additional vocabulary, and defines taxonomies, properties, relations and constraints. The fundamental concept for context organisation in context-aware systems is W4H [40]: who, where, when and what, which are according to context entity, place, time and activity, respectively, and how (the devices used). Essentially, W4H contains all the questions one might ask about the context. The 'where', that is, location, has seen much attention in terms of its use and acquisition. This has resulted in great successes with most current context-aware applications in fact being location-aware applications. 'When' is, time zones apart, fairly trivial. Both 'Who' and 'How' are often captured by people and resource profiles—however, these two ideas can be more complex. Our model, not surprisingly, includes classes for all these concepts. However, with the devices getting more personal, we decided to consider how and who together as part of our entity class. Thus, we identify four main high-level components of context for our model:

Entity (who and how): An entity that performs some action and can be an actor, an organisation or an object (machine or another service, etc.).

Place (where): A place is the physical location in a pervasive environment.

Time (when): The time is usually tagged against interactions and observations in the system.

Activity (what): Any information that allows the capture of the user's activity occurs in this category.

We are most interested in the activity context here. However, this is interlinked with other areas of context. For example, two people might be meeting in an office or in a bar—obviously, in both cases, they have a meeting (or a discussion); however, in the former case, it is most likely work related, while in the latter, it is more likely to be of a social nature (especially if it is after work hours). So, the location and time lets us infer the nature of the activity. In the light of this, we cannot just look at a subsection of the context model, but rather require an overview of the whole model.

10.3.2 Context Model

An overview of the proposed context model based on the considerations above and inspirations from the literature is presented in Figure 10.3. The main inspiration of our context model is based on the ontology by Gu et al. [35], which provides a vocabulary for representing knowledge about a domain and for describing specific situations in a domain. However, their structure addresses taxonomies that are closely related with the more traditional physical sensors and hence provide a gap for our research. As we are dealing here with software sensors, we have adapted some links to suit our needs more closely and expanded specific areas of the ontology. Other inspiration came from models such as the one targeting hardware sensors in CONON [41] or a specific domain in inContext [16].

The model defines a common vocabulary and structure of the context, allowing for sharing context-related information between users and services to enable operations. It includes machine-processable definitions of the basic concepts in the domain and range and relations among them and reasoning become possible by explicit definition. This model represents the main concept and subclasses that we require for our work in a convenient way—as always with ontologies one could debate little aspects here and there, and we are not claiming that this model is superior to existing ones; it simply places focus on the elements that we require and presents them in a convenient way while being generic at the same time and thus allowing reuse in other domains and applications. Of course, we are reusing ontologies where appropriate, such as for people (FOAF—friend of a friend), which is not detailed in Figure 10.3. To keep the figure readable, and to not overload the paper with detailed figures, we have not included all possible links (for example, all entities will have a location; some links might be created through attributes, others through reasoning rules) and also not all details of each class. We will now discuss each main class in a bit more detail and also hint at the kind of information that can be gleaned by combining data. The focus will be on the `Activity` ontology.

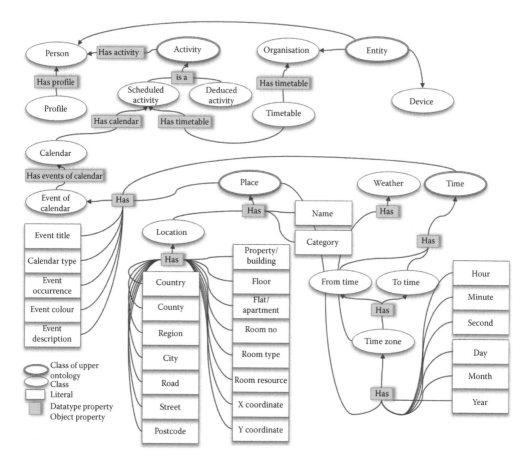

FIGURE 10.3
The context model centred on activity.

10.3.2.1 Entity Ontology (Person, Organisation, Device)

The entity ontology captures information about people, organisations and devices. Typically, such information might include names and other descriptive data, but often also skills and capabilities—the latter showing what can be offered by the entity. For example, an output device such as a screen would have resolutions and colour profiles as capabilities; a person might be a programmer capable of using specific technologies or languages. Most of this context information is semi-stable (it is less dynamic as learning new skills takes time and devices will not be constantly upgraded).

The social connection between entities must also be considered, and social connections are dynamic (people might have different relations to each other in different contexts). Devices might be (and predominately are) personal, but they might also share resources or be accessible by a number of people. People might work for several organisations, for example, in consultancy roles, and worse, there might not even be a very clear time separation between such activities as a person might be on the premise of company X while dealing with a call from company Y. Also, the relations between people are complex: a certain person might be the subordinate of another person in the company hierarchy, but for a specific project he might be the project leader and thus, for the project, outrank his superior.

Considering such relations and roles, one could derive conclusions about activities: a lecturer in a room with 50 students on a university campus is most likely giving a class—if the room contains many fixed computer devices, it might even be a computer lab and the class would be a laboratory class.

10.3.2.2 Place Ontology

Places may have names, are assigned to categories (e.g., restaurant), and will have physical locations (i.e., address, coordinates) and often associated resources (computers, tables, chairs, etc.), all providing useful context information.

The presence of an entity at different places lets us deduce much about their activities. At the most trivial level, people who are in their office are probably working, while at home they are probably not working (unless we know that they are a mobile worker and the time of the observation tells us it is working hours). Of course, the devices that they use and what they do with them influence this: someone using their mobile phone to undertake a call to their partner from the office is not working at that moment. In the context of a university environment, rooms have specific purposes and these let us deduce further information. A staff member alone in their office is probably preparing teaching material or undertaking research; while if they are in the office with a PhD student, they are probably in a supervision meeting.

10.3.2.3 Time Ontology

Events occur at specific times or might be scheduled to have start and end times. Sometimes, time can only be provided as an interval enclosing the actual event time. A time instance is a point or an interval on the universal timeline [40].

Time is obviously often crucial when considering relations between facts—recall the example under the entity ontology where we loosely stated that the lecturer and the students are in the same room. Obviously, a precise definition of that would need to refer to the time interval for the class and make precise that the students and lecturer are in the room at the same time.

Time plays a second very important role; not all context information will be immediately evaluated. Some might be stored for later use, and here it is important to use the time to decide on the lifetime of an item of information. This is a significant problem in itself and is an item for future research. Consider, for example, a GPS coordinate for a user who is travelling in a car—this will very quickly be incorrect; however, the city or country that he finds himself in will be more long-lived (unless we only identified the location just before he tried to cross a border).

10.3.2.4 Activity Ontology

The activity ontology considers two primary types of activities: scheduled and deduced. The former represents activities that are planned for a specific time and location with a specific purpose (e.g., a project meeting or a class or flight) and usually can be identified from calendar entries or teaching timetables or airline schedules. The latter represents the activities that occur in a more informal manner and can be inferred by combining context data through rules.

For example, one might deduce that a user is missing his flight if he was scheduled to fly in 10 min, the airline flight information states that the flights are on time and the user is known to still be in his office away from the airport.

10.3.3 Inference Procedures for Context Reasoning

To derive new facts by applying rules on existing facts in the knowledge base is called inference mechanism. The most trivial inference is, of course, that where no items are combined (essentially a lookup of some information). For example, a scheduled activity can simply be looked up. Much more interesting is, of course, the use of several bits of information that together let us arrive at some new knowledge: knowledge gained by non-trivial inference. Such inference is guided by rules. For example, we can deduce an activity by combining two or more items of context data, such as by using a staff member's work calendar, their role and location and the organisation's timetable, we might derive that a certain professor is indeed giving a class. We can argue that the more information we combine, the better will be our judgement—in some way, considering each piece of information as witnesses. Note that this does not come without downsides, as some information might be more reliable or up to date; we will return to this point later on when we briefly consider degrees of confidence in the scenario. Obviously, we can store derived or inferred information back into the context model, on the one hand to save the effort of re-computing it every time, and on the other to ease its use in further deductions.

Simple inferences use the hierarchical structure of the ontologies; typically, the sub- and super-class relationships. The most common of such forms is transitivity ($A = B \wedge B = C \Rightarrow A = C$): if a person is physically present in Leicester and we know that Leicester is a city in the United Kingdom, we can conclude that the person is in the United Kingdom.

More complex reasoning will involve domain-specific rules that specify more intricate relationships between context items, which cannot be directly observed from the ontology. However, these can be quite generic in their nature, such as the impact of weather conditions on travel conditions and, by extension, the presence of people in certain places.

A typical rule is (more examples are shown later when we consider an example scenario):

$$Timetable(f, o) \wedge Weather(l) teachingClass(af, c) \qquad (10.1)$$

where f is a faculty member, o is the organisational calendar (timetable), l is a location, af is another faculty member (distinct from f) and c is a postgraduate class.

In general, we could argue that if a person is currently in location X, is expected to be in location Y in 20 min and the normal time of transport between the two locations is 20 min, then the person should be able to fulfil the expectation. If we now also observe traffic delays or adverse weather conditions, we can deduce that the travel time will not be satisfied and hence the person will be delayed.

Considering the above rule, we say that if a staff member is foreseen to teach a certain class by the organisation's timetable, but the weather conditions are adverse, an alternative faculty member might conduct that class (or at least make an announcement about the delay).

10.3.4 Information Retrieval

SPARQL [42] (SPARQL protocol and RDF query language) is a query language that allows basic queries over RDF, such as conjunctive queries, UNIONs of queries (i.e., disjunction), OPTIONAL query parts, and filtering solutions by means of a set of basic filter expressions.

For instance, let us assume one wants to answer the following query: 'Check whether a room is double-booked during a certain period of time; if so, get the list of all conflicted scheduled activities along with the attendants involved'. This could be expressed in

SPARQL by means of a simple conjunctive query (assuming some technicalities, namely, that we have 'ctx' as default prefix http://www.cs.le.ac.uk/ontology/contextmodel# and that some OWL properties are defined in the ontology, such as hasVenue, hasTimespan, startsAt,finishesBy, supervisorOf):

```
PREFIX rdf: <http://www.w3.org/1999/02/22-rdf-syntax-ns#>
PREFIX ctx: <http://www.cs.le.ac.uk/ontology/contextmodel#>

SELECT ?act1 ?act2 ?p1 ?p2 WHERE {

    ?p1 ctx:hasActivity ?act1.
    ?p2 ctx:hasActivity ?act2.

    ?act1 rdf:type ctx:ScheduledActivity.
    ?act2 rdf:type ctx:ScheduledActivity.

    ?act1 ctx:hasVenue ?loc.
    ?act2 ctx:hasVenue ?loc.

    ?loc rdf:type ct:Room

    ?act1 ctx:hasTimespan ?ts1.
    ?act2 ctx:hasTimespan ?ts2.

    ?ts1 ctx:startsAt ?start_time_act1.
    ?ts1 ctx:finishesBy ?end_time_act1.

    ?ts2 ctx:startsAt ?start_time_act2.
    ?ts2 ctx:finishesBy ?end_time_act2.

    #check overlaps
    FILTER ((?start_time_act1>=?start_time_act2 && ?start_time_act1
    <=end_time_act_2)||
    (?start_time_act2>=?start_time_act1 && ?start_time_act2<=end_time_
    act_1))

}
```

10.4 Context Acquisition

From the architecture (see Figure 10.1.), it is apparent that the sensors are placed in between the user and the services. These sensors capture the raw SOAP messages that are being used to invoke the web services, observing communications at the protocol level at the point where messages are reaching or leaving the web services server. SOAP [43] is an XML-based communication protocol to allow exchange of structured information between web services and applications over HTTP. SOAP is a textual protocol. By being payload to HTTP messages, SOAP messages are able to get around firewalls (a great advantage from practical points of view, but of course also a huge concern to system security experts). However, SOAP is a W3C recommendation and the standard protocol for web service interactions.

```xml
<soap:Envelope xmlns:soap="http://www.w3.org/2003/05/soap-envelope" xmlns:ser="http://service.calendar.google.com">
  <soap:Header/>
  <soap:Body>
    <ser:getTitle>
        <!--Optional:-->
        <ser:title>Meeting</ser:title>
    </ser:getTitle>

        <ser:getFromDay>
                <!--Optional:-->
                <ser:fromDay>10/07/2012</ser:fromDay>
        </ser:getFromDay>

        <ser:getFromTime>
                <!--Optional:-->
                <ser:fromTime>18:00:00</ser:fromTime>
        </ser:getFromTime>

        <ser:getToDay>
                <!--Optional:-->
                <ser:toDay>10/07/2012</ser:toDay>
        </ser:getToDay>

        <ser:getToTime>
                <!--Optional:-->
                <ser:toTime>19:00:00</ser:toTime>
        </ser:getToTime>

        <ser:getLocation>
                <!--Optional:-->
                <ser:location>Leicester</ser:location>
        </ser:getLocation>

        <ser:getDescription>
                <!--Optional:-->
                <ser:description>This is a meeting with a colleague</ser:description>
        </ser:getDescription>

  </soap:Body>
</soap:Envelope>
```

FIGURE 10.4
A typical SOAP request message.

Observing SOAP messages is quite simple, as handlers residing on the client and/or the server side during a service invocation allow the interception of SOAP messages. These handlers can capture a SOAP request/response message before it is sent from or returned to the client. The handlers forward the information from the SOAP exchange to the sensors. The handlers are part of the standard SOAP server architecture, although by default they are disabled for security reasons. The handlers can be activated for the client side (in the web.xml configuration file) and/or for the server side (in the webservices.xml configuration file).

For example, while creating a new entry in the calendar service, a user inputs data that will then travel from the client to the web service using the SOAP protocol. These SOAP messages carry the arguments the user passes to create an event in the calendar, such as Title, FromDay, FromTime, ToDay, ToTime, Location and Description in the request. Figure 10.4 shows a typical SOAP message for the calendar service.

We need to observe both requests and responses with our sensors, as they will carry different information. In the specific case here, the response will only be the acknowledgement stating that the entry is added into the calendar and thus not be very interesting in providing context information. For example, for a weather service, both messages will be important, as the request will specify a location and the response will inform about the actual weather.

10.5 Mapping

This section considers the mapping of XML data extracted from SOAP messages to data in the context model.

The input to the mapping task is raw XML data and the output would be a context model populated with the new data gathered. XML files contain tags and values and their relation expressed through nesting and the general tree structure, while in the OWL context model, we have classes, instance data and explicit relations expressed through properties. There are obvious mappings from tags and values to classes and instances at a conceptual level (we discussed this earlier in Section 10.2); however, at the concrete data level considered here, we require novel techniques as we indicated earlier.

To automate this matching, we use a syntactical comparison, enriched with semantic information. We can distinguish four cases, based on how well the tags and ontology instances align. The simplest scenario is where the XML tag name is identical to the instance name in the ontology (note that we assume that syntactically identical terms are referring to identical concepts in this domain; however, the context (such as super- or sub-tags) in which the tag occurs in the XML file could provide further insight). This case will be referred to as a direct match and we will formalise this later. The more interesting cases are those where no direct match is found between the tag and the ontology. One way to address this could be to create new instances and extend the context model; however, having a stable structure is more suitable for reasoning, so we are looking at methods of identifying existing instances with which the new data can be directly associated. We identify two possibilities of creating this association: a synonym of the tag matches with the ontology concept (a synonym match) or the hierarchical context of the word contains the ontology concept (a hypernym match). The fourth case covers all those concepts that occur in terms and cannot meaningfully be matched to the ontology.

10.5.1 Methodology

Our methodology considers the cases in turn and inserts the value from the XML document into the most appropriate matched ontology instance as shown in Figure 10.5. Briefly, the approach goes through three steps:

1. Attempt to match the tag with the instance; if found, insert the value as a literal and finish. If not found, go to the next step.
2. Identify synonyms for the tag and match each synonym with the instance; if found, insert the value as a literal and finish. If not found, go to the next step.
3. Find the tag in a lexical database (such as WordNet) providing a hierarchy (that is, hypernyms). Match each hypernym with the instance; if found, insert the value as a literal and finish. If not found, go to the next step.

Note that for step 3 we could consider an alternative of enhancing the ontology with elements along a path in the hierarchy. While this would require extra effort in reasoning, we at least know that the expansion is conducted in a structured way and the structure of the lexical database could be used as an extra factor in reasoning. We will now define the four levels of match in detail.

10.5.2 Direct Match

The easiest kind of match is one where the tag matches the instance of the ontology (see also Figure 10.6), and in this case we wish to insert the value from the tag directly as literal associated to the instance.

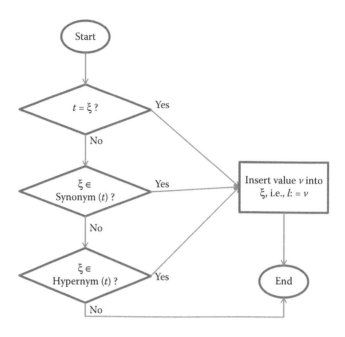

FIGURE 10.5
Overview of XML to OWL mapping methodology.

FIGURE 10.6
Direct match.

Definition: Direct Match
Let

- I be an instance of an ontology O
- l be a literal of instance I
- t be a tag in an XML schema Ξ
- v be a value for t

then

$$l = v \text{ iff } I = t.$$

10.5.3 Synonym Match

The second type of match is where we find that a synonym of the tag (e.g., location and placement can be seen as synonymous in this context, see Figure 10.7) matches the instance in the ontology.

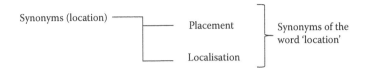

FIGURE 10.7
Synonym match.

Definition: Synonym Match
Let

- I be an instance of an ontology O
- *l* be a literal of instance I
- *t* be a tag in an XML schema Ξ
- *v* be a value for *t*
- *synonym(t)* be a set of synonyms for *t* as obtained from a lexical database

then

$$l = v \text{ iff } I \in synonym(t).$$

10.5.4 Hypernym Match

The third type of match is where no direct match and no synonym match can be found. Existing work merges tags by looking at the super-tags and inserting that value (thus assuming a semantic relation between the super- and sub-tag). For example, if a person tag has a position sub-tag and the position would not exist, the value would be inserted with a person class, rather than with location as would be desirable. The alternative is to create a new ontology model.

Both are not suitable here—the former for reasons indicated above and the latter because we have a fixed context model and we consider the structural model to fulfil the needs of a context-aware system making use of software sensors, and hence we need to instantiate the model.

The proposed hypernym match (see Figure 10.8) is generic and could be used for any purpose where XML data needs to be inserted into fixed ontologies.

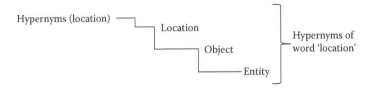

FIGURE 10.8
Hypernym match.

Definition: Hypernym Match
Let

- I be an instance of an ontology O
- *l* be a literal of instance I
- *t* be a tag in an XML schema Ξ
- *v* be a value for *t*
- *hypernym(t)* be a set of hypernyms for *t* (that is, the branch of the tree in which *t* is found) as obtained from a lexical database

then

$$l = v \text{ iff } I \in hypernym(t).$$

The hypernyms are obtained from a lexical database (e.g., WordNet) and checked for match with the instance; if a match is found, the value is inserted. The hypernyms are processed in order, going upwards in the hierarchy of words and the first match will be chosen (as clearly that will be closest in meaning to the tag).

10.5.5 No Match

We can arrive at a situation where none of the above steps can achieve a match for the data item from the XML file. While it seems desirable to match as much information as possible, there might be data that simply does not add to the context, and hence it is safe to ignore that. Furthermore, even if a data item could be adding to the context, one should keep in mind that any context data gathered is often not 100% accurate and could be contradictory to information simultaneously gathered, changes quickly and hence has a limited period of validity and is possibly made more or less reliable through reasoning anyhow. In light of these missing factors, some items can be seen as a negligible problem.

10.6 Sample Scenario

A research student wants to meet with his supervisor to discuss his work without prior arrangement but does not know about the availability of the supervisor. He is, of course, aware that the supervisor is busy, with involvement in research projects and teaching as well as administrative tasks.

Meanwhile, the supervisor is working on a computer on one of his research projects and does not want to be interrupted; however, he has not updated his current context.

In these kinds of real-world situations, one is not only concerned if the person is available—actually available can have many meanings (someone is 'in', 'free', 'could be disturbed in an emergency', etc.). One could update one's current context, which is tedious and needs to be repeated every time the context is changed (and hence probably is not done at all), or provide some means through which the context is automatically updated and can be queried by authorised users—this work clearly supports the latter.

10.6.1 Queries

We have discussed the context model and the insertion of context earlier, so here we are looking at making use of it. A typical question to ask in the context of the scenario might be *What is my supervisor doing on 22nd Feb 2011 from 09:00 to 11:00?*. We assume here that the person asking, that is, the research student, is authorised to ask about the teacher.

Formally, as we are using an RDF-based ontology, we can ask questions such as this through the SPARQL query language:

```
PREFIX rdf: <http://www.w3.org/1999/02/22-rdf-syntax-ns#>
PREFIX ctx: <http://www.cs.le.ac.uk/ontology/contextmodel#>

Select ?faculty ?act WHERE{

#all scheduled activities falls within the time span

  ?faculty ctx:hasActivity ?act.
  ?faculty ctx:supervisorOf ctx:a_student.
  ?act ctx:hasTimespan ?ts.
  ?ts ctx:startsAt ?start_time.
  ?ts ctx:finishesBy ?end_time.

  FILTER ((?start_time >="2011-22-02T9:00:00"^^xsd:dateTime
       && ?start_time <="2011-22-02T11:00:00"^^xsd:dateTime)
    || ("2011-22-02T09:00:00"^^xsd:dateTime >=?start_time
       && ?"2011-22-02T09:00:00"^^xsd:dateTime <=?end_time))

}
```

This SPARQL query looks very much like an SQL query, and asks about a faculty member's activity at a particular time—some of the arguments (values after the ?) could be instantiated with specific values, directly filtering the results to those matching.

10.6.2 Rules

Simple queries asking about the data stored in the model are easy to formulate and ask, but do not allow one to harness the full power of the semantic technologies at hand. We can enhance the richness of questions to be asked by allowing the use of additional reasoning rules that can combine existing facts to derive new facts. Many of these rules can be quite specific to certain domains, reflecting the interpretation of data in those domains. For example, in the university, there might be sufficient information on rooms and schedules available to derive that a professor being in a teaching room in the postgraduate block is probably teaching a postgrad class. Here are some of the rules that we used in the context of the scenario (we have seen rule 3 earlier):

$$hasPlace(?fm, ?pl) \wedge Time(?t) \rightarrow isTeachingClass(?fm, ?pc) \qquad (10.2)$$

where *fm* is a faculty member, *pl* is place name or name of a building, *t* is the time zone and *pc* is a postgraduate class.

$$hasTimetable(?fm, ?ocal) \wedge hasCalendar(?fm, ?pcal) \rightarrow isTeachingClass(?fm, ?pc) \qquad (10.3)$$

where *fm* is a faculty member, *ocal* is the organisational calendar (including a timetable), *pcal* is the personal calendar and *pc* is a postgraduate class.

$$hasTimetable(?fm, ?ocal) \wedge Weather(?loc) \rightarrow isTeachingClass(?afm, ?pc) \quad (10.4)$$

where *fm* is a faculty member, *ocal* is the organisational calendar (timetable), *loc* is a location, *afm* is another faculty member (distinct from *fm*) and *pc* is a postgraduate class.

10.6.3 Degree of Confidence

Having gathered data from different sources and being able to derive new conclusions on them, we should turn our focus to the issue of how certain we can be about the information derived. We term this concept *degree of confidence* as it should show how certain an enquiring user can be that the information she receives truly reflects the activity of the entity enquired about. This could, for example, be quite crucial if billing of consultants' time or travel arrangements to meet someone will be based on such activity sensing.

Returning to our example, a faculty member might have private and work calendars. If they are inserting events in their work calendar, which reflects working time activities, these are probably accurate for day time. However, social activities would usually take place outside working hours, and hence the private calendar might be more accurate for such activities. In large organisations, there might be further departmental and institutional calendars to be considered. So, to identify what a user is doing, it might be worthwhile considering which data source is most appropriate and should take priority in case of events being in each of them.

At this stage, we propose a simple ranking of data sources, which is dynamic over time in that different preferences are given at different times.

Following our example, assuming the user to have an organisation calendar (*OC*), a work calendar (*WC*) and a personal calendar (*PC*), we get the following ranking:

```
If (time is 8:00am and 5:30pm) and (day is Mon to Fri) then
OC > WC > PC;
Else
PC > WC > OC
```

This states that during normal working hours the organisation and work calendars are more reliable than the private calendar, while out of working hours the calendars that are under the user's control gain in relevance.

One could consider a more fine-grained approach, and indeed this is one of our aspects of further work. For example, we could calculate the confidence of having correctly identified activities during mapping from XML to OWL, based on specific instances and also on how certain we are about the match that has been made when inserting data.

10.7 Conclusion and Future Work

In this chapter, we have introduced software sensors as an inexpensive alternative to hardware sensors. They are particularly well suited to detecting user activities as they can exploit the multitude of data exchanged between users and the web-enabled services they use. Hardware sensors will still have their place and are orthogonal to software sensors in that sense: they are very well suited to obtain environment data such as temperature or physical location measurements. The exchanged data enrich the information available about user activities and hence will allow for more effective context-aware systems.

In the chapter, we introduced the fields of sensors for context-aware systems and context modelling. We presented our context model (an adaptation of ideas from the literature that suits our needs well). We discuss how SOAP messages can be extracted from the exchanged network messages as a first stage in sensing through software sensors. We have presented a solution to the problem of associating data from SOAP messages with elements in the context model—thus allowing the generation of knowledge to be used in the context-aware system. This technique for mapping data from the XML payload in SOAP messages into an existing OWL model to then allow for reasoning rules to use that data to answer user queries is a revised version from that presented in Ref. [8].

The current work is concerned with refining the implementation of the test system and conducting a more realistic case study. Future work will consider whether we can also use the data that are observed but are not captured by the current mapping process.

Two areas of study deserve more attention, which will be more important for software sensors, but also have implications for hardware sensors: (1) confidence in observations and (2) lifetime of information.

The former needs to study how much we can trust the information we have gained, that is, how reliable is the sensor in reporting high-quality information? This is very crucial, as reasoning about the data influences its reliability, and can do so by making it more reliable (two indicators showing the same fact are better than one), but can of course also degrade the quality (if we give high weight to bad information)—we are sure legal systems around the world cope with this problem too! Of course, the problem is aggravated for software sensors (measuring, e.g., the temperature using a hardware sensor is well understood), as we need to make decisions on data items and allocate them to specific categories.

The latter is also very interesting, as context data change, but they do not do so at the same rate for all elements, and even for a specific element, they might change at different rates: both a timetable and a calendar will provide insight into scheduled meetings; however, the former is usually more stable mostly for practical reasons in that it is shared by more people.

The final issue for future work—and something we deliberately kept out of the scope of our current work—is the matter of privacy. Obviously, any context-aware system will violate privacy in some way by tracking what people do. While many people will feel that this is objectionable, there are also a large number of people who make such information freely available on Twitter or Facebook. Coming full circle, it is again location that is at the forefront of the discussion in this area [44], but this discussion will need to be expanded in the future to cover activity sensing too.

References

1. M. Baldauf, S. Dustdar, F. Rosenberg: A survey on context-aware systems. *International Journal of Ad Hoc and Ubiquitous Computing*, 2: 263–277. 2007.
2. J. Budzik and K.J. Hammond: User interactions with everyday applications as context for just-in-time information access. In *Proceedings of the 5th International Conference on Intelligent User Interfaces (IUI '00)*. ACM, New Orleans, LA, pp. 44–51. 2000.

3. L. Finkelstein, E. Gabrilovich, Y. Matias, E. Rivlin, Z. Solan, G. Wolfman, E. Ruppin: Placing search in context: The concept revisited. *ACM Transactions on Information Systems.* 20(1): 116–131. 2002.
4. H.L. Truong, C. Dorn, G. Casella, A. Polleres, S. Reiff-Marganiec, S. Dustdar, S.: inContext: On coupling and sharing context for collaborative teams. In *Proceedings of the 14th International Conference on Concurrent Enterprising (ICE2008).* pp. 225–232. 2008.
5. Gmail. https://mail.google.com/
6. J.S. Wilson: *Sensor Technology Handbook.* Newnes, Burlington, MA, 2005.
7. S. Loke: *Context-Aware Pervasive Systems: Architectures for a New Breed of Applications.* Auerbach Pub, Boca Raton, FL, 2006.
8. K.T. Pathan, S. Reiff-Marganiec, Y. Hong: Mapping for activity recognition in the context-aware systems using software sensors. In *Proceedings of the 2011 IEEE Ninth International Conference on Dependable, Autonomic and Secure Computing (DASC '11).* IEEE Computer Society, pp. 215–221. 2011.
9. W. Mark: The computer for the 21st century. *Scientific American,* 265, 94–104. 1991.
10. M. Satyanarayanan: Pervasive computing: Vision and challenges. Personal communications 8(4). *IEEE,* 10–17. 2001.
11. A. Hopper: The Clifford Paterson lecture. Sentient computing. *Philosophical Transactions of the Royal Society A: Mathematical, Physical and Engineering Sciences,* 358: 2349–2358, 2000.
12. M. Raento, A. Oulasvirta, R. Petit, H. Toivonen. Contextphone: A prototyping platform for context-aware mobile applications. *IEEE Pervasive Computing,* 4(2): 51–59, 2005.
13. M. Solarski, L. Strick, K. Motonaga, C. Noda, W. Kellerer. Flexible middleware support for future mobile services and their context-aware adaptation. In F. A. Aagesen, C. Anutariya and V. Wuwongse, editors, *INTELLCOMM,* Vol. 3283 of LNCS, pp. 281–292. Springer, Heidelberg, Germany, 2004.
14. G.D. Abowd, A. K. Dey, P. J. Brown, N. Davies, M. Smith, P. Steggles. Towards a better understanding of context and context-awareness. In *HUC '99: Proceedings of the 1st International Symposium on Handheld and Ubiquitous Computing.* Springer, pp. 304–307. 1999.
15. ECOSPACE: eProfessionals Collaboration Space, http://www.ip-ecospace.org.
16. H.L. Truong, S. Dustdar, D. Baggio, S. Corlosquet, C. Dorn, G. Giuliani, R. Gombotz et al. inContext: A pervasive and collaborative working environment for emerging team forms. In *International Symposium on Applications and the Internet (SAINT 2008).* IEEE, Turku, Finland, pp. 118–125. 2008.
17. L.A. Klein, Sensor and Data Fusion Concepts and Applications, 2nd ed. SPIE Press, Bellingham, WA, 1999.
18. A. Ferscha, S. Vogl, W. Beer: Ubiquitous context sensing in wireless environments. In *Workshop on Distributed and Parallel Systems (DAPSYS).* Kluwer, pp. 98–108. 2002.
19. D. Luckham, *The Power of Events: An Introduction to Complex Event Processing in Distributed Enterprise Systems.* Addison-Wesley Longman, Amsterdam, 2002.
20. M. Tilly and S. Reiff-Marganiec. Matching customer requests to service offerings in real-time. In *Proceedings of the 2011 ACM Symposium on Applied Computing (SAC '11).* ACM, Taichung, Taiwan, pp. 456–461. 2011.
21. A. Schmidt, K.A. Aidoo, A. Takaluoma, U. Tuomela, K. Van Laerhoven, W. Van de Velde. Advanced interaction inContext. In *Proceedings of the 1st International Symposium on Handheld and Ubiquitous Computing (HUC '99).* Springer, pp. 89–101. 1999.
22. B.E. Brewington and G. Cybenko. Keeping up with the changing Web. *Computer* 33(5): 52–58, 2000.
23. M. Tilly, S. Reiff-Marganiec, H. Janicke. *Efficient Data Processing for Large Scale Cloud Services.* Services 2012–FM-S&C. IEEE, Honolulu, HI, 2012.
24. A. Harter, A. Hopper, P. Steggles, A. Ward, P. Webster. The Anatomy of a Context-aware Application. In *Proceedings of the 5th Annual ACM/IEEE International Conference on Mobile Computing and Networking (MobiCom '99).* ACM, pp. 59–68. 1999.
25. B.P. Clarkson: *Life Patterns: Structure from Wearable Sensors.* MIT, Cambridge, MA, 2003.

26. G.M. Voelker and B.N. Bershad: Mobisaic: An information system for a mobile wireless computing environment. *Kluwer International Series in Engineering and Computer Science*, 375–394. 1996.
27. B. Schilit, N. Adams, R. Want: Context-aware computing applications. In *Proceedings of the 1994 First Workshop on Mobile Computing Systems and Applications (WMCSA '94)*. IEEE Computer Society, pp. 85–90. 1994.
28. J. Bauer: *Identification and Modeling of Contexts for Different Information Scenarios in Air Traffic*. Technische Universität Berlin, Diplomarbeit. 2003.
29. K. Henricksen, J. Indulska, A. Rakotonirainy: Generating context management infrastructure from context models. In *4th International Conference on Mobile Data Management (MDM)*, Melbourne, Australia. pp. 1–6. 2003.
30. K. Henricksen, J. Indulska, A. Rakotonirainy: Modeling context information in pervasive computing systems. In *Proceedings of the First International Conference on Pervasive Computing*. Springer, pp. 167–180. 2002.
31. J. McCarthy: Notes on formalizing context. In *Proceedings of the 13th International Joint Conference on Artificial Intelligence–Volume 1 (IJCAI'93)*. Morgan Kaufmann Publishers Inc., pp. 555–560. 1993.
32. T.R. Gruber: A translation approach to portable ontology specifications. *Knowledge Acquisition*, 5(2): 199–220, 1993.
33. P. Otzturk and A. Aamodt: Towards a model of context for case-based diagnostic problem solving. In *Proceedings of Incontext 1997*, pp. 198–208. 1997.
34. H. Chen, T. Finin, A. Joshi: Using OWL in a pervasive computing broker. Defense Technical Information Center. 2003.
35. T. Gu, X.H. Wang, H.K. Pung, D.Q. Zhang: An ontology-based context model in intelligent environments. In *Proceedings of Communication Networks and Distributed Systems Modelling and Simulation Conference*, San Diego, CA. pp. 270–275. 2004.
36. H. Bohring and S. Auer: Mapping XML to OWL ontologies. *Leipziger Informatik-Tage*, LNI Vol 72, 147–156, 2005.
37. R. Ghawi and N. Cullot: Building ontologies from XML data sources. In *20th International Workshop on Database and Expert Systems Application, 2009. DEXA '09*. pp. 480–484, 2009.
38. T. Rodrigues, P. Rosa, J. Cardoso: Mapping XML to existing OWL ontologies. *International Conference WWW/Internet 2006*. pp. 72–77. 2006.
39. Fellbaum, C.: *WordNet: An Electronic Lexical Database*. MIT Press, Cambridge, MA, 1998.
40. R. de Freitas Bulcao Neto and M. da Graca Campos Pimentel: Toward a domain-independent semantic model for context-aware computing. In *Proceedings of the Third Latin American Web Congress(LA-Web'05)*. IEEE. 2005.
41. T. Gu, H.K. Pung, D.Q. Zhang: Toward an OSGi-based infrastructure for context-aware applications. *IEEE Pervasive Computing* 3(4): IEEE, 66–74, 2004.
42. E. Prud'hommeaux and A. Seaborne: SPARQL Query Language for RDF. http://www.w3.org/TR/rdf-sparql-query/. W3C. 2008.
43. N. Mitra and Y. Lafontitle: Soap Version 1.2 Part 0: Primer. http://www.w3.org/TR/soap12-part0/. W3C. 2007.
44. Stephen B. Wicker. The loss of location privacy in the cellular age. *Communications of the ACM* 55(8): 60–68, 2012.

11

Multi-Agent Framework for Distributed Leasing-Based Injection Mould Remanufacturing

Bo Xing, Wen-Jing Gao and Tshilidzi Marwala

CONTENTS

11.1 Introduction 267
11.2 Motivations 269
    11.2.1 Leasing 269
        11.2.1.1 Is Leasing 'Greener' than Selling? 269
        11.2.1.2 Eco-Leasing 270
    11.2.2 Remanufacturing 270
        11.2.2.1 What Is Remanufacturing? 270
        11.2.2.2 Why Remanufacturing? 272
        11.2.2.3 Obstacles to Remanufacturing 272
        11.2.2.4 Who Is Remanufacturing? 274
        11.2.2.5 Case Study: Remanufacturing Strategy Embedded in Leasing Products 276
    11.2.3 Role of Mould Leasing and Remanufacturing 276
        11.2.3.1 Process Description of Leasing-Based Injection Mould Remanufacturing 276
11.3 Multi-Agent Framework for Leasing-Based Injection Mould Remanufacturing 278
    11.3.1 Multi-Agent System 278
    11.3.2 Electronic Commerce 279
    11.3.3 Proposed Multi-Agent Framework 279
        11.3.3.1 Description of Participants 280
        11.3.3.2 Description of Business Unit 280
        11.3.3.3 Description of Production Planning and Scheduling Unit 281
        11.3.3.4 Description of Warehousing Unit 281
        11.3.3.5 Description of Supplier Unit 282
    11.3.4 Cooperation and Interaction Mechanism 282
11.4 Collaborative Leasing-Based Injection Mould Remanufacturing Process Analysis..... 283
11.5 Conclusions 285
References 286

11.1 Introduction

The last few decades have seen a significant improvement in the quality of life through mass production and advanced technologies. However, this has led to a sharp increase in the strain on the environment in the meantime because though the phenomenon of industrialisation increases the quality of life for many, the costs of supporting it include the mass consumption of materials and energy, which not only pushes the exploitation of natural resources and the capability of biosphere to the limit but also directly leads to the mass disposal of retired products as waste when they do not generate any value for users. Consequently, the traditional ways of business thinking, which is through sell/purchase and then 'forget', meaning that the seller's responsibility ends at the point of sale, are questioned. There is a need to assess both the economic and environmental values and impacts of business decisions through identification of the 'whole of the life' costs and services associated with the products sold/leased [1,2].

To address the growing problems of diminished natural resources and increased waste, governments around the world have established or proposed stricter legislation to prevent the open-loop 'sell and forget' mode of transacting for a producer. That is, producers are required to take responsibility for their products at the end-of-life (EoL). Specific examples of closing such waste-loops cited by Ref. [3] include the take-back obligations on the waste electronic and electrical equipment (WEEE) directive in Europe, extended producer responsibility (EPR) in Japan and Taiwan, and voluntary product stewardship in Australia. These external influences make the companies change their business models in which they replace product ownership with creative service offerings [4]. In this view, products should not be valued for their material make-up but for the services that they provide to the user [5].

One example of this thinking is when the product is linked to a functional sale (e.g., leasing) offer. Research has found that usage-phase impacts play a major role in determining the advantages realised by leasing [6]. In 2002, the equipment leasing association estimated that of the total investment in business equipment by companies, nearly one-third, or $204 billion, would be financed through leasing [3]. Products such as electronic equipment that continually consume energy negatively impact the environment much more during their use than during manufacturing. Proponents of leasing claim that by maintaining ownership of the product the manufacturer can successfully put in place a service strategy to preserve EoL value and a product recovery strategy consisting of reuse, remanufacturing and recycling.

In this chapter, we look at the mould industry to illustrate the challenges of the emerging business model (i.e., leasing-based model) that can be found in many other industries as well. The design and manufacturing of moulds represent a significant link in the entire production chain, especially in the automotive industry, consumer electronics and consumer goods industries because nearly all produced discrete parts are formed using production processes that employ moulds [7]. Moulds, similar to machine tools, may represent a small investment compared to the overall value of an entire production programme. However, the maximum value of firms depends on the characteristics of moulds as the quality and lead times of moulds affect the economics of producing a very large number of component, subassemblies and assemblies. Nowadays, there are very few industries that can justify the purchase of a modern moulding machine to manufacture the same part over its entire lifetime. The largest consideration of a mould to survive for thousands of moulding cycles is the activities involved in mould maintenance and modifications [7]. Under these circumstances, mould leasing, as an alternative solution for mould maintenance and modifications, is gaining more and more attention during the past few years.

Briefly, this chapter is organised as follows. Section 11.2 includes a background and literature review about leasing, and gives examples of papers within the context of remanufacturing. In Section 11.3, we introduce the problem and the model formulation. Then, we outline the implement environment in Section 11.4. Finally, Section 11.5 draws conclusions of the work presented in this chapter.

11.2 Motivations

11.2.1 Leasing

11.2.1.1 Is Leasing 'Greener' than Selling?

In addition to consumer goods, capital goods and durable goods also offer a great potential for ecological and economic optimisations. It can be described as a lessee acquiring the use of an asset from a lessor for a period of time in exchange for a regular lease payment. In addition, the lessor, for legal as well as financial reasons, remains the owner of the leasing object, and takes over maintenance, service, as well as guarantee for the leased-out object. Examples of these include daily necessities such as cars, furniture, computers and other electronic appliances.

Normally, leases are often categorised as operating leases or financing leases. The former refers to arrangements where the term of the lease is substantially less than the useful life of the product and the cost of the lease is much less than the full acquisition cost of that product. The latter refers to an intermediate or long-term commitment by both parties to the lease. Full-payout leases, in which the period of the lease is structured so that the lease ends when the product is at the EoL, are not typical.

Over the years, a growing number of advocates have claimed adopting leasing as a business model as being more 'green'. The practice of leasing products, rather than selling them, is a strategy for increasing resource productivity by moving to a pattern of closed-loop material use by manufacturers [8,9]. The authors of Ref. [10] analysed the problems associated with marketing a durable through leases and sales. Their goal was to understand the strategic issues associated with concurrently leasing and selling a product and determining the conditions under which this concurrent strategy was optimal. Later, the same authors [11] examined competition in a duopoly; they investigated a firm's rationale in choosing an optimal mix of leasing and selling and to understand how it was affected by the nature of competition in the market and the embedded quality in the product. They argued that a competitive environment forced firms to adopt strategies where they only sell their products or use a combination of leasing and selling. In addition, the authors of Ref. [12] identified how a durable goods manufacturer can use a combination of leasing and selling to balance its strategic commitment across both its own market and the complementary market.

However, there have also been claims contrary to this, and numerous complications to leasing exist that may hinder its potential environmental improvements [13]. Thus, the answer to the question 'Is leasing greener than selling?' is not clear cut. At least, leasing does not automatically bring about the environmental goals sought by EPR. The success or failure of leasing as a business depends critically on the leasing parties' ability to forecast the residual values of the leased products [14]. Furthermore, understanding how and when leasing may contribute to reduced energy consumption of resource use is vital for businesses to reduce their overall impact. Meanwhile, knowing how certain product characteristics affect the manner in which leasing may influence energy usage is also important.

11.2.1.2 Eco-Leasing

Recently, there is a new terminology called 'eco-leasing' presented in Ref. [15]. The authors point out that 'eco-leasing' needs to be distinguished from traditional leasing in at least two ways. First, there should be no option of final purchase of the product so that the responsibility for the product always stays with the provider. Second, there needs to be a close relationship between the producer and customer for repair and service requirements. In light of this statement, remanufacturing within the eco-leasing concept represents a valuable opportunity to overwhelm such challenges.

Xerox provides an example. Since 2001, when Xerox started to provide leasing services, the company made several structural organisational changes, such as having its asset recovery engineers work directly with product designers, developing a hierarchy of options for managing EoL products that places reuse and remanufacturing ahead of recycling, and building EoL factors into criteria for new product design. These changes led the company to a 'win–win' position. Overall, the green manufacturing programme saves the company 40–65% in manufacturing costs through the reuse of parts and materials [16]. Similar activities are undertaken by Hewlett Packard Corporation for computers and peripherals and by Canon for print and copy cartridges.

Overall, the economic and environmental benefits of this concept can be summarised as follows:

- Benefits to the leasing/remanufacturing company such as additional sales of interior upgrade parts or replacement interiors; promotes ongoing relationships and brand loyalty; closes material loop to lower material costs and increases profit margin; steady income stream solidifies budget forecasting; and eco-friendly equity.
- Benefits to the customer such as less responsibility; option to replace broken parts or upgrade to new interior; eco-friendly satisfaction; service available, more materially intensive product, better durability; and operating leases can have accounting and tax advantages for some customers by shifting costs to the operating budget from the capital budget, conserving cash.

11.2.2 Remanufacturing

11.2.2.1 What Is Remanufacturing?

Remanufacturing—the industrial process that returns non-functional products or modules to 'like new' conditions—is part of the sustainability equation because it can lend itself to sustainability practices by giving products an extended life cycle via incremental upgrades. Recently, remanufacturing has been receiving growing attention for various reasons such as consumer awareness, environmental concerns, economic benefit and legislative pressure. Remanufactured products include photocopiers, computers, cellular phones, automotive parts, office furniture, tyres and machine tools/equipments. For example, a worn-out vending machine can be remanufactured to a better-than-new condition by replacing worn parts and equipping it with the latest circuit boards, coin-detection technologies and exterior graphic panels for a fraction of the price of a new unit. This process can be repeated as necessary.

Currently, there is no single standard definition of remanufacturing existing in the literature and we have listed some of the definitions as follows:

- *Definition 1*: 'Remanufacturing is an industrial process whereby used products referred to as cores are restored to useful life. During this process, the core passes through a number of remanufacturing operations, e.g. inspection, disassembly, component reprocessing, reassembly, and testing to ensure it meets the desired product standards. This could sometimes mean that the cores need to be upgraded and modernised according to the customer requirements' [17].

- *Definition 2*: 'Remanufacturing is an end-of-life strategy that reduces the use of raw materials and saves energy while preserving the value added during the design and manufacturing processes' [18].
- *Definition 3*: 'Remanufacturing can be seen then as an advantageous product recovery option. Not only is it the case, as it is with other options e.g. recycling, that less waste must be landfilled and less virgin material consumed in manufacturing but also the value added in the manufacturing of the components is also "recovered". It also saves the energy needed to transform and sort the material in recycling products' [19].
- *Definition 4*: 'Remanufacturing is the ultimate form of recycling. It conserves not only the raw material content but also much of the value added during the processes required to manufacture new products' [20].
- *Definition 5*: 'Remanufacturing is particularly well suited for end-of-life products that include components characterised by long technology cycles and low technological obsolescence, and when ex ante uncertainty regarding usage intensity results in "over-engineering for certain user groups in order to meet the needs of other user groups"' [21].
- *Definition 6*: 'Remanufacturing is a basis for an innovative paradigm shift in the manufacturing industry—from selling physical products to supplying services through product systems' [22].
- *Definition 7*: 'Remanufacturing has been considered as the transformation of used products, consisting of components and parts, into units which satisfy exactly the same quality and other standards as new products' [23].
- *Definition 8*: 'Remanufacturing is rapidly emerging as an important form of waste prevention and environmentally conscious manufacturing. Firms are discovering it to be a profitable approach while at the same time enhancing their image as environmentally responsible, for a wide range of products' [24].
- *Definition 9*: 'Remanufacturing is an efficient environmental programme, not just as a cost-effective means to reduce waste, but as an integral part of the firm's manufacturing and marketing strategy' [25].
- *Definition 10*: 'Remanufacturing preserves the product's identity and performs the required disassembly, sorting, refurbishing and assembly operations in order to bring the products to a desired level of quality' [26].

Though remanufacturing has been interpreted from various viewpoints, these definitions all share some similarities: First, most researchers agree that used products are the main input for remanufacturing. They play the role of raw materials in traditional manufacturing. Second, except going through some traditional manufacturing shop floor procedures such as assembly and testing, unlike normal raw materials, some additional operations such as disassembly, cleaning, inspection and component reprocessing must be done on used products before they can be put into production. Third, the majority of definitions conclude that remanufacturing is not performed at the expense of losing product quality and this unique characteristic distinguishes remanufacturing from other confusing terms such as reconditioning, repair and rebuild.

11.2.2.2 Why Remanufacturing?

The reasons for adopting remanufacturing are manifold:

- On enterprise side: Green image can help remanufacturing companies to distinguish themselves from their competitors; valuable data can be gathered through remanufacturing so that the original product design and functionality can be improved; new business opportunities are created for the after sale service market by offering customers new low-cost solutions with remanufactured products.
- On customer side: A lower price, typically 40–60% less than similar new products, is a great enjoyment for customers to embrace remanufacturing [27].
- On community side: Owing to its labour-intensive nature, remanufacturing can create more job positions for the employment market; meanwhile, it also serves as a forum for workers' problem-solving skills, being more rewarding than traditional production-line jobs.
- On policymaker side: Several recently passed EPR directives such as WEEE and end-of-life vehicle (ELV) have heralded the start of a new era of waste management policy for durable goods worldwide. In this context, remanufacturing, as a means of meeting these legislations' requirements, may help governments to gain insights of the early impacts of EPR directives.
- On environment side: Across all life cycle stages (e.g., beginning-of-life, middle-of-life and end-of-life), design decisions influence the resulting cost and environmental impact of a product. The reason that used product remanufacturing is a meaningful subject lies in the fact that this strategy removes some burden from the life cycle cost and environmental impact by eliminating the need for new materials and components for future products.

11.2.2.3 Obstacles to Remanufacturing

Although there are good reasons to get involved in the remanufacturing practice, at the same time, there are many obstacles to the development of remanufacturing, which limit their implementation. One of the major issues faced by the firms involved in remanufacturing is taking back used products before the end of their useful life so that some revenue can be generated by remanufacturing or reusing them. Another concern is the uncertainty in the quantity, timing and quality of returned products [4].

11.2.2.3.1 Economic Viability of Remanufacturing

Certainly, determining if remanufacturing is even economically viable is important. The authors of Refs. [16,28,29] sought to develop mathematical models for determining the profitability of supply chains with remanufacturing. Focus is on the relationship between 'pricing decisions in the forward channel and the incentives to collect used products under different reverse channel structures'. Other scholars have attempted to use game theory/competition concept to model remanufacturing decisions in order to affect the profitability of remanufacturing. For example, the authors of Ref. [30] examined how third-party remanufacturing can induce competitive behaviour when the recycled products cannibalised the demand for the original products. Numerous scenarios were considered in this paper. The most general results agreed with the argument for remanufacturing as a means of increasing a firm's revenue by opening new secondary markets.

In the same vein, Ref. [31] also focuses on this phenomenon. The authors investigated the effects remanufacturing has on the sale of new products, a reason why many firms have chosen not to remanufacture their products. In order to do that, they developed a model to determine which conditions were necessary for remanufacturing to be profitable, despite the possible loss of sale of new products. The key results determined that it is more profitable to enter the remanufactured products market as this acts as a deterrent for new entrants into the used goods market. Also, the lower-cost remanufactured goods can open a secondary market for buyers not willing to pay for full-priced new goods. Perhaps most interestingly, the authors found that 'when collection is the major portion of the total remanufacturing fixed and/or variable cost, the OEM is better off remanufacturing'.

In addition, Ref. [32] discussed the factors that influence the remanufacturing environment, focusing on the managerial importance. The authors continued the discussion of the effect of remanufactured goods on the sale of new goods by using the term 'market segmentation'. Numerous important factors were taken into account in their model, including consumer's preference for new goods over used goods, market competition and their party remanufacturers, and the OEM's control over the product design for remanufacture. The goods were assumed to exist for one remanufacturing cycle only, and the cost to do so was constant over time. With proper designing, a good can be made to be reused numerous times, and improved technology can make remanufacturing less costly over time as well. Given this analysis, the argument for remanufacturing was very strong.

11.2.2.3.2 Managing the Uncertainty of Returned Products
Most of the literature focusing on supply chains and remanufacturing has assumed product returns to be an exogenous process, meaning that the company has no control over the quality and quantity of returned products [33,34]. One of the first papers to investigate the accuracy of this assumption is in Ref. [35]. The authors argued that companies can actively manage product returns, rather than accept them passively. Successfully controlling the quality of the products returned can be done using a market-driven approach in which the end-users are motivated to return EoL products by financial incentives, including deposits, credit towards a remanufactured or new unit, cash for product returns and leasing [36]. This assertion is expanded on a follow-up paper [37] by the same authors. They developed a framework for determining optimal prices and the corresponding profitability when a firm can proactively influence product returns.

Nowadays, several firms that are actively involved in remanufacturing recover their used products by leasing their new products. Product leasing results in the most predictable return stream and is a popular option for firms that also remanufacture [38]. That means, leasing helps a firm in getting a consistent flow of used products for remanufacturing, which in fact reduces the uncertainty in the quantity and timing for returns [39]. In addition, product characteristics affect the viability of a leasing. The authors of Ref. [10] contended that the relative profitability of leasing and selling hinged on the depreciation rate of the asset. A lower depreciation rate in leasing produced financial benefits as it prevented a product from losing significant reuse or remanufacturing value at the end of its life. This means an increased residual value, resulting in an increased recovery value or resell/release value for the product.

So, at this point, leasing turns out to be a viable strategy that helps to better manage the return process and prevent the residue value as well.

11.2.2.4 Who Is Remanufacturing?

Despite the above-mentioned barriers, remanufacturing, as an environmentally benign alternative, is still experiencing a rapid development during the past decades. Two major forms of remanufacturing firms exist: third-party remanufacturers and OEM remanufacturers. Third-party remanufacturers, also known as independent remanufacturers, do not manufacture the original product. Remanufactured products are usually sold to the replacement parts stores or are contracted by OEMs to remanufacture replacement parts and act as suppliers for OEMs who sell remanufactured items through their existing dealer networks (Bras, 2007).

Xerox [30,36,40–42] is a global company offering products and services for printing, publishing, copying, storing and sharing documents. These include copiers, printers, scanners, fax machines and document management software. It is interesting that despite numerous ownership and management changes, the company's commitment to remanufacturing has remained strong since the 1960s. Therefore, Xerox is often quoted as an excellent example of an environmentally and economically successful remanufacturing system. Xerox's remanufacturing system is considered by some to be the 'modern classic' example of remanufacturing, which is 'by far the most advanced'. By remanufacturing used copiers, Xerox has saved millions of dollars in raw material and waste disposal costs. Some analysts claim that Xerox's success is due to the fact that its products are robust, large, easy to disassemble and valuable when remanufactured.

Single-use camera remanufacturing [43,44] is a frequently mentioned example. In Japan, Fuji Film [21,41,45,46] developed a remanufacturing system for single-use cameras. Although film cameras have been replaced by digital cameras in recent years, the case provides useful information. The company developed a fully automated system that disassembles collected used cameras, inspects and cleans the parts, and reassembles the cameras. The flash, battery, plastic and mechanical parts are reused. The company reports that more than 82% of all camera components collected, by camera weight, is reused or recycled. The products were designed to make automated remanufacturing possible.

Caterpillar's global remanufacturing business [47–49] is currently one of the largest in the world in terms of volume, recycling more than 50,000 tonnes of products (over 2.2 million EoL units) each year. Through their large-scale activities, Caterpillar has helped change the business of remanufacturing. Caterpillar first entered remanufacturing in 1972 and now, Caterpillar remanufacturing services is one of its fastest-growing divisions—annual revenue is over $1 billion and is reputedly growing at 20% a year. In 2005, Caterpillar spent $1.5 billion on purchasing remanufacturing facilities around the world so that Caterpillar could remanufacture in those markets. Caterpillar now has remanufacturing facilities across the globe, including Shrewsbury, UK and Nuevo Laredo, Mexico; Shanghai, China was the 14th facility and was opened in early 2006.

Swedish forklift truck manufacturer BT Products (BT) [17,50] is another example. BT has built a system of contracts where the customer can choose from different kinds of rental programmes according to their needs. For these programmes, the customer never owns the forklift trucks; at the end of the contract, the forklift is returned to BT for remanufacturing and eventually service for a new customer in a new contract. According to BT, remanufacturing volumes have been doubled during the past few years and they currently exceed the number of forklift trucks being newly produced in the ordinary manufacturing facility.

In 1850, the Flen plant [51–53] started to manufacture harvest machines and nearly 100 years later, in 1960, the production was replaced by the remanufacturing of automotive and marine engines. As of 1998, Volvo Flen is a fraction of Volvo Parts, a business unit of the Volvo Group. Volvo Parts supports the six business areas: Volvo trucks, Renault trucks, Mack trucks, Volvo Penta, Volvo buses and Volvo Construction Equipment. It provides services and tools for the aftermarket throughout the whole supply chain. The vision of the Volvo Parts division is to be number one in the after sales market and to be perceived as easy to do business with. Volvo Parts Flen AB has 220 employees and total sales are approximately 55 million Euros a year. The main activities are remanufacturing of petrol and diesel engines for trucks, buses and cars where remanufacturing of bus, truck and car engines account for half of the total sales and the remaining part is made up of remanufacturing and manufacturing water pumps and packaging of cylinder liner kits.

RetreadCo [54] is a wholly owned subsidiary of NewTireCo, a major European-based new tyre manufacturer. RetreadCo has operations in France, Germany, Belgium and Luxemburg. Its main plant retreads more than 4500 tyres per day and employs 700 people in manufacturing or commercial functions. RetreadCo retreads passenger car, van, heavy truck and earthmover tyres and produces retread rubber for export. RetreadCo retreads used tyres of almost every brand and uses a licensed technology for used NewTireCo tyres. With the acquisition of RetreadCo, NewTireCo obtained a strong position in the European retread market. Similarly, in North America, NewTireCo entered the retread market by acquiring an independent tyre dealer with retread operations.

In Germany, Mercedes-Benz (MB) [21,55] offers the owners of MB cars, vans or trucks the option of replacing their present engine with a remanufactured engine, of the same or different type, with the same quality as a new engine, but for a price 20–30% lower than the price of a similar new engine. This offer holds for at least 20 years after a new car, van or truck has been purchased. MB offers similar options for water pumps, crank cases, crank shafts and other parts produced by MB itself. Meanwhile, from the mid-1960s, BMW initiated a programme to recondition and remanufacture high-value used components for resale as used parts. The resale of these remanufactured parts is at 50–80% the price of new parts, and includes a notable profit margin for the company (taking into account the costs of labour).

In South Africa, RemTec [56] is a leading remanufacturer of petrol and diesel engines for light commercial and passenger vehicles. Originally established in 1963 as Volkswagen Remanufacturing to remanufacture engines for VW SA, the facility was purchased by a Port Elizabeth-based concern in July 1995 and relocated from Uitenhage to Port Elizabeth. RemTec is currently the exclusive, approved supplier of remanufactured engines to Volkswagen SA, General Motors SA, Ford SA, Nissan SA and Land Rover SA. Its Port Elizabeth plant was the first official large-scale remanufacturer in South Africa and occupies 4000 square metres. The layout has been specifically designed to optimise 40 years of experience in the remanufacturing business and has been used by other original equipment remanufacturing plants throughout the country. A RemTec premium engine carries a benchmark 'best in industry' 100,000 km/12 month warranty, which is the same as that of a brand new engine. When compared with a new engine, a RemTec premium engine is much more affordable—at roughly half the price.

In March, 2011, Mercedes-Benz South Africa (MBSA) [57,58] launched a programme that would offer remanufactured truck and car parts to customers at a 5–30% discount compared with new parts, but with the same 12-month warranty currently available on new parts. 'Quality is the same and the warranty is the same, but the price is lower' is the main characteristic of this programme. There were 102 part numbers available in the initially launched remanufacturing programme such as starter motors and air pumps and all of them are sourced from Germany. The aim of this programme is to offer South African customers a great, affordable alternative to new parts, while in the meantime, to protect them against dangerous grey parts.

11.2.2.5 Case Study: Remanufacturing Strategy Embedded in Leasing Products

The authors of Ref. [59] provided an example of Pitney Bowes, which manufactures mailing equipment that matches customised documents to envelopes, weighs the parcel, prints the γ age and sorts mail by zip code. The company leases about 90% of its new product manufacturing, and sells the remainder. At the end of a leasing contract (normally for four years), customers often upgrade to newer-generation equipment if it is available. In these cases, the customer returns the used equipment to the company, which tests and evaluates the condition of the used machine, and makes a disposition decision: scrap for material recovery (recycling), which is done for the worst-quality returns; dismantle for spare parts harvesting, which is done for medium-quality returns; or (potential) remanufacturing, which is done for the best-quality returns.

In a similar vein, the authors of Ref. [60] proposed a business model that integrated leasing and remanufacturing strategies for baby pram manufacturers. They argued that, with appropriate design changes that enable easy and cheaper remanufacturing, utilising the leasing option was expected to generate long-term sustainable profits, which otherwise would be lost to the second-hand market. Furthermore, the model proposed by [61] allowed electronic equipment leasing companies to simultaneously make optimal decisions about lease lengths, product flows and EoL product disposal. And the authors of Ref. [39] considered a company that leased new products and also sold remanufactured versions of the new product that become available at the end of their lease periods.

11.2.3 Role of Mould Leasing and Remanufacturing

Moulding plays a paramount role in our daily life. This extremely popular mass production tool is widely relied upon because of its ability to produce dimensionally stable parts for many cycles and to do so at very little cost per unit. Traditional moulding is often dedicated and expensive. Any design changes lead to the tooling becoming unsuitable for its specific use, and a new one has to be made. Consequently, the increased variety of moulds makes them less affordable for some customers and alternatives to the ownership-based use of moulds might be developed. In view of this fact, recently, a new trend of focusing to create a marketable reuse/remanufacturing of moulds by using modular design and organisationally closed distribution concepts has become more and more popular. This is supported by research conducted by INFORM, which argues that leasing can be used to manage the reverse logistics involved with remanufacturing [8]. The report on leasing found that leases 'can increase the probability that a company will own and be responsible for managing its products at the EoL and internalise the costs of doing so'. Leasing provides both a motive for product take-back and a structure to aid in the logistics involved with closed material loops.

11.2.3.1 Process Description of Leasing-Based Injection Mould Remanufacturing

Injection moulding, where injection moulds are widely utilised, is a manufacturing method by which large quantities of small- to medium-sized objects can be produced from low-melting-point materials.

From food containers to computer cases, many of the plastic products that we use today are manufactured by injection moulding. Liquid material (usually plastic, zinc, brass or aluminium) is injected into a mould constructed from steel or aluminium. These moulds contain the liquid material while it solidifies within a cavity. After the part has cooled, the mould opens up and the part is ejected [62]. The harvested parts can be either final products or components of assembled products. Typically, mould design and fabrication consists of a set of key stages that usually take several months to complete. A great deal of design knowledge and manufacturing know-how always complicates the production of a new mould.

In an ideal condition, an injection mould would be discarded when it completes its design lifetime. However, owing to the requirement of mass customisation, an injection mould always becomes obsolete earlier than it is expected. For instance, in some cases, the original product design is largely retained and only minor changes are needed, say the position of the company's logo. Under these circumstances, only the two blocks of blank mould need to be remachined while the other parts of the mould can be retained. This is one of several scenarios where a remanufacturing strategy can exactly fit into. For the rest of this section, we will describe how our example mould is remanufactured.

The mould remanufacturing process consists of reverse logistics (i.e., used mould collection), reprocessing (e.g., disassembly, cleaning, inspection, remachining, reassembly and testing) and remanufactured mould redistribution. First, after the injection mould is returned to the remanufacturer through the reverse logistics system, a partial or complete disassembly will be performed.

Disassembly is normally motivated by two goals—to obtain pure secondary materials and to isolate environmentally relevant materials from other materials. The disassembly process needs facilities and is normally done either manually or automatically. This process can be viewed as new technologies on the industrial level, where the industry itself has a pioneering role in setting new standards and creating new solutions. Meanwhile, they require know-how to maintain high productivity. To decide which mould is to be assigned to which process, certain constraints, for example, the availability of the demand for components, the cost for and number of remanufacturing resources need to be considered. Furthermore, the material that is not disassembled can be recycled through mechanical processing lines. Then, the disassembled components will be cleaned. According to Ref. [63], the cleaning operation is the most time consuming and entails much more than just removing dirt; it also means de-greasing, de-oiling, de-rusting and freeing components from paint.

Right after the cleaning stage, a thorough inspection will be performed to identify the potential damages to the key components. For example, owing to the impact of high working temperature and high working pressure, the strength of some components in an injection mould (e.g., carrier ejector and sprue side in our case) might be degraded. The remanufacturing plan of two blocks of blank mould will also be determined at this step (e.g., purchasing raw material to build new blocks or using state-of-the-art technology to fix them). This inspection process results in different sequences and times for the reprocessing procedure.

After the remanufacturing plan has been confirmed, remanufacturable components will be remachined to restore to its desired working condition. Geometrical change of the components through metal cutting like grinding will change the dimensions. Sometimes, after remachining, for example, a highly worn-out product will not match the standard tolerance, such as the diameter of a pin, and must be scrapped. The replacement components are supplied by either external procurement or internal retrieval of components from other used injection moulds. In the case where parts have to be purchased, some additional issues might be generated due to reasons such as (1) long lead times for purchased products, (2) single supplier for parts or components, (3) poor visibility of requirements, (4) small purchase quantities leading to unresponsive vendors, (5) parts no longer in production and (6) vendor's minimal purchase requirements.

Once the remachining operations are done, the injection mould will be reassembled. The reassembly of components is often done with power tools and assembly equipment as in new product assembly. Testing operations are typically executed after the reassembly stage to make sure that the remanufactured injection mould meets newly manufactured product standards and customer requirements. In fact, during the reprocessing stage, the component's quality is continuously assured through applied measurements [63]. A remanufactured injection mould is said to be ready for delivery after passing through all necessary testing experiments.

As noted earlier, the remanufacturing process is labour intensive and highly variable, with many parties involved. These participants play an essential role in distributing and delivering merchandise to the consumers. Therefore, real-time interaction and coordination among remanufacturers, suppliers and customers are necessary for leasing-based mould remanufacturing in response to customer requests.

11.3 Multi-Agent Framework for Leasing-Based Injection Mould Remanufacturing

Obviously, if managers cannot quantify the potential financial (and non-financial) benefits, they are unlikely to consider return flows as anything other than a nuisance, for which they must minimise losses. With this recognition, we adopt a new business model (i.e., a leasing-based mould remanufacturing model) that shows top managers how to release the enormous value that is currently unrecognised and unappreciated. Meanwhile, we also develop new operational models to help them manage the day-to-day tactical elements so that they can realise their business objectives as well.

In general, the purpose of this section is to present a distributed agent-based mould remanufacturing framework (d-AMR) for moulds that are leased to customers and that can be remanufactured upon their return. A multi-agent system (MAS) provides natural key metaphors that facilitate a high-level and understandable description of the problem domain and the aspired solution. The objective of d-AMR is to increase the usability and prolong the life span of moulds through the organised rebuilding and remanufacturing of the mould design as well as the refurbishment and reuse of their components. Therefore, it is necessary to develop organisational and methodical strategies to initiate new distribution and business models between the supplier and user of moulds to realise a workable remanufacturing concept.

11.3.1 Multi-Agent System

An MAS is a system that applies various autonomous agents to accomplish some specified goals. Such a system is suitable for resource allocation problems since the nature of resource trading requires multiple agents to request for geographically dispersed heterogeneous resources. In order to support such a dynamic situation, an MAS must be extended to accommodate the agent with a high negotiation skill capability, which is the ability of learning from the requirements of resources advertised by the involving agents during each interaction. One way of achieving this is to analyse the negotiation tactics and strategies in every counter-offer. The typical MAS scenario involves a set of autonomously situated entities interacting with each other and exploiting resources in the environment to achieve a common goal [64,65]. Those entities can be formulated as an abstraction of an agent, an object or, in general, a (social) communication partner that contains both the code and data necessary to perform some well-defined functions. Agents interact differently depending on situations. For instance, agents work cooperatively to achieve a common goal, while they behave competitively when their goals are conflicting.

11.3.2 Electronic Commerce

Electronic commerce (e-commerce) had an immense effect on the manner in which businesses order goods and have them transported with the major portion of e-commerce transactions being in the form of business-to-business (B2B). At large, e-commerce is defined as sharing business information, maintaining business relationships, operating business negotiations, settling and executing agreements by means of telecommunication networks, often the Internet, in order to achieve business transactions [66]. One hope of e-commerce was that direct merchants would profit by the elimination of middlemen—those traders who are simply go-betweens from the producer to the consumer. On the basis of that, advances in e-commerce have unveiled new opportunities for the management of supply chain networks by increasing the mutual visibility of consumers and suppliers, and by raising the possibility that some of their trading processes may be automated [67].

Nowadays, e-commerce sites increasingly use agent-based systems for providing goods and services to their customers, which are time consuming and laborious [68]. In the literature, there are many agent-based e-commerce systems. One of the famous examples is the trading agent competition (TAC) [69], which is a test bed for intelligent software agents that interact simultaneously to obtain services for customers. In addition, some of them have also studied product returns in the e-commerce context [70–72]. In light of this context, e-commerce is examined in terms of technologies and emerging services, which are used to improve trading of used products and parts, including marketing, purchasing, sales and post sales.

11.3.3 Proposed Multi-Agent Framework

The idealistic model for our remanufacturing-centred process should be a three-layered structure only involving customers, lessor/remanufacturers and suppliers. The non-value-added transferring process should be cut off to the maximum extent in order to achieve optimal efficiency. In the complex structure of d-AMR, by making the most of the information flow and cooperation mechanism with the back up of business units, production planning and scheduling (PPS) units, warehousing units and supplier units, a conceptual supply chain model can be abstracted from the network system.

In the proposed conceptual model, abstracted from the e-commerce structure, all the middle structures, such as wholesaler, distributor and retailer, have been replaced by the backup system of four units. This approach will enable the remanufacturer to identify the end-customer's demand directly and embrace a more efficient marketing and delivery channel. Like the other applications, agent technology provides higher flexibility, mobility and intelligence to the entire system. The framework allows different remanufacturing sectors within each unit to plan and correct deviance independently according to their own needs, while not sacrificing the feasibility and synchronisation by collaborating with internal or external partners. The functionalities of each unit and its corresponding remanufacturing sectors are detailed as follows.

11.3.3.1 Description of Participants

The system is designed for the following three participants to meet their operational requirements and to enhance their business competitiveness:

- *Customer/lessee*: Expecting their requests for products or services from distributors and remanufacturers to be completed on time
- *Lessor/remanufacturer*: Receiving orders from customers and efficiently fulfilling them by cooperating with the internal operation departments and logistic service providers
- *Logistic service provider/supplier*: Providing logistic services with a full range of information support provided to both buyers and sellers in order to enhance the business efficiency of the customer/lessee

11.3.3.2 Description of Business Unit

In our design, there are five types of agents working for a business unit: administration agent, reception agent, finance agent, knowledge base agent and yellow pages agent. The functional behaviours of these agents are defined as follows:

- *Administration agent*: This agent obtains the information from the customer and then performs coordination between other agents to fulfil the customer's request. It is responsible for gathering all the incoming orders and eliciting relevant information regarding customer choices, preferences, specifications and handling customer requests for order modification or cancellation. Customer preferences and profiles are stored in a database for further analysis to detect trends and develop customer relationship.
- *Reception agent*: This agent processes order information coming from the administration agent. After consulting with the finance agent about the credit of such customers, a customer demand will then transfer the requirement to the corresponding agent in the PPS unit. After receiving a reply from the PPS unit, the reception agent will finalise a preliminary feedback report including cost, lead time and so on and send it back to the administration agent. This loop might go several rounds until an agreement between the administration agent and the customer is achieved.
- *Finance agent*: It is responsible for controlling the lessee's contract payments.
- *Knowledge base agent*: Imagine an agent whose task is to search for the internal structure of the mould designs, given some keywords and constraints. For this task, prior knowledge (provided by the literature or mould-making companies) of the methods file, and of the alternative mould features by the method database and the mould feature geometry database is normally required. This may be acceptable for static environments, where the agent has permanent access to a database in its local network or the Internet. But in a dynamic environment, the addresses have to be acquired during runtime using some discovery mechanism.

For example, if the agent is executed on a mobile device, the model may restrict the ways to gathering information to save transmission costs or even delay the request until an appropriate and low-priced Internet access is available.
- *Yellow pages agent*: This agent provides all required information about registered suppliers. For example, when the procurement agent invites some suppliers to join parts supply bidding task, the yellow pages agent will be notified to connect the appropriate suppliers.

11.3.3.3 Description of Production Planning and Scheduling Unit

Three types of agents are designed for the PPS unit and the functionalities of these agents are defined as follows:

- *Production planning agent*: Upon receiving a request from the reception agent, the production planning agent will set out to obtain the information needed. Through considering local constraints (e.g., number of breakdown machines and current production schedule) and verifying the availability of the potential required parts with the warehousing unit, the production planning agent generates production plans and then gives the reception agent a notification.
- *Remanufacturing management agent*: This agent executes the remanufacturing processes to satisfy orders for production. It collects the remanufacturing capacity and situation information from workshops (including the disassembly agent, cleaning agent, inspection agent, machining agent, reassembly agent and testing agent) and communicates with the production planning agent to smooth the production plan, and then allocates the production tasks to the related workshops.
- *Workshop agents*: Each workshop agent (including disassembly, cleaning, inspection, machining, reassembly and testing) has available objects that can be modified by local actions or actions from other agents. Actions are made possible by task flows, which are sequences of tasks and called the planning protocol. Obviously, the planning protocol is triggered upon reception of a new demand plan from a customer.

11.3.3.4 Description of Warehousing Unit

Three types of agents are designed for a warehousing unit and the functionalities of these agents are defined as follows:

- *Warehousing agent*: This agent handles the logistic information from the inventory agent, the remanufacturing management agent and the procurement agent. It is responsible for coordinating the logistic process in the enterprise domain to achieve the best possible results in terms of on-time delivery, cost minimisation and so forth.
- *Inventory agent*: It deals with the inventory-related tasks, such as keeping track of materials, work-in-process and finished products inventory records, to meet the production requirement, determining the reordering point and responding to the inventory availability queries from the production planning agent.
- *Procurement agent*: The procurement agent analyses the whole production plan and decomposes the material requirements into specific categories in order to communicate the outgoing order information with certain suppliers.

11.3.3.5 Description of Supplier Unit

Three types of agents are designed for the supplier unit, and the functionalities of these agents are defined as follows:

- *Supplier interface agent*: It provides an intelligent interface for the suppliers to bid for the outgoing orders of material requirement. It is responsible for inviting supplier biddings to fulfil the material requirements and proclaiming the award of a bid.
- *Supplier management agent*: It gives out supplier bidding rules and analyses the supplier bidding information from the supplier interface agent. It is responsible for managing supplier information and choosing the most suitable and reliable suppliers based on material requirements in the production process. The related supplier's information is stored in the supplier profile database as cases to facilitate future procurement process.
- *Delivery agent*: The delivery agent gathers the material flow information inside the remanufacturer entity to facilitate the inside logistics and generate the outside logistics information. It is responsible for coordinating plans and suppliers in the enterprise domains to achieve the best possible results in terms of the goals of the closed-loop supply chain, including on-time delivery, cost minimisation and so forth.

11.3.4 Cooperation and Interaction Mechanism

In network economy, cooperation between supply chain partners plays an important role in the more dynamic, interactive and distributed supply chain. The following part describes the interactions among these agents.

The remanufacturer receives an incoming order from its customer. The incoming order flows to the administration agent. During the planning phase, the administration agent sends a demand plan to the reception agent to consult with the finance agent and then to the knowledge base agent to check the mould features. After identification, the knowledge base agent sends the needed parts/components to the warehouse agent to verify if those are in stock. For non-available products, it sends a demand plan to the production planning agent. Using this demand plan, along with resource constraints and lead times, the production planning agent builds its plan considering infinite supply and transmits it to the procurement agent.

The procurement agent specifies the demand plan into different categories of material sources and generates the outgoing orders. The procurement agent first reaches the supplier profile database, if there are suitable suppliers based on the previous trading information; the outgoing orders can be sent directly to the delivery agent by the supplier management agent. If there are not suitable suppliers in the database, a round of bidding should be carried out. The suppliers bid on the platform of the supplier interface agent, and the bidding information subsequently goes to the supplier management agent. After receiving the applications of bid, the supplier management agent will give a comprehensive evaluation of the suppliers in consideration of quality, price, due date, service and so on, and then choose the most suitable suppliers according to the evaluation results. Finally, the supplier management agent transmits its preferred plan to the delivery agent to complete the transport task. When suppliers answer the demand plans, the administration agent receives a supply plan and communicates with the customer. Meanwhile, the yellow pages agent will notice and update all required information about the suppliers.

In addition to argumentation generation, selection is the key step of argumentation. Normally, a promising solution, that is, a bid of the highest promising value, will be selected. However, a promising solution to a remanufacturer may not necessarily be a good choice from the viewpoint of a global solution. A practical remedy to overcome the limitation is to diversify the search in the vicinity of local optimal via random constructions; this attempts to avoid local optimal by jumping out of them early in the computation. In this way, the production planning agent may achieve coordination and coherence among the decisions of the remanufacturing management agent through a series of negotiations and adjustments with workshop agents that are made individually, but in a coordinated fashion.

If an event occurs in the internal supply chain operations, any agent can initiate collaboration with its internal clients and suppliers by sending a revised demand or supply plan. This can be triggered by an agent who needs some products to fulfil inventory, lost production or new demand. This explains why agents are also responsible for continuously monitoring their environment and reacting to disturbances. Owing to the interaction context, an agent's environment is also made up of all messages received from other agents specifying a new or modified requirement or replenishment, a contingency situation or a high-priority requirement to process.

11.4 Collaborative Leasing-Based Injection Mould Remanufacturing Process Analysis

The goal of the process analysis is to identify the key processes and to reduce the overall cycle time and process cost. We explain the cooperative bidding mechanism of a multi-agent supply chain in Figure 11.1 as an example.

Numbers 1 to 9 represent the communication sequence between supply chain partners. Cooperation can be triggered by advertising orders of the remanufacturer.

1. The procurement agent first reaches the supplier profile database.
2. If there are suitable suppliers based on the previous trading information, the outgoing orders can be sent directly to the delivery agent by the supplier management agent.
3. If not, the procurement agent advertises its outgoing orders through the supplier interface agent to all the potential suppliers.
4. After receiving the advertising orders, the potential suppliers makes the decision of bidding.
5. If the supplier decides to bid, it will plan the application of a bid.
6. The supplier bids on the platform of the supplier interface agent.
7. After receiving the applications of a bid, the supplier management agent will give a comprehensive evaluation of the suppliers in consideration of quality, price, due

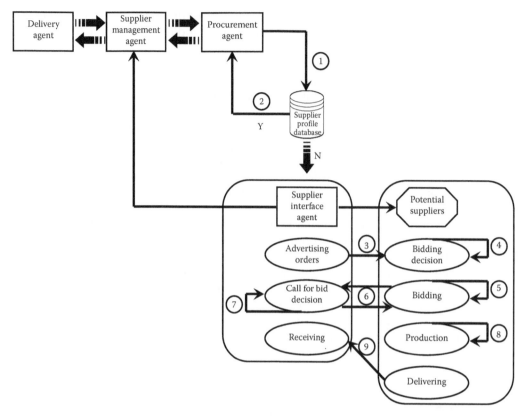

FIGURE 11.1
Cooperation mechanism of multi-agent supply chain.

date, service and so on, and then choose the most suitable suppliers according to the evaluation results; the agent meanwhile gives necessary replies to the not-chosen suppliers through its supplier interface agent.

8. The supplier who has won the bid carries out production to fulfil the incoming orders.
9. The supplier delivers the finished products to the delivery agent.

To test the practicality of the proposed multi-agent framework, all the interactions among participating agents in our distributed injection mould remanufacturing scenario are suggested to be implemented in Java agent development framework (JADE), the most widespread agent-oriented middleware in use today, which is completely in compliance with the foundation for intelligent, physical agents (FIPA) standards. In addition to establishing a low-cost and effective communication model among partners to support process tracking, the d-AMR system adopts the radio frequency identification (RFID) technology.

As can be seen in Figure 11.2, the agent community is split into four different sites. The communication protocol between different agents is through the JADE platform by the Internet inter-ORB protocol [73] and focuses on JADE to JADE.

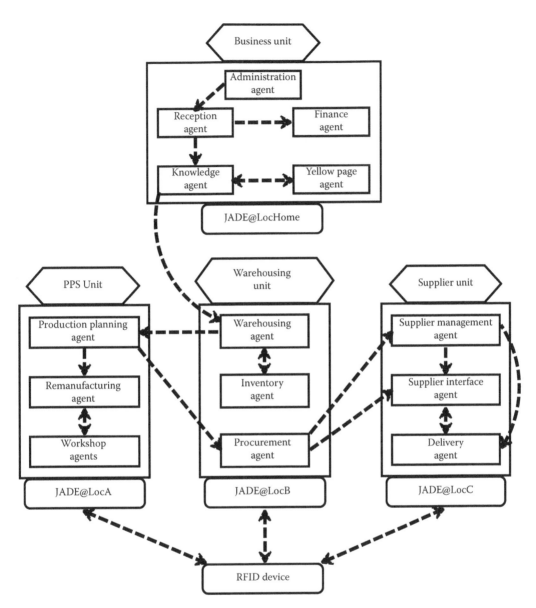

FIGURE 11.2
Deployment view of agents on d-AMR system.

11.5 Conclusions

In this chapter, a distributed leasing-based mould remanufacturing model, which utilises the advantage of MAS such as reactivity, utility evaluation, anticipation, scalability and negotiation, is introduced. The proposed framework can be a powerful tool to reach appreciated gains when implemented in a multi-agent programming environment such as JADE. Following the design of model architecture, conceptualisation of the required agent's intelligent behaviours and the description of their implementation environment, future work can be pursued with the following prospects in mind: First, different agent configurations should be experimented against real-world mould remanufacturing scenarios to identify the different situations where agents can perform well or not. The reasons need to be determined for further improvement of our model design. Second, it will be of great interest to increase the agent's learning competency. An agent geared with certain learning capabilities would be able to update its utility functions and thus to modify its preference for an action. This should lead to an even more agile and performing multi-agent framework. Last but not least, the target application scenario is on mould remanufacturing; nevertheless, other remanufacturing processes share many similarities. Therefore by recognising the dissimilarities between mould remanufacturing and other types of remanufacturing activities, our proposed framework should be extended to other remanufacturing cases. This is highly promising and can thus form a more generalised distributed leasing-based remanufacturing process based on MAS.

References

1. Maxwell, D., Sheate, W. and Vorst, R.v.d. 2006. Functional and systems aspects of the sustainable product and service development approach for industry. *Journal of Cleaner Production 14*, 1466–1479.
2. Rose, C.M. 2000. Design for environment: A method for formulating product end-of-life strategies. Doctoral Thesis. (Stanford University).
3. Sharma, M., Ammons, J.C. and Hartman, J.C. 2007. Asset management with reverse product flows and environmental considerations. *Computers & Operations Research 34*, 464–486.
4. Guide, V.D.R., Harrison, T.P. and Wassenhove, L.N.V. 2003. The challenge of closed-loop supply chains. *Interfaces 33*, 3–6.
5. Stahel, W.R. 1997. The service economy: 'Wealth without resource consumption'? *Philosophical Transactions: Mathematical, Physical and Engineering Sciences 355*, 1309–1319.
6. Intlekofer, K. 2010. Environmental implications of leasing. In school of mechanical engineering, Master's Thesis. (Georgia Institute of Technology).
7. Altan, T., Lilly, B. and Yen, Y.C. 2001. Manufacturing of dies and molds. *CIRP Annals–Manufacturing Technology 50*, 404–422.
8. Fishbein, B.K., McGarry, L.S. and Dillion, P.S. 2000. Leasing: A step towards producer responsibility. (New York: Technical Report, INFORM, Inc.).
9. Mont, O.K. 2002. Clarifying the concept of product–service system. *Journal of Cleaner Production 10*, 237–245.
10. Desai, P. and Purohit, D. 1998. Leasing and selling: Optimal marketing strategies for a durable goods firm. *Management Science 44*, S19-S34.
11. Desai, P.S. and Purohit, D. 1999. Competition in durable goods markets: The strategic consequences of leasing and selling. *Marketing Science 18*, 42–58.
12. Bhaskaran, S.R. and Gilbert, S.M. 2005. Selling and leasing strategies for durable goods with complementary products. *Management Science 51*, 1278–1290.
13. Desai, P. and Purohit, D. 1998. Leasing and selling: Optimal marketing strategies for a durable goods firm. *Management Science 44*(11), S19–S34.
14. Lifset, R. and Lindhqvist, T. 2000. Does leasing improve end of product life management. *Journal of Industrial Ecology 3*, 10–13.
15. Cooper, T. and Evans, S. 2000. *Products to Services*. (UK: Sheffield Hallam University).
16. Savaskan, R.C., Bhattacharya, S. and Wassenhove, L.N.V. 2004. Closed-loop supply chain models with product remanufacturing. *Management Science 50*, 239–252.

17. Östlin, J., Sundin, E. and Björkman, M. 2008. Importance of closed-loop supply chain relationships for product remanufacturing. *International Journal of Production Economics* 115, 336–348.
18. Zwolinski, P., Lopez-Ontiveros, M.-A. and Brissaud, D. 2006. Integrated design of remanufacturable products based on product profiles. *Journal of Cleaner Production* 14, 1333–1345.
19. Langella, I.M. 2007. Planning demand-driven disassembly for remanufacturing. Doctoral Thesis. (Magdeburg, Germany: Universität Magdeburg).
20. Giuntini, R. and Gaudette, K. 2003. Remanufacturing: The next great opportunity for boosting US productivity. *Business Horizons* 46(6), 41–48.
21. Toffel, M.W. 2004. Strategic management of product recovery. *California Management Review* 46, 120–141.
22. Sundin, E. and Bras, B. 2005. Making functional sales environmentally and economically beneficial through product remanufacturing. *Journal of Cleaner Production* 13, 913–925.
23. Guide, V.D.R., Srivastava, R. and Kraus, M.E. 1998. Proactive expediting policies for recoverable manufacturing. *Journal of the Operational Research Society* 49, 479–491.
24. Guide, V.D.R., Jayaraman, V. and Srivastava, R. 1999. Production planning and control for remanufacturing: A state-of-the-art survey. *Robotics and Computer Integrated Manufacturing* 15, 221–230.
25. Ferrer, G. 2003. Yield information and supplier responsiveness in remanufacturing operations. *European Journal of Operational Research* 149, 540–556.
26. Sasikumar, P. and Kannan, G. 2008. Issues in reverse supply chains, part I: End-of-life product recovery and inventory management–an overview. *International Journal of Sustainable Engineering* 1, 154–172.
27. Sahni, S., Boustani, A., Gutowski, T. and Graves, S. 2010. *Engine Remanufacturing and Energy Savings*. (Cambridge, MA: Massachusetts Institute of Technology, Report No.: MITEI-1-d-2010).
28. Zhu, Q. and Sarkis, J. 2004. Relationships between operational practices and performance among early adopters of green supply chain management practices in Chinese manufacturing enterprises. *Journal of Operations Management* 22, 265–289.
29. Kumar, S. and Putnam, V. 2008. Cradle to cradle: Reverse logistics strategies and opportunities across three industry sectors. *International Journal of Production Economics* 115, 305–315.
30. Majumder, P. and Groenevelt, H. 2001. Competition in remanufacturing. *Production and Operations Management* 10, 125–141.
31. Ferguson, M.E. and Toktay, L.B. 2006. The effect of competition on recovery strategies. *Production and Operations Management* 15, 351–368.
32. Debo, L.G., Toktay, L.B. and Wassenhove, L.N.V. 2005. Market segmentation and product technology selection for remanufacturable products. *Management Science* 51, 1193–1205.
33. Guide, V.D.R., Srivastava, R. and Kraus, M.E. 1997. Product structure complexity and scheduling of operations in recoverable manufacturing. *International Journal of Production Research* 35, 3179–3199.
34. Guide, V.D.R. and Srivastava, R. 1998. Inventory buffers in recoverable manufacturing. *Journal of Operations Management* 16, 551–568.
35. Guide, V.D.R. and Wassenhove, L.N.V. 2001. Managing product returns for remanufacturing. *Production and Operations Management* 10, 142–155.
36. Guide, V.D.R., Jayaraman, V., Srivastava, R. and Benton, W.C. 2000. Supply-chain management for recoverable manufacturing systems. *Interfaces* 30, 125–142.
37. Guide, V.D.R., Teunter, R.H. and Wassenhove, L.N.V. 2003. Matching demand and supply to maximize profits from remanufacturing. *Manufacturing & Service Operations Management* 5, 303–316.
38. Denizel, M., Ferguson, M. and Souza, G.G.C. 2010. Multiperiod remanufacturing planning with uncertain quality of inputs. *IEEE Transactions on Engineering Management* 57, 394–404.
39. Aras, N., Güllü, R. and Yürülmez, S. 2011. Optimal inventory and pricing policies for remanufacturable leased products. *International Journal of Production Economics* 133, 262–271.
40. Berko-Boateng, V.J., Azar, J., Jong, E.D. and Yander, G.A. 1993. Asset recycle management— A total approach to product design for the environment. In *Proceedings of the International Symposium on Electronics and the Environment*. (Arlington, Virginia: IEEE).

41. Kerr, W. 1999. Remanufacturing and eco-efficiency: A case study of photocopier remanufacturing at Fuji Xerox Australia. In *International Institute for Industrial Environmental Economics (IIEEE), Volume Master of Science in Environmental Management and Policy*. (Lund, Sweden: Lund University).
42. Ahn, H. 2009. On the profit gains of competing reverse channels for remanufacturing of refillable containers. *Journal of Service Science 1*, 147–190.
43. Grant, D.B. and Banomyong, R. 2010. Design of closed-loop supply chain and product recovery management for fast-moving consumer goods: The case of a single-use camera. *Asia Pacific Journal of Marketing and Logistics 22*, 232–246.
44. Metta, H. 2011. A multi-stage decision support model for coordinated sustainable product and supply chain design. College of Engineering, Doctoral Thesis. (Lexington, Kentucky: University of Kentucky).
45. Kerr, W. and Ryan, C. 2001. Eco-efficiency gains from remanufacturing: A case study of photocopier remanufacturing at Fuji Xerox Australia. *Journal of Cleaner Production 9*, 75–81.
46. Matsumoto, M. 2009. Business frameworks for sustainable society: A case study on reuse industries in Japan. *Journal of Cleaner Production 17*, 1547–1555.
47. Allen, D., Bauer, D., Bras, B., Gutowski, T., Murphy, C., Piwonka, T., Sheng, P., Sutherland, J., Thurston, D. and Wolff, E. 2002. Environmentally benign manufacturing: Trends in Europe, Japan, and the USA. *Journal of Manufacturing Science and Engineering 124*, 908–920.
48. Brat, I. 2006. Caterpillar gets bugs out of old equipment: Growing remanufacturing division is central to earnings-stabilization plan. *The Wall Street Journal*, 5 July 2006, p. 1.
49. Gray, C. and Charter, M. 2007. Remanufacturing and product design: Designing for the 7th generation. (The Centre for Sustainable Design, University College for the Creative Arts, Farnham, UK), p. 77.
50. Sundin, E., Lindahl, M. and Ijomah, W. 2009. Product design for product/service systems: Design experiences from Swedish industry. *Journal of Manufacturing Technology Management 20*, 723–753.
51. Mähl, M. and Östlin, J. 2006. *Lean Remanufacturing—Material Flows at Volvo Parts Flen*. Department of Business Studies. Master Thesis. (Uppsala, Sweden: Uppsala University).
52. Sandvall, F. and Stelin, C. 2006. *The Remanufacturing Offer: A Case Study of Volvo Construction Equipment Implementing and Expanding Reman in Russia*. School of Business, Volume Bachelor of Science. (Stockholm: Stockholm University).
53. Khalil, L. and Olofsson, M. 2007. *Reverse Logistics Study at Volvo CE CST Europe*. Institution for Innovation, Design and Product Development, Volume Master of Science. (Eskilstuna: Mälardalen University).
54. Debo, L.G. and Wassenhove, L.N.V. 2005. Tire recovery: The RetreadCo case. In: *Managing Closed-Loop Supply Chains*, S.D.P. Flapper, J.A.E.E.v. Nunen and L.N.V. Wassenhove, eds. (Berlin Heidelberg: Springer-Verlag), pp. 119–128.
55. Driesch, H.-M., Oyen, H.E.v. and Flapper, S.D.P. 2005. Recovery of car engines: The Mercedes-Benz case. In: *Managing Closed-Loop Supply Chains*, S.D.P. Flapper, J.A.E.E.v. Nunen and L.N.V. Wassenhove, eds. (Berlin Heidelberg: Springer-Verlag), pp. 157–166.
56. RemTec_SA http://www.remtec.co.za (assessed on 07 October 2012).
57. Mercedes-Benz_SA http://www.mercedes-benz.co.za (assessed on 07 October 2012).
58. Mercedes-Benz_SA 2009. Genuine Mercedes-Benz remanufacture parts catalog. p. 81.
59. Souza, G. C. 2010. Production planning and control for remanufacturing. *Closed-Loop Supply Chains New Developments to Improve the Sustainability of Business Practices*, M.E. Ferguson and G.C . Souza, eds. (Boca Raton, FL: Taylor & Francis), pp. 119–130.
60. Mont, O., Dalhammar, C. and Jacobsson, N. 2006. A new business model for baby prams based on leasing and product remanufacturing. *Journal of Cleaner Production 14*, 1509–1518.
61. Sharma, M. 2004. Reverse logistics and environmental considerations in equipment leasing and asset management. Doctoral Thesis. (Georgia Institute of Technology).
62. Buchok, A. 2008. The design, manufacturing and use of economically friendly injection molds. Department of Materials Science and Engineering, Bachelor Thesis. (Massachusetts Institute of Technology).

63. Steinhilper, R. 1998. *Remanufacturing: The Ultimate Form of Recycling.* (Stuttgart, Germany: Fraunhofer IRB Verlag).
64. Wooldridge, M. and Jennings, N.R. 1995. Intelligent agents: Theory and practice. *The Knowledge Engineering Review 10*, 115–152.
65. Wooldridge, M. 2009. *An Introduction to Multiagent Systems*, 2nd Edition. (West Sussex, England: John Wiley & Sons, Ltd).
66. Kokkinaki, A.I., Dekker, R., Nunen, J.v. and Pappis, C. 1999. *An Exploratory Study on Electronic Commerce for Reverse Logistics.* (Netherlands: Econometric Institute, Erasmus University Rotterdam).
67. Chopra, S., Dougan, D. and Taylor, G. 2001. B2B: E-commerce opportunities. *Supply Chain Management Review*, May/June, 50–58.
68. Mohanty, B.K. and Passi, K. 2010. Agent based e-commerce systems that react to buyers' feedbacks—A fuzzy approach. *International Journal of Approximate Reasoning 51*, 948–963.
69. Collins, J. and Sadeh, N. 2009. Guest Editors' introduction to special section: Supply chain trading agent research. *Electronic Commerce Research and Applications 8*, 61–62.
70. Ryan, S.M., Min, K.J. and Ólafsson, S. 2002. Experimental study of reverse logistics e-commerce. In *Proceedings of the IEEE International Symposium on Electronics and the Environment (ISEE)*, pp. 218–223. (IEEE).
71. Choi, T.-M., Li, D. and Yan, H. 2004. Optimal returns policy for supply chain with e-marketplace. *International Journal of Production Economics 8*, 205–227.
72. Ramanathan, R. 2011. An empirical analysis on the influence of risk on relationships between handling of product returns and customer loyalty in e-commerce. *International Journal of Production Economics 130*, 255–261.
73. Bellifemine, F., Caire, G. and Greenwood, D. 2007. *Developing Multi-Agent Systems with JADE.* (West Sussex, England: John Wiley & Sons, Ltd).

12

The Smart Operating Room: smartOR

Marcus Köny*, Julia Benzko*, Michael Czaplik*, Björn Marschollek*, Marian Walter, Rolf Rossaint, Klaus Radermacher and Steffen Leonhardt

CONTENTS

12.1 Introduction 292
    12.1.1 Motivation 292
    12.1.2 Technical Challenges 293
    12.1.3 smartOR Project 293
12.2 Background 294
    12.2.1 Medical Background 294
    12.2.2 SOA-Based Integration for the Operating Room 295
    12.2.3 Open Surgical Communication Bus 296
    12.2.4 Network Interfaces and Protocols 297
    12.2.5 M^2IO Switch 298
    12.2.6 Related Works 299
12.3 Concept of the smartOR 300
    12.3.1 System Overview 300
    12.3.2 Nomenclature and Interface Description 301
    12.3.3 Interaction with the Hospital IT/Post-Operative Care 302
    12.3.4 Security Aspects and Patient Safety 302
    12.3.5 Risk Management 304
12.4 User Interaction: Workstations 304
    12.4.1 Surgical Workstation 304
    12.4.2 Anaesthesia Workstation 306
        12.4.2.1 Overview 306
        12.4.2.2 SomnuCare Software Concept 307
        12.4.2.3 Interfacing Medical Devices 308
        12.4.2.4 Memory Mapped File Engine 309
        12.4.2.5 Innovative Clinical Applications 309
12.5 Use Case Laparoscopic Intervention 311
12.6 Conclusion and Discussion 313
Acknowledgements 314
References 314

---

* These authors contributed equally.

12.1 Introduction

12.1.1 Motivation

During the last few decades, the usage of high-tech devices has increased intensively in daily routine as well as in patient care. Certainly, technical equipment has made a considerable contribution to improving the quality of clinical treatment. In particular, anaesthesia and critical care have profited remarkably by the engineering progress. Patient monitors, highly integrated microscopes and endoscopes, tracking and navigation tools, anaesthesia machines, electronic syringe pumps and infusion pumps are worth mentioning, as well as machines such as the extracorporeal oxygenation apparatus, heart assist devices and intraoperative cell salvage machines, to name just a few (Figure 12.1). Besides, advances in software engineering have led to more and more sophisticated applications providing image processing, medical analysis and decision support. Not least, the potentialities for telemedicine have risen from technologisation for the benefit of treatment quality and resource conservation.

Altogether, outstanding advances and progresses have been made in medical treatment that can be improved further by networking. Currently, devices mainly operate independently except for some standard couplings such as between the anaesthesia machine and the patient monitor. However, by interconnection, innovative and smart algorithms can be established such as improved intelligent alarm concepts, decision support facilities and automation procedures since interdependencies and synergies between devices exist.

A prerequisite, however, is that medical devices and their combinations are approved, safe, robust and easy to handle even in critical and stressful situations. Therefore, one has to take into account that medical staff, such as surgeons and anaesthetists, in our case, have to focus on the patient and not do error correcting or undertake workarounds. Moreover, surgeons and anaesthetists have different points of views and demands. One surgeon focuses on the surgical intervention, caring for optimal conditions regarding overview

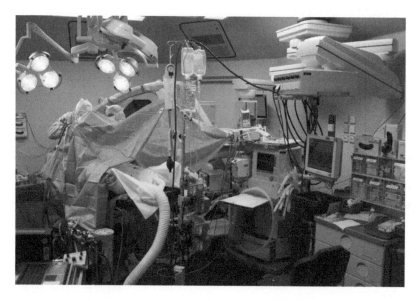

FIGURE 12.1
Typical scenario in the operating room.

and underlying anatomy, whereas another manages a reliable narcosis under stable cardiopulmonary conditions. By doing so, more or less technical effort is required, depending on the surgical procedure and the patient. Of course, there is a difference between a laparoscopic appendectomy, a neurosurgical intervention and a heart transplantation.

In general, medical staff is adapted to certain devices and user interfaces so that a cutover to other frontends in principle involves a risk. Therefore, cross-functional cooperation is crucial for defining requirements and specifications. Obviously, the current situation is characterised by a huge variety of devices and machines with different control concepts and displays (analog, digital, LCD, touchscreen, etc.), which is confusing and risky for patient safety.

12.1.2 Technical Challenges

There are several technical challenges in Section 12.1.1. The need for clearly arranged user interfaces makes standardisation necessary. New standards and methods for information presentation and device operation must be developed. Standardised interfaces for displaying information and operating devices would help the staff in the operating room (OR) to handle the multitude of different devices. Unfortunately, every device would still have its own displays and an interaction between the devices is still missing. Therefore, every device should be connected to a network. This would enable devices to exchange information with other devices in the network and profit thereby. Furthermore, the information and alarms of every device can be submitted to central workstations—one for the anaesthesia workplace and one for the surgical workplace. These workstations are able to evaluate information received from the devices in the OR and present optimised information for the surgical and anaesthesia workplace. Furthermore, the workstations offer a central user interface for operating devices in the OR. All these challenges must be handled while ensuring a safe operation of all devices in a safety-critical environment.

To guarantee a safe operation of medical devices, every device on the market must comply with different standards and laws. This is comparable to the certification of devices used in safety-critical applications such as airplanes or robotics. Before being brought to the market, every medical product must undergo a risk analysis according to DIN EN ISO 14971 or the future ISO 80001 first. These standards describe the risk management process of medical devices to guarantee a safe operation. By enumerating risks and possible impacts a medical device causes, different counteractions must be evaluated [2]. Furthermore, every device must be compliant with the criteria of electrical safety according to standards such as ISO 60601 and the medical device directive.

Risk management and certification must be done for every medical device. Connecting two medical devices, such as, for example, a patient monitor and an anaesthesia machine, requires a new risk management and certification because new aspects that the connection generates must be considered. As can be imagined, supporting a flexible, modular and plug-and-play connection between a multitude of devices [3] makes the assessment of risks, especially for one vendor, extremely difficult. Therefore, new standards and methods for risk management must be developed based on a state-of-the-art hardware and software protocol.

12.1.3 smartOR Project

To facilitate the smooth handling of medical devices in the OR more and more, the so-called integrated operating room systems are used to optimise the human–machine interaction and enable data exchange between systems. Often, these are proprietary and closed systems, so that the integration of components of third-party vendors is not possible or possible only with great effort. Therefore, the smartOR project (www.smartOR.de) has developed a manufacturer-independent framework for networking medical devices. Currently available standards and concepts cover the requirements for a modular, flexible and plug-and-play integration rather poorly. To overcome these obstacles, an integration concept based on a service-oriented architecture (SOA) is presented, which is an enhancement of the developed integration framework in the OrthoMIT project (www.orthomit.de). The framework is based on well-known networking and Internet technologies such as Ethernet, TCP/UDP and web services. The web service standard guarantees a version-independent compatibility between the devices and allows a flexible handling of the communication regardless of the operating system. Medical devices are available as services and are handled by central management components. Standardised interfaces are developed within the project to enable the data and command exchange of individual systems. Furthermore, aspects of risk management and certification are regarded during the complete design process of described concepts and protocols.

Therefore, first, the basic technologies and modules used in the smartOR project are discussed in Section 12.2. Expanding on those, aspects of interfaces, nomenclature, patient safety and risk management are described in Section 12.3. In Section 12.4, the focus is on user interaction. Two workstations, one for the surgical workplace and the other for the anaesthesia workplace, are described, which represent the central display and operation panels for all devices in the operating room. Finally, this chapter closes with a use case as proof of concept and evaluation of the protocols in Section 12.5, followed by a conclusion and discussion in Section 12.6.

12.2 Background

12.2.1 Medical Background

Although distributed network intelligence plays a cross-departmental role in several situations, we want to focus on networking in operation rooms and subsequent post-operative care such as intensive care or post-anaesthesia care units. Regarding surgical interventions, it is necessary to differentiate between elective and non-elective cases. Since non-elective or emergency procedures are critical, saving time is an important goal, sometimes at the expense of costly arrangement and pre-examinations. Of course, for these cases, requirements for technical equipment or software differ completely from elective surgical procedures. Nevertheless, the patient's state is dynamic and not always predictable since complications or adverse events can occur at any time.

In general, patient safety is the main focus. Another aspect is the increase of efficiency. Hence, patient monitoring and the application of sophisticated technical equipment should be targeted. The kind and severity of surgical operation is one factor; the pre-medical history and the current health state of the concerned patient is another factor. Anyway, utilised devices have to be robust and reliable. Additional functions and features that are optional and 'nice-to-have' must not affect basic functions or complicate handling. Nevertheless, the monitoring of several parameters and data is crucial for performing a safe anaesthesia and therewith a successful operation.

Usually, the sole analysis of one single parameter is neither reasonable nor target-aimed. Besides, its evaluation and interpretation are dependent on the current surgical progress and the patient's physical condition. Thus, it has to be related to the current context and its plausibility should be checked with other measured data and the clinical situation. Moreover, it is hardly surprising that the involved medical staff has various demands and requires specific information. Anyway, the work routines of a surgical nurse, an assistant doctor, a surgeon, an anaesthetist and an anaesthesia nurse, to mention the most common attendants, are completely different. Since monitoring of the physical condition during anaesthesia is the central task of the anaesthetist, the surgeon only needs very brief and compact information about the current medical situation. Mostly, for that, periodic oral information by the anaesthetist is sufficient and adequate. Apart from that, for the surgeon, other data are of peculiar interest such as the preliminary medical findings and images such as x-rays, CT data or MRI. Furthermore, some very specific tools and technical devices such as navigation tools, cameras or further intraoperative medical imaging, are used as appropriate.

Collecting all available information, consolidation and evaluation in consideration of the current circumstances and surgical progress is crucial for controlling the vital functions and the physical condition of the patient. The enormous number of diverse graphical user interfaces and displays from multiple devices and machines is sometimes awkward to handle. Thus, a manufacturer-spanning overview of context-related consolidated information would be beneficial. This could also be the basis for more sophisticated supporting systems. But, to date, alerting systems are predominately fixed on one single parameter. The plausibility check and evaluation of the clinical relevance of an initiated alarm is carried out with the help of further parameters, the clinical sight and experience of the concerned physician. At this point, the usage of adapted hardware and software could provide benefits. Furthermore, additional information about the patient's pre-medical history or preliminary medical findings (test results, images, ECG, etc.) matters, so an easy-to-handle recall of appropriate data should be possible for every concerned team member.

Finally, information about the performed operation, including surgical and anaesthetic details, should be consolidated, processed and transferred to the hospital information system automatically. In this manner, surrender of relevant data that are significant for further post-operative care (on the post-anaesthesia care unit, intensive care unit or even a general ward) could be carried out.

12.2.2 SOA-Based Integration for the Operating Room

For the integration of medical devices in the operating room, an SOA framework has been chosen in the smartOR project. SOA is a paradigm for organising and utilising distributed functionality and resources [4]. The functionalities and resources may be under different ownerships and are implemented as reusable and distributed services. They are technically and functionally loosely coupled and used by other components or services. Access is realised over the so-called interface descriptions, which specify how a service consumer has to interact with a service provider. SOA is technology-independent; it describes a paradigm instead of a precise realisation. Therefore, it is an optimal solution connecting completely different devices from various manufacturers and production dates. The applicability of SOA for a modular, flexible and integrated operating room system architecture has been proved and can be seen in Ref. [3].

To implement SOA, different technologies can be used. In the smartOR project, the proposed integration framework is implemented by building a central communication network called open surgical communication bus (OSCB). Furthermore, technologies such as web services, service oriented access protocol (SOAP), web service description language (WSDL), extensible markup language (XML) and devices profile for web services (DPWS), which are described in this section, are used.

12.2.3 Open Surgical Communication Bus

OSCB is the central transportation medium and the common communication principle. It is realised over an IP-based network, which can be optionally expanded with a central communication server. To establish communication between medical devices over OSCB, open and standardised interfaces are necessary. Figure 12.2 shows the SOA-based integration framework for the operating room with the following safety and management components:

- Service manager
- Monitoring system
- Event manager
- Workflow engine
- Communication server
- Gateway

Medical devices can be a service consumer and/or service provider. A service provider is a device that offers its capabilities encapsulated in a service. A service consumer uses services provided by a service provider. The system is managed by the service manager. Every service provider has to register the offered service to the service manager. The registration consists of the publication of the interface description and some meta-information such as vendor, device description, device type and so on. Service consumers can ask the service manager for the needed services. For safety, security and reliability of the whole system, the service manager provides further services that are also important for

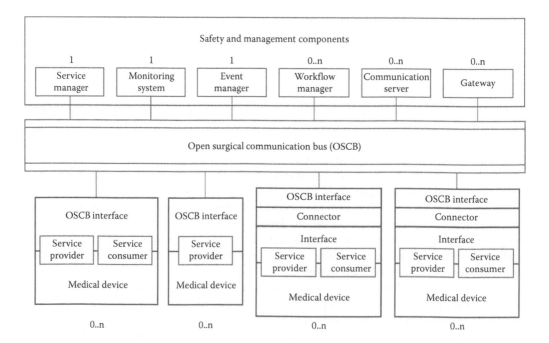

FIGURE 12.2
SOA-based integration framework for OR.

risk management. These services are initialisation routines and access control. The event manager is an optional component whose function is the central management of all events occurring on the network. Therefore, service providers have to register their events, while service consumers can subscribe for events and are then notified upon request of the occurrence of an event. The communication server/connector is used to transform data formats and to integrate components without a standardised format. The gateway in the network connects the OR network (OSCB) with the hospital IT network. The two primary functions are the logical separation of the two networks plus the function as a protocol converter and security system. An optional component is the monitoring system for monitoring the network and integrated devices. It supports the service manager in detecting faulty equipment or services and reports problems via the event manager.

12.2.4 Network Interfaces and Protocols

The physical layer of OSCB uses Ethernet. It provides high bandwidth and low operation costs due to its prevalence in the most clinics. It is a best-effort solution; for example, there is no guarantee about the disposition of the data (for further reading, refer to Refs. [5–9]). Ethernet covers the physical and data link layer in the ISO/OSI reference model, and thus forms the basis for the IP protocol used on the network layer. The transport layer is completed by TCP (transmission control protocol) and UDP (user datagram protocol), depending on the application. TCP establishes reliable communication between two end points and addresses the alleged drawback of sporadic non-disposition of network packets where necessary. With respect to the underlying techniques, OSCB can only be defined for device identification, device management, exchange of parameters and commands with no hard real-time demands. Large amounts of data and real-time data, such as high-resolution video, must be transmitted using other protocols. Protocol negotiation can be done using OSCB, though. Additionally, there can be other physical channels, such as the use of an matrix switch (see Section 12.2.5) or an additional Ethernet channel.

In the upper layers, web service technology is used. Web services are defined by the World Wide Web Consortium (W3C) as services that may support the cooperation between different applications on different platforms [10]. Each service is characterised by a uniform resource identifier (URI) and thus is uniquely addressable [11]. Its interface is human and machine readable and can be described by WSDL. Web services are autonomous and cannot be influenced by whether and how a message is processed by them. Message exchanges between service providers and service consumers are based on XML. XML is a meta-language that allows hierarchically structured data to be used using text documents. By the exclusive representation of text, it is platform-independent and hence is ideal for the message payload. Beyond this, the W3C recommendation XML schema is used to define the structure of an XML document [12]. XML schema provides a set of basic data types, supports complex types by restricting or expanding existing types and allows the definition and structuring of the message format using the XML-based SOAP network protocol [13]. SOAP is used for exchanging messages between two devices on the network and specifies the structure of the messages and the representation of the data it contains. Hence, the sender and recipient are able to define and interpret the messages correctly. SOAP does not define the semantics of message contents. Although remote procedure calls are available via SOAP, they are not used at the smartOR project. SOAP messages consist of an optional header and a body element, which are enclosed in an envelope. To transmit a message, the SOAP protocol data must be embedded in other protocols from the application and transport layer. The most popular and widely used example is HTTP over TCP. Currently, XML, XML Schema and SOAP have the status of a W3C recommendation.

In summary, communication partners use the SOAP protocol to agree on a message format. In this case, the message data are always going to be encapsulated in an XML document, which itself is structurally defined by an XML schema. As the SOAP protocol can only define a message structure, it must employ other methods or protocols to perform the actual message transport and delivery. While smartOR relies on the widely used HTTP, other protocols such as SMTP or FTP are possible.

For a detailed description of the offered services, devices use the XML-based WSDL. WSDL is independent of protocols, programming languages and platforms and hence meets the requirements set by a heterogeneous clinical network. In particular, machine readability is guaranteed by the XML format. This is important inasmuch as every service consumer can now access information on the syntactic form in which service calls are to be made. SmartOR web services provide a detailed interface description to consumers using WSDL.

For web services, there are many additional modular specifications in the context of SOAP and WSDL use, all named with the prefix 'WS-' and collectively referred to as 'WS-*'. WS-Discovery is used for the discovery of devices and their services in the network (either a certain type or all), while WS-Addressing provides permanent and unique addressing; WS-Security describes how the integrity and confidentiality of messages can be ensured. The WS-* specifications extend the functionality of SOAP and WSDL and are heavily used in the smartOR project.

To meet the requirements of resource-limited devices, of which there are typically plenty in OR networks, web services must be carefully implemented. For a secure communication via web services, for facilitating the identification and description of devices and services and for eventing mechanisms, DPWS is used. DPWS describes series-to-use WS-* standards and presents specifications for implementing. In smartOR, in particular, WS-Discovery, WS-Addressing and WS-Security are used besides other WS-* specifications.

12.2.5 M²IO Switch

Despite a steadily higher throughput and lower latency for Ethernet connections, video or audio signals cannot always be transmitted reliably or with sufficient quality. In these cases, conventional KVM switches (keyboard–video–mouse) are primarily used for direct user interaction. Therefore, a KVM switch is integrated in the smartOR concept. Devices such as the endoscope, which need to display a high-quality video signal on the surgical workstation, can set up the KVM switch using the OSCB and route the video and audio signals to the workstation or other devices (Figure 12.3). The KVM switch is called the matrix switch (man–machine input/output) and has been defined to meet the following requirements:

- OSCB basic functions:
    - Read the current configuration
    - Switching all inputs and outputs in a matrix
- Input/output types: HDMI/DVI, VGA and S-video (input only), USB keyboard, mouse and touch screen, analog audio (duplex), RS232
- Operation with extenders to improve cable length problem

The Smart Operating Room

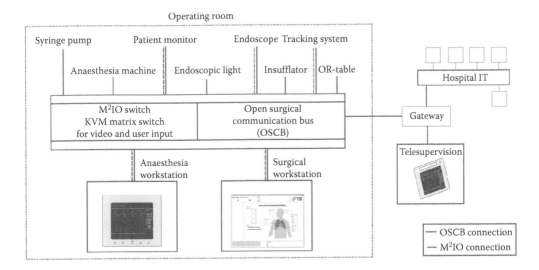

FIGURE 12.3
Overview of the smartOR.

- Configurable picture-in-picture for simultaneous display of the planning software and control
- Emulation of USB protocols of various screens with different resolutions
- Conformity to standards: DIN60601-1 (electrical safety), DIN 62304 (software) and so on.

Thereby, several smartOR project objectives should be achieved. It is possible to operate the switch functions via web services and OSCB. The matrix switch can also provide an easy-to-use dynamic user interface for the devices registered over OSCB. If error messages are sent that conform to the protocol, a notification can be delivered to the user by the switch.

12.2.6 Related Works

During the last few years, many approaches have been developed to build integrated networked operating rooms and optimised human–machine interaction concepts. For example, there are many integrated operating rooms from Richard Wolf GmbH, Karl Storz or Olympus for minimal invasive endoscope-based surgery. All these integrated operating room solutions are based on networked devices to exchange information between medical devices. For example, the endoscope can submit the current exposure to the endoscopic light to optimise the exposition. Furthermore, there are central operating panels visualising the current state of surgical devices and allowing the user to control different devices from one panel. Additionally, these devices can be operated by voice activation. Unfortunately, these solutions are built by only one manufacturer and only devices from licensed vendors can be integrated in these solutions. Furthermore, the interaction and exchange of vital signs between the anaesthesia workplace and the surgical workplace are missing. To support a manufacturer-independent device interoperability, SOA on the basis of web services has been used in previous interdisciplinary European research projects such as, for example, the OrthoMIT [14,15] and FUSION [16] projects. During these projects, the web service-based SOA architecture could be proved. Furthermore, similar concepts have been developed in the context of MDPnP [17,18], projects by Julian Goldman in the United States. Finally, the smartOR project aims to merge the previously named aspects of a web service-based SOA architecture to allow a manufacturer-independent device interoperability, allowing information exchange between medical devices and central optimised human–machine interaction using central optimised display units. Finally, the concept is proofed in a demonstrator and the developed protocols are going to be standardised.

12.3 Concept of the smartOR

12.3.1 System Overview

The overview of an anaesthesia-dominated smartOR instance is shown in Figure 12.3, with essential technical components used for most surgical interventions.

The main component of the integrated OR is OSCB, which is the central transportation medium and the common communication principle. All devices in OR are connected to OSCB. To establish communication between the medical devices over OSCB, open and standardised interfaces are necessary, which are described in detail later. At the surgical workplace, the surgeon has a central display and control component, which can be controlled during a surgical intervention using touchless input methods by himself or the assistants. All available and controllable devices by the surgeon can be found on the graphical user interface of the surgical workstation. It presents context-related information for the surgical team, such as vital signs and relevant alarms, which potentially require further interaction, as well as the patient's current health state. Accordingly, the anaesthesia workstation provides all information about the patient that is relevant for the anaesthesiologist by consolidating the information of patient monitor, anaesthesia machine and syringe pumps during the intervention. Additionally, a telesupervision system [1] is integrated.

Moreover, several partners of the smartOR project have implemented the OSCB support and integrated their devices; for example, the integration of a tracking system, an insufflator, an endoscope with endoscopic light and an OR table have been demonstrated. All devices offer their functionality using web service encapsulation and communicate conformably to the OSCB protocol.

Optionally, a protocol converter can be used to integrate devices that do not implement the OSCB protocol. The converter can be a medical approved PC, such as, for example, the surgical or anaesthesia workstation, or an embedded system with the capability to run the OSCB protocol and connect to the medical device. The gateway is used for connecting the medical IT network in the OR to the hospital IT network. The two primary functions are the logical separation of the two networks plus the function as protocol converter and security system. The gateway filters potentially unsafe traffic from the hospital IT network and sensitive data by an integrated firewall. Furthermore, the gateway forwards data to the telesupervision system to the hospital IT, which is virtually connected to the operating room.

With this concept, all components can be substituted by others as long as they use the defined nomenclature and interface description that is characterised in the next section.

12.3.2 Nomenclature and Interface Description

To ensure the sustainability of the smartOR project, the consortium developed proposals for the standardisation of the protocol and the interfaces. These proposals are submitted to national and international bodies. So far, for the OSCB protocol, nomenclatures are defined for the devices that are used in the demonstrator (see Section 12.5). They are developed based on ISO 11073-10101 [19], which are used for semantic descriptions of devices, measurements, units, events and so on; see also Ref. [20].

Nomenclatures include a unique reference ID in the entire network, for example, for groups and types of devices, physical units or events. Other data can be assigned to this ID, depending on this determination. Data that can be assigned to a parameter or attribute could be, for example, a description, an understandable plain-text name, a type and the unit of the represented value. For an event, it may suffice if only the reference ID is transmitted. For example, if a device goes into standby mode, the sender and the reference ID is sufficient for the receiver of the message, since the information is immediately clear from the occurred event.

The designation is built using specified rules, which are based on the standard 11073. For example, the designation of an insufflator device would be MDC_DEV_INSUFFLATOR. MDC stands for medical device component; the second part specifies if it is a device (DEV), a command (CMD) or an event (EVT). Parameters are named using the device name as the second part. For example, MDC_ECG_HEART_RATE stands for the ECG heart rate. Examples of events that can be sent from an OR table are: MDC_EVT_ORTABLE_CHARGING, MDC_EVT_ORTABLE_BATT_OPERATED, MDC_EVT_ORTABLE_BATT_CHARGING, MDC_EVT_ORTABLE_ BATT_LOW. They describe the used battery and its charge state. A selection of possible parameters of an OR table, as well as their definitions and format, is listed in Table 12.1. The strength of the open protocol is that manufacturers can expand the parameters and functions according to their personal needs.

The interface definition is the core of OSCB. The goal is an encapsulation of the devices and their functions realised over DPWS and web services as well as the achievement of an efficient SOA-based machine–machine communication. The main focus lies on interoperability and openness to ensure a best-possible cross-vendor interaction of the participating devices. Furthermore, the discovery of other devices, their registration and deregistration plus the description of services are unified. As a model for the descriptive services, WSDL is used in version 1.1. WSDL describes the interface and specifies the characteristics and protocols for the transmission of messages. In OSCB, SOAP is implemented over HTTP in conjunction with TCP.

TABLE 12.1

Parameters of an OR Table

Reference ID	Definition	Type	Format	Unit of Measure
MDC_ORTABLE_POS_HEIGHT	Heights of OR-Table	Numeric	1000	mm
MDC_ORTABLE_POS_ LONGITUDINAL	Longitudinal position of OR-Table top	Numeric	1000	mm
MDC_ORTABLE_POS_BACK	Angle of the back	Numeric	1.1	degree
MDC_ORTABLE_POS_LEG_BOTH	Angle of leg	Numeric	1.1	degree
MDC_ORTABLE_POS_LEG_LEFT	Angle of left leg	Numeric	1.1	degree
MDC_ORTABLE_POS_LEG_RIGHT	Angle of right leg	Numeric	1.1	degree
MDC_ORTABLE_POS_TREND	Trendelenburg angle	Numeric	1.1	degree
MDC_ORTABLE_POS_TILT	Tilt angle	Numeric	1.1	degree

Services will be kept as general as possible to keep OSCB flexible and clear. There are two categories of services: required services, which must be implemented by any device, and individual services, which are dependent on the particular device and usually encapsulate a function. The mandatory identification service has to transmit the device's skills plus supported parameters and attributes upon request. Typically, this request is made by the service manager immediately after finding the device. To identify the devices, medical device profiles can be used. A mandatory control service allows one to set and read parameters. The reference ID of each parameter and the used command are passed. If the parameter should be set to a fixed value, the value must be submitted. To proceed on an OR table one step up, a call could look like this in pseudo-code:

setParameter(MDC_ORTABLE_POS_HEIGHT, MDC_CMD_PLUS, null);

Here, the increment command MDC_CMD_PLUS is used and the third value passed is ignored. Alternatively, this could be a specific target value for the parameter. In this case, the command would be MDC_CMD_VALUE instead of MDC_CMD_PLUS.

Feedback is made asynchronously via events. Syntactic and semantic errors are supported, along with the achievement of an intermediate or final value for a parameter. If a service control assigns an insufflator to set a certain pressure, the insufflator informs the requesting device regularly on intermediate values or any errors that may occur due to external circumstances.

12.3.3 Interaction with the Hospital IT/Post-Operative Care

Information needed during intervention, such as the patient's anamnesis or digital imaging and communications in medicine (DICOM) images from radiology, can be transferred from the hospital IT through the gateway to concerned devices in the operating room. For example, anamnesis is relevant for the surgeon and the anaesthesiologist for an optimised therapy. The DICOM images can be used by the navigation system for brain surgery.

After the intervention, an anaesthesia and surgery report can be transferred to the hospital information system. These reports are necessary for treatment after intervention in post-operative care or future therapies.

Devices exchanging such information need to communicate with the gateway to perform a data transfer. This is necessary due to the security restriction described in Section 12.3.4, to filter information for potential security risks for the smartOR (OSCB) network.

12.3.4 Security Aspects and Patient Safety

The implementation of the proposed SOA-based integration approach to a clinically applicable reliable system requires further steps. By adding networking functionality and an open bus to common standalone devices, these systems are suddenly exposed to malware threats and network attacks. The IEC 80001-1 risk management standard defines security goals for medical networks: patient protection, device and data safety as well as communication process stability.

Patient protection particularly includes the prevention of misbehaviour of medical devices that could physically harm the patient. Device and data safety refers to the availability of medical devices in a proper state and the confidentiality of sensitive data such as patient identity, diagnoses, medical history and medication. A lack of confidentiality can cause legal problems to the responsible organisation. Communication process stability ensures that devices can use available services and exchange data in the network.

Important hazardous situations are malware-infected systems in the same network and hacker attacks from outside the network. Moreover, devices must only accept commands from eligible service consumers at any time.

Approaches using centralised components bear the risk of creating one or more single points of failure. Components such as the service manager and the event manager are essential for the operation of the OSCB-based network. Failure of one of the central components may cause the breakdown of the OSCB network and make service providers unavailable for service consumers, which again can result in an insufficient patient-centred care. Hence, it is necessary to create redundancy of the important hardware and software and implement an additional decentralised communication method that can be fallen back upon an emergency situation.

Problems resulting from deficient information security can directly influence patient safety. For example, a syringe pump or an anaesthesia machine may cause serious harm to the patient by applying an incorrect dosage of anaesthesia medication. By allowing devices on the network to change these settings, there exists a new way of accessing them. It must be guaranteed that devices only accept authenticated, authorised and validated commands from service consumers.

The first step to protect the smartOR network is to separate it from the remaining (medical and non-medical) clinic IT. This adds a layer of security because an attacker needs physical access to at least one device in an access-controlled environment. However, communication with other networks can be desirable as electronic patient records can be stored, for example, by PACS systems outside the smartOR network and could provide essential data for the surgical workflow. To address this issue, a dedicated gateway is used as the only interface to the clinic IT network. The gateway is protected by a firewall and only permits non-malicious network traffic. This, however, only protects from attacks originating from the easily accessible network. Other devices in the smartOR network may come with a separate Internet connection and are at risk of being infected with malware, which makes them a possible threat to all units in the network. Hence, malware protection must be implemented on all devices. The proper usage of antivirus software or other counter-mechanisms, including related risk management, is left to the respective medical device manufacturer.

To meet the requirements for information security in IT networks [21], several approaches can be used. The Organization for the Advancement of Structured Information Standards (OASIS) published an extension to SOAP called WS-Security. The extension aims to provide confidentiality and integrity using the W3C recommendations XML Encryption and XML Signature. Possible techniques that can be applied are X.509 certificates or protocols following the Kerberos paradigm.

From a medical device perspective, it is necessary to use access control mechanisms to make sure that only commands from eligible service consumers are accepted. All smartOR devices and systems feature an access control list (ACL) that defines which device types (e.g., cameras and OR tables) are eligible. In addition, the responsible operator IT risk manager is able to define exceptions and additions to ACL for every possible device combination and thereby overrule the vendor-supplied ACL. These exceptions and additions are stored in a central database and are transferred to the smartOR devices in defined intervals or on manual request (e.g., immediately before starting surgery). Decisions on allowance or denial are made based on ACL and the local copy of the exception list is available on the device.

At least for medical devices that are critical for the patient's health status, this necessitates the prevention of impersonation attacks. To be able to authenticate other devices, digital signatures and certificates are needed and this requires the use of asymmetric cryptography and the implementation of a public key infrastructure (PKI). Despite there being the option to employ a central smartOR PKI that issues certificates for every smartOR device, the current OSCB specification stipulates an extra PKI to be run by each responsible organisation. Hence, it is the risk manager's task to issue a certificate for each device and to store the facility's public key on the device during the first initialisation.

A further aspect in the field of security is updates that are released for operating systems in use. Owing to newly discovered threats (caused by security flaws in the operating system) or new versions of computer viruses, this update may be essential for the operating system security [22]. However, these updates may not always be immediately transferred to a running system because, under certain circumstances, it is not directly clear that there are no unwanted side effects on the existing functionality of the medical device.

12.3.5 Risk Management

The IEC 80001-1 (application of risk management for IT networks incorporating medical devices) will provide a basis for modular, multi-vendor integration of medical devices in clinical IT networks in the future [23]. It defines the roles and responsibilities and describes the risk management measures in the overall context.

The technical risk analysis performed by Ibach for the SOA-based framework has shown that a flexible, modular integration with a central user interface is possible [3]. Regarding the human–machine interaction, a first risk analysis has been performed [24]. For this, the endoscopic use case from the smartOR project has been abstracted, that is, medical devices and input/output devices and users are considered. To reflect the different processes, three modes of communication have been defined: user–system communication, system–user communication and system–system communication. The mAIXuse approach [25], which has been developed at the Chair of Medical Engineering at RWTH Aachen University, has been applied as part of the first analysis and then examined for potential errors in the information process. The analysis identifies potential risks in the following aspects:

- Process of perception (visual, auditory, tactile)
- Cognitive processing (sensorimotor, rule-based and knowledge-based regulatory action level)
- Context of motor action (defect classification according to the external appearance)

For example, the initial risk analysis has shown that before surgery takes place, a system initialisation has to be done. At this point, it can be examined whether there are suitable input/output devices present for the medical equipment used. To reduce or avoid possible errors, the creation of situation awareness should be a main goal of system and GUI design.

12.4 User Interaction: Workstations

12.4.1 Surgical Workstation

In Figure 12.4, a screenshot of the surgical workstation can be seen with the consolidated patient's state on top and activated OR-table control panel. On the left side, all devices

The Smart Operating Room

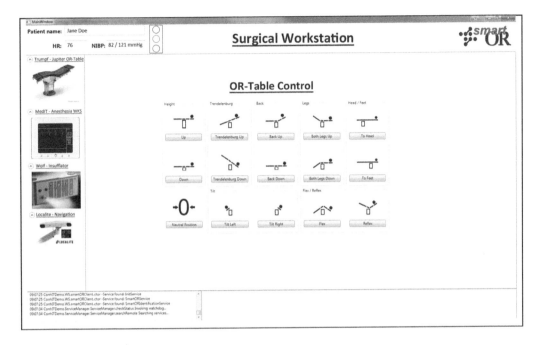

FIGURE 12.4
Surgical workstation with activated OR table control.

connected to OSCB are listed. Detailed information and options for device control can be accessed by selecting a device. In the detail frame, individual information according to the device is displayed. This detailed information, for example, a control panel, is delivered by the device vendor and can be designed to present individual device control functions and properties. Such customised control panels for device operations are described in a platform-independent XML-based language. The control panels must be requested by the workstation from the devices over the OSCB.

The functionality of the surgical workstation is optimised for the workplace of the surgeon. For example, during a standard intervention, the surgical team only needs the most important information about the patient, such as heart rate, blood pressure and a traffic light showing the patient's state, which are displayed in a condensed form on top of the application. However, the devices used by the surgeon, such as, for example, the OR table control, an endoscope or the tracking system, can be controlled by the surgeon or an assistant. Furthermore, the information of the surgical workflow is displayed. Using the information of the workflow, the surgeon can be informed of future steps of the intervention. For example, devices can automatically be prepared for use in the following step of the intervention.

In case of an alarm on a specific device, the device is highlighted in the list on the left side. In order not to interrupt possible operation by the surgeon, the panel specifying the alarm is not shown automatically. However, detailed information can be retrieved by selecting the according device manually. For example, detailed information about a blood pressure alarm can be retrieved by activating the anaesthesia workstation's panel. Furthermore, detailed information about the patient's state and vital signs can be obtained from the anaesthesia workstation.

12.4.2 Anaesthesia Workstation

12.4.2.1 Overview

Typically, the anaesthesiologist's workplace consists of a patient monitor, an anaesthesia machine and syringe pumps for drug application. These components are essential and necessary for monitoring the patient's condition and performing the narcosis. The workstation retrieves vital signs and alarms for further processing and combines them to be presented on a central display unit on the workstation. Furthermore, an intelligent alarm concept based on the patient's vital signs combined with information from the smartOR network supports the anaesthesiologist. Additionally, a telesupervision system is integrated for training and teleconsultation purposes.

In Figure 12.3, the anaesthesia workstation and the position in the smartOR network can be seen. In Figure 12.5, the typical components of the anaesthesiologist's workplace are shown. The workstation is the central component of the anaesthesia workplace. It receives data such as the ECG, blood pressure, breathing parameter and applied drugs from the patient monitor, anaesthesia machine and syringe pumps, processes these data and presents them in a consolidated uniform way. Furthermore, the anaesthesia workstation uses data received over OSCB from surgical devices such as the OR table or an insufflator, combines them with the vital signs of the patient and generates context-sensitive alarms and patient information. During the intervention, the workstation sends consolidated information about the patient to the surgical workstation, as described in Section 12.4.1. In case of a critical situation, detailed information about the patient's state and additional vital signs are sent to the surgical workstation on request, using an additional real-time connection.

The main component of the anaesthesia workstation is a common medically approved PC in combination with special software written in C++/Qt called *SomnuCare*.

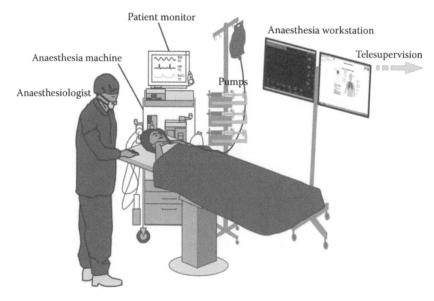

FIGURE 12.5
Overview of the anaesthesia workplace.

12.4.2.2 SomnuCare Software Concept

The SomnuCare software has three main tasks. The first is collecting the information from all connected devices using different interfaces and storing vital signs, commands and alarms in a central storage unit. The second task is to analyse these data to present the information in an efficient and adapted way. Furthermore, this information is used to build a support system for the anaesthesiologist. The third task is to provide vital signs, alarms and consolidated information to OSCB. Additionally, it acts as a server for the telesupervision system. Thus, the workstation features a video and audio channel to provide further information about the current situation in the OR to the supervisor.

In Figure 12.6, the main components of the SomnuCare software are shown. The SomnuCare software uses a modular architecture. This allows the programmer to easily integrate new components such as interfaces for medical devices or processing medical data stream. The two central modules are the memory mapped file (MMF) storage and the SomnuCare interfaces. The interfaces retrieve data from connected devices using different hardware and software interfaces and store the data in the MMF module. To easily integrate new devices into SomnuCare, the interfaces offer a uniform application programming Interface (API). Data stored in the MMF module can be easily accessed by other instances within the software, as a GUI, which presents vital signs and additional information generated by SomnuCare, the telesupervision server or OSCB.

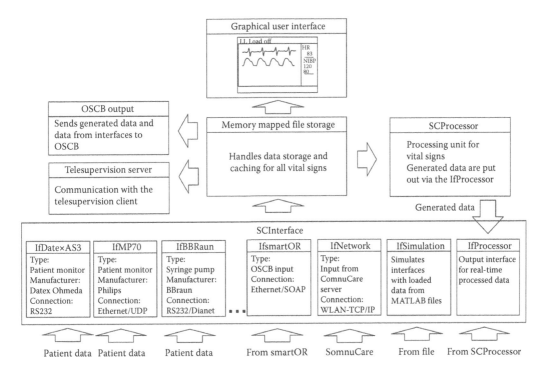

FIGURE 12.6
Overview of SomnuCare software. (Modified after M. Köny et al., *The Third International Conference on Emerging Network Intelligence*, Lisbon, Portugal, 2011.)

12.4.2.3 Interfacing Medical Devices

Medical devices can be connected to SomnuCare using different methods and protocols. Every device supporting the OSCB protocol can be connected to the anaesthesia workstation regardless of the type and manufacturer.

Devices using proprietary protocols or protocols that are not compliant with OSCB can be connected through generic interfaces in SomnuCare. Generic interfaces are software modules that can be accessed by SomnuCare using a uniform software API, which is shown in Figure 12.7. Communication with the medical device itself is performed in the interface itself. This modular design enables a fast and easy integration of new devices.

As the connection to medical devices can consist of different physical interfaces and software protocols, the implementation of the interface itself is left to the programmer. For example, a simple serial RS232 protocol can be implemented directly in the interface, whereas a complex protocol based on UDP multicasts can be implemented in an additional thread that communicates with the interface.

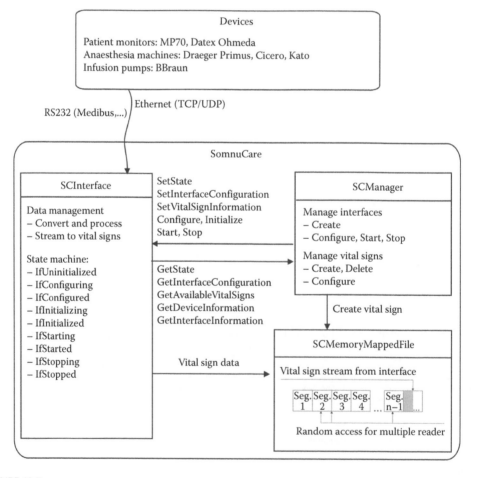

FIGURE 12.7
SomnuCare interface API.

12.4.2.4 Memory Mapped File Engine

The MMF engine is one of the central modules in SomnuCare, since it stores all data, such as vital signs and alarms, from the generic interfaces. Every vital sign is represented by two MMFs, one for the data and the other for managing the data in MMF. Data within the MMFs are stored in a binary format that enables, in connection with the control segment, one writer (the interface) to store data in MMF, while multiple readers are allowed to read data from MMF, as can be seen in Figure 12.8. The writer is only allowed to append data because MMF acts as a log file. However, multiple readers can access the file randomly. Furthermore, a reader can register its instance and read only appended data since the last read operation. MMF itself handles the coordination of the write and read pointers.

Multiple high data rate vital sign streams, such as ECG or EEG, need a lot of memory. Hence, the whole MMF is not mapped into the memory. Only the current write segment and segments being currently read are being mapped into the memory. This enables the workstation to record vital signs over hours without increasing the used memory. This feature is described in detail in Ref. [26].

12.4.2.5 Innovative Clinical Applications

One of the most important aspects of human–machine interaction is the efficient and safe presentation of patient alarms and decision support in anaesthesia. Patient alarms inform the physician about a possible critical state of the patient. For example, a sudden increase of the blood pressure can, depending on the situation, be associated with an insufficient depth of narcosis. However, there are many other influences that can cause such an increase of blood pressure. State-of-the-art assistance systems such as the Draeger Smart Pilot View (www.draeger.com) support the anaesthesiologist by predicting the depth of narcosis depending on the currently applied drugs and vital signs. Comparable to the anaesthesia workstation, this system consolidates information from connected devices such as the patient monitor, the anaesthesia machine and syringe pumps to perform the prediction. As already discussed, these integrated solutions are mostly supported by only one manufacturer or licensed partners. The system presented in this section is used in the use case discussed in Section 12.5 and combines vital signs from OSCB, intelligent algorithms and events from used surgical devices to an intelligent decision support system, which can be used in the operating room and by the telesupervision system introduced in Section 12.3.1.

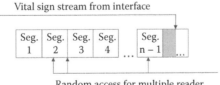

FIGURE 12.8
Segmentation principle of the memory mapped file. (Adapted from M. Köny et al. *The Third International Conference on Emerging Network Intelligence*, Lisbon, Portugal, 2011.)

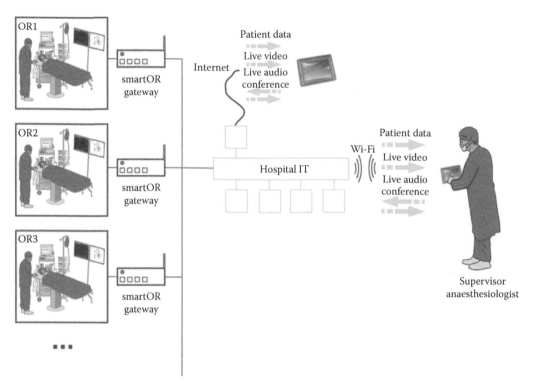

FIGURE 12.9
Typical telesupervision scenario.

A typical telesupervision scenario is shown in Figure 12.9. Vital signs and alarms from the patient are transmitted to a remote device, such as a tablet PC or a mobile phone. An optional video and audio stream complements the information of the respective situation in the OR.

There are two application fields for telesupervision. The first is to support an anaesthesiologist in education. During the education of an anaesthesiologist, the trainee performs multiple different types of narcosis. An experienced senior physician supervises multiple interventions. The trainee can receive support from the supervisor. The telesupervision system assists the supervisor to provide this support, without being present in the operating room. Since the client side presents exactly the same information that the anaesthesiologist in place sees, the supervisor can quickly get an overview, especially in critical situations. In this application field, the supervisor stays in the hospital and the connection to the operating room is realised via the hospital IT infrastructure to the smartOR network using the smartOR gateway. The second application area is similar to teleconsultation. An anaesthesiologist in the operating room retrieves support by a colleague in another hospital. Similar to the first application field, data from the OR must pass the gateway. After that, data must pass through the hospital firewall to the target. Depending on the Internet connection, there may only be a reduced bandwidth available, and only a reduced set of vital signs can be transmitted with a limited video quality.

12.5 Use Case Laparoscopic Intervention

For the evaluation of the presented concepts and protocols, a demonstration scenario has been created. The demonstration OR consists of the components described in Section 12.3.1 and Figure 12.3.

The subject of the use case is a gynaecological laparoscopic intervention of a 35-year-old woman. It demonstrates the interaction of the devices used by the concerned medical staff. In the simplified use case, the following devices are relevant:

- The *surgical workstation* provides relevant displays, controls and user interfaces for the surgical team (nurses, assistants, surgeon).
- The *operating room table* is controllable to achieve an appropriate position and decline for the intervention.
- The *insufflator* is used by the surgeon to blow CO_2 into the abdomen for the intervention.
- The *anaesthesia workstation* with a central display shows consolidated information from the patient monitor, anaesthesia machine and syringe pumps. Furthermore, it generates context-dependent alarms and controls devices used by the surgical team under certain conditions.
- The task of the *event manager* is to interchange events and alarms of medical devices among each other.

These devices are connected to OSCB and make use of information provided by other devices in the network. The central user interface components are the surgical and the anaesthesia workstation. The displays of the two workstations present optimised and context-sensitive information, in particular, during critical situations. To show the patient's state, both workstations show a traffic light, where green means the patient is healthy, yellow means the patient's state is alarming and red means the patient's state is critical.

The scenario starts after induction of anaesthesia and initiation of the volume-controlled mechanical ventilation. The patient's state is healthy and all vital signs are within the alarm limits. For laparoscopy, a caponoperitoneum is needed. Therefore, CO_2 is used to fill the abdominal cavity to improve the conditions of visibility during endoscopy. After the initiation of gas insufflation, the anaesthesia machine raises an alarm because the maximum tolerable breathing pressure must be increased to overcome the abdominal counter-pressure. After resolving this situation by changing the appropriate alarm limits with the anaesthesiologic workstation, the alarm is cancelled. The patient's state is shown as 'green' again. For the subsequent surgical steps, a declined position of the table is required—bringing the upper body to a degraded position. Therefore, the surgical workstation is used to control the operating table. Because of the further increased intra-thoracic pressure caused by the abdominal organs, the anaesthesia workstation raises two alarms. One alarm represents the breathing pressure, which overstepped the defined limit; the other alerts due to the too small amount of applied tidal volume. The patient's state is now reclassified as critical and is transferred to the surgical workstation. The anaesthetist requests to reduce the declination angle of the OR table and changes the OR table position accordingly. Thus, the physical state of the patient is 'green' again.

The mentioned steps of the use case with the underlying network communication are displayed in the following table:

1. Phase after induction of anaesthesia:

    The anaesthesia workstation submits consolidated information to the surgical workstation about the state of the patient (green = healthy)
    The anaesthesia workstation submits continuous information about the patient's state and selected parameters to all participants on OSCB

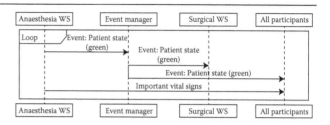

2. Start of the intervention: (CO_2 insufflation into the abdomen)

    The insufflator submits the start of the CO_2 insufflation to all participants

3. Alarm at the anaesthesia workstation:

    Owing to the increased abdominal pressure, the breathing pressure exceeds the predefined alarm limits and triggers an alarm that is forwarded through OSCB. Additionally, more detailed context information (CO_2 insufflation) will be provided
    The anaesthesia workstation submits consolidated information to the surgical workstation about the patient's state (yellow = alarming)

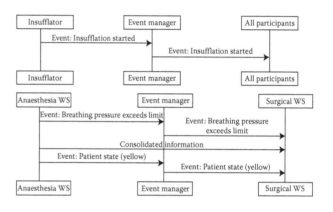

4. The anaesthesiologist modifies the alarm limit at the anaesthesia workstation:

    The alarm limit of maximum breathing pressure is increased
    The anaesthesia workstation cancels the alarm and sends the information to all participants
    The anaesthesia workstation submits consolidated information to the surgical workstation about the patient's state (green = healthy)

5. The surgeon changes the OR table position:

    The breathing pressure exceeds the new set-up limit and the tidal volume cannot be applied
    The anaesthesia workstation sends breathing pressure and tidal volume alarms to the event manager
    The anaesthesia workstation submits consolidated information to the surgical workstation about the updated patient's state (red = critical)

6. The anaesthesiologist changes the OR table position by the anaesthesia workstation:

   The anaesthesia workstation requests the authorisation from the surgical workstation

   The surgeon grants the request and the surgical workstation sends the permission for controlling the OR table

   The anaesthesia workstation transmits the new OR table position to the OR table

7. Situation is solved:

   The anaesthesia workstation submits consolidated information to the surgical workstation about the patient's state (green = healthy)

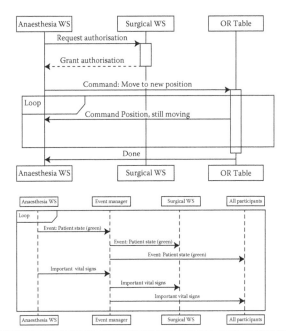

12.6 Conclusion and Discussion

The development of the smartOR standard is driven by a medical motivation for a standardised human–machine interaction and the need for a manufacturer-independent network in the operating room. Therefore, the development has two main focuses: the OSCB protocol, which aims to build a modular and flexible protocol for manufacturer-independent networking, and new innovative concepts for human–machine interaction in the operating room.

OSCB is based on an SOA, which consists of service providers and consumers. Normally medical devices represent both, a service provider and a service consumer. The communication channel Ethernet is used, with the IP-based TCP and UDP protocol in the upper layers. The application layer is based on DPWS-compatible frameworks, which offer web services for every device. Using web service technology, different devices and protocol versions are able to communicate with each other. A unified nomenclature is defined by implementing the ISO 11073 standard for exchanged parameters, commands and events (alarms). To conform to the criteria for medical devices and include risk through the network, risk management according to ISO 11971 and ISO 80001 is included during the design process. During the design process, the developed protocols and standards have been tested in a standard surgical use case.

The human interaction concepts for operating medical devices make use of the connection between all technical medical devices in the operating room. Based on two workstations, one for the surgical and one for the anaesthesia workplace, intelligent visualisation and operation concepts have been developed. The surgical workstation presents optimised information for the surgical workplace, such as the consolidated patient's state and parameters of surgical devices. Furthermore, all surgical devices in the operating room can be operated by the surgical workstation. A surgical workflow system determines the current state and the possible next steps of the intervention-determining device usage and actions. The anaesthesia workstation is the central display component of the anaesthesiologist's workplace. Vital signs and alarms from the patient monitor, anaesthesia machine and syringe pumps are collected and presented in a uniform patient monitor style. Furthermore, the anaesthesia workstation collects data from OSCB, such as information provided by the applied surgical devices. To provide a decision support for the anaesthesiologist, collected data are evaluated in real time. For example, to estimate the patient's condition, multiple vital signs and alarms are analysed and the state is presented in a traffic light scheme. Depending on the used surgical devices, actual vital signs and alarms, intelligent alarms dependent on the context of the intervention can support the anaesthesiologist during critical situations.

In future, the smartOR architecture, for example, the OSCB protocol, aims to be an open manufacturer-independent standard for medical device networking. This standard enables every manufacturer, even smaller ones, to integrate their devices into the new OR architecture. Using appropriate methods, additional risks resulting from the networked connection are reduced to a minimum. Therefore, a safe operation is guaranteed. Furthermore, patient safety and treatment can be improved using modern, clearly arranged user interfaces and supporting technologies.

Acknowledgements

The smartOR project is supported by the German Ministry of Economics and Technology and has the support code 01MMA09041A.

The authors thank the following smartOR project partners for their contribution to the presented collaborative work:

- Innovation Center for Computer Assisted Surgery (ICCAS) of the University of Leipzig
- LOCALITE GmbH
- Richard Wolf GmbH
- SurgiTAIX AG
- Synagon Gmbh
- VDE/DGBMT

Furthermore, the project is support by clinics of different branches.

References

1. M. Walter, Telesupervision und automatisierung in der anästhesie, in *VDE Kongress 2006 Aachen*, Aachen, Germany, 2006.
2. J. P. Bläsing, *Medizinprodukte: Risikomanagement im Lebenszyklusmodell*. Ulm, Germany: TQU VERLAG, 2008.

3. B. Ibach, *Konzeption und Entwicklung einer serviceorientierten Integrationsarchitektur für die Vernetzung von Medizinprodukten im Operationssaal*, 8th ed., Aachener Beiträge zur Medizintechnik, Ed.: Shaker Verlag, 2001.
4. C. M. MacKenzie, K. Laskey, F. McCabe, P. Brown and R. Metz, Reference model for service oriented architecture 1.0., in *OASIS Committee Draft*, 2006.
5. S. Poehlsen, *Entwicklung einer Service-orientierten Architektur zur vernetzten Kommunikation zwischen medizinischen Geräten, Systemen und Applikationen*, PhD thesis, Ed. Universtität zu Lübeck, 2010.
6. O. Holmeide and T. Skeie, VoIP drives realtime Ethernet, 2001.
7. D., Paula, Introduction to real-time Ethernet I, in *Circuits and Systems Research Centre*. Ireland, University of Limerick, 2004.
8. J. Jasperneite, P. Neumann, M. Theis and K. Watson, Deterministic real-time communication with switched Ethernet, in *4th IEEE International Workshop on Factory Communication Systems*, 2002, pp. 11–18.
9. J. Loeser and J. Wolter, Scheduling support for Hard Real–Time Ethernet Networking, in *Proceedings of the Workshop on Architectures for Cooperative Embedded Real-Time Systems (WACERTS'04)*, Lisbon, Portugal, 2004.
10. H. Haas and A. Brown, W3C-Web Services Glossary. 2004, February. [Online]. http://www.w3.org/TR/ws-gloss/.
11. D. Kossmann and F. Leymann, Web Services, *Informatik-Spektrum*, 27(2), 117–128, 2004.
12. M. Skulschus and M. Wiederstein, *XML Schema*. Bonn: Gallileo Press, 2004.
13. N. Mitra et al. SOAP Version 1.2 Part 0: Primer/W3C. [Online]. http://www.w3.org/TR/2007/REC-soap12-part0-20070427/, 2007.
14. J. A. K. Ohnsorge, K. Radermacher and F. U. N. C. Buschmann, The orthoMIT project. Gentle surgery using innovative technology, *Bundesgesundheitsblatt*, 52(3), 287–296, 2009.
15. orthoMIT–Minimal-Invasive Orthopädische Therapie. 2012. [Online]. http://www.orthomit.de/.
16. FUSION–Future Environment for Gentle Liver Surgery Using Image-Guided Planning and Intro-Operative Navigation. 2012, Oct. [Online]. http://www.somit-fusion.de.
17. Medical Device Plug and Play. 2012, Oct. [Online]. http://www.mdpnp.org/.
18. J. Goldman, Advancing the adoption of medical device plug-and-play interoperability to improve patient safety and healthcare efficiency, Medical Device 'Plug-and-Play' Interoperability Program, 2000.
19. ISO/IEEE 11073-10101:2004, Health Informatics—Point-of-care MedicalDevice Communication—Part 10101: Nomenclature, 2004.
20. J. Benzko, A. Janß and K. Radermacher, smartOR–plug&play in the OR, *International Journal of Computer Assisted Radiology and Surgery (CARS)*, June 2012.
21. M. Kaeo, Designing network security, Second Edition, in *Dedicated minimally invasive surgery suites increase operating room efficiency*, D.R. Urbach, J.B. Speer, B. Waterman-Hukari, G.F. Foraker, P.D. Hansen, L.L. Swanström and T.A.G. Kenyon, Eds.: Cisco Press.
22. T. Jaeger, Operating system security, *Synthesis Lectures on Information Security, Privacy, and Trust*, 1(1), 2008.
23. IEC 80001-1:2010, Application of risk management for IT-networks incorporating medical devices—Part1: Roles responsibilities and activities, 2010.
24. J. Benzko, A. Janß and K. Radermacher, Bewertung und Gestaltung der Mensch-Maschine-Schnittstellen in integrierten OP-Systemen, *Fortschritt-Berichte VDI Reihe 17 Biotechnik/Medizintechnik*, vol. Automatisierungsverfahren für die Medizin, 17, 29–30, 2012.
25. A. Janß, W. Lauer, F. Chuembou Pekam and K. Radermacher, Using new model-based techniques for the user interface design of medical devices and systems, In: *Human Centered design of e-health technologies: Concepts, methods and applications*, Roecker, C., Ziefle, M. Eds. IGI Group or Medical Information Science Reference, Hershey, pp. 234–251, 2011.
26. M. Köny, M. Czaplik, M. Walter, R. Rossaint and S. Leonhardt, A new telesupervision system integrated in an intelligent networked operating room, in *The Third International Conference on Emerging Network Intelligence*, Lisbon, Portugal, 2011.

13

Distributed Online Safety Monitor Based on Multi-Agent System and AADL Safety Assessment Model

Amer Dheedan

CONTENTS

13.1 Introduction 317
    13.1.1 Fault Detection and Diagnosis 317
    13.1.2 Alarm Annunciation 318
    13.1.3 Fault Controlling 319
    13.1.4 Monolithic and Distributed Online Safety Monitors 319
    13.1.5 Motivation 320
13.2 The Monitored System 321
13.3 Distributed Online Safety Monitor 323
    13.3.1 Distributed Monitoring Model 325
    13.3.2 Multi-Agent System 330
13.4 Case Study: ABS 335
13.5 Discussion, Validation and Conclusion 341
References 341

13.1 Introduction

Dating back to the early 1980s, research effort has focused on the development of advanced computer-based monitors. Since then, computerised online safety monitors started to appear as computer systems are installed in the control rooms of plants and flight decks of aircraft [1,2]. Computerised monitors have been dealt with differently in terms of (a) their capacity to deliver three safety tasks: fault detection and diagnosis, alarm annunciation and fault controlling; (b) their architectural nature: monitors could be developed from multi-agent (distributed) or monolithic (centralised) reasoning.

13.1.1 Fault Detection and Diagnosis

Fault detection and diagnosis techniques are typically developed as model-based and data-based techniques [3,4]. The distinction between these techniques lies in the way of deriving the knowledge that informs the real-time reasoning. Specifically, knowledge of model-based techniques is derived from offline design models, such as data flow diagrams (DFD), functional flow block diagrams (FFBD) or, more recently, from models defined in the unified modelling language (UML). Knowledge about the normal behaviour of the monitored system can be obtained directly from these models. To obtain knowledge about abnormal behaviour, analysis techniques such as HAZard and OPerability study (HAZOP), functional failure analysis (FFA) and failure mode and effect analysis (FMEA) are used to analyse the design models [5].

Knowledge of data-based techniques, on the other hand, is derived from the online context of the monitored system. Knowledge about normal behaviour is obtained by empirical experiment of fault-free operation of the monitored system. To derive knowledge about abnormal behaviour, the possible faults of the basic components are identified (by applying FMEA to the basic components) and injected experimentally in the operational context. The resulting symptoms and ultimate effects on the functionality of the system are then modelled [6].

In both model-based and data-based techniques, monitoring knowledge is presented to real-time reasoning in executable format as monitoring models. To deliver fault detection and diagnosis, a monitoring algorithm executes the monitoring model by instantiating, evaluating and verifying modelled conditions with real-time sensory data.

Model-based techniques have exploited a wide range of monitoring models, such as Goal Tree Success Tree (GTST) [7–9], fault trees [10–12], signed direct graph [13,14], diagnostic observers [15,16] and parity equations [17–20].

Similar variety can be seen with the data-based techniques. Consider, for example, rule-based expert systems [21–24] and qualitative trends analysis [25–27], artificial neural networks [28], principal component analysis [29,30] and partial least squares [31].

13.1.2 Alarm Annunciation

Alarm is the key means to bring the occurrence of faults to the attention of the operators [32]. Developing an alarm technique involves the consideration of alarm definition, alarm processing and alarm prioritisation and availability [33,34].

Alarm definition concerns the definition of mode dependency, which is required to establish a distinction between events that occur due to normal operation and others that occur due to faults, so that confusing alarms can be eliminated. State-machines [11], operational sequence diagrams [35] and system control charts [32] are among the models that have been exploited to address this issue. Alarm definition also concerns the definition of an effective threshold, the violation of which would result in verifying the occurrence of an event. Thresholds should not be too sensitive and result in false verification, and at the same time, not too relaxed, which would result in late verification and depriving the operators of knowledge about the actual conditions [34].

In alarm processing, distinction among genuine, consequent and false alarms should be achieved. While genuine alarms should be released, consequent and false alarms should be filtered out to avoid confusing alarm avalanches. Cause–consequent analysis of the design models can establish the distinction between causal alarms that concern the maintenance operators and consequent alarms that concern the pilot operators [36,37]. Sensory measurement validation can eliminate the potential for false alarms. Recent techniques achieve validation through analytical redundancy among sensors, for example, see [38–41]. On the other hand, earlier techniques depended on hardware redundancy [42], that is, redundant sensors. Although hardware redundancy techniques offer adequate robustness, their applicability is limited since they demand increases in cost, weight and volume.

Alarm prioritisation and availability is a process in which alarms are given priorities according to their importance, so they are selected and announced accordingly [34]. The highest priority is always given to safety consequences [43]. Dynamic and group presentation are two strategies to prioritise alarms. In dynamic prioritisation, alarms might be prioritised by (a) different colours (red, amber, magenta) [34]; (b) different severities, such as catastrophic, critical, marginal and insignificant [44]; (c) presenting the highest priority alarms and hiding and facilitating optional access to the less important ones [34]. Group presentation takes advantage of the screen display (LCD) to present alarm information in windows according to the hierarchical architecture of the monitored process and the importance of the relevant functionality [45]. Windows may allow operator interaction through facilitating the silencing of the alarm sound or suppressing the illuminated alarm lights [46].

13.1.3 Fault Controlling

Practically, fault controlling is considered in parallel with the controlling process. Fault controlling is implemented in two different approaches. The first is by manual interference of the system's operators, in which further to the need of an advanced alarm technique, the operators should also be trained and provided with guidance on controlling faults [47–49].

The other approach is achieved automatically by a computerised controller, which is commonly called a fault-tolerant control system (FTCS) [50,51]. FTCSs, in turn, are classified into active fault-tolerant controlling (AFTC) and passive fault-tolerant controlling (PFTC) [47].

Research on AFTC has been motivated by the aircraft flight control system [51]. Faults are controlled by selecting and applying the corresponding corrective procedure. An engine fault of a two-engine aircraft, for example, requires a procedure of: (a) cutting off fuel flow to the faulty engine; (b) the achievement of cross feed from the tanks that were feeding the faulty engine; (c) applying the corresponding command movements to control the surface and compensational instructions to the operative engine [52].

PFTC relies mainly on redundant components, such as multiple control computers and backup sensors and actuators [53,54]. Typically, provision of redundant components is implemented by hot or cold standby redundancy. In hot standby redundancy, the system is provided with parallel redundant components that operate simultaneously (powered up). Each component monitors the output of the other(s). Should any of them fail, the others take over. In cold standby redundancy, only one component is online (powered up) and other copies are on standby (powered down). Should the online component fail, it is powered down and one of the standby components is powered up by a controller [55].

13.1.4 Monolithic and Distributed Online Safety Monitors

Monolithic and multi-agent are two common classes of computerised monitors. The monolithic monitor in Ref. [11] has been developed from a monitoring model derived from the application of the Hierarchically Performed Hazard Origin and Propagation Studies (HiP-HOPS) safety assessment technique [56]. The model consists of a hierarchy of state-machines (as a behavioural model) that record the behaviour of the monitored system and its sub-systems and a number of fault trees as diagnostic models that relate detected faults to their underlying causes. The concept was motivated by observation of the fact that immense offline knowledge ceases its benefit and is rendered useless after certifying the safe deployment of critical systems. The exploitation of that knowledge in the context of the online monitoring results accordingly in an effective and cost-effective monitoring model.

A quite similar monolithic monitor is developed in Ref. [12]. The only difference is that the hierarchy of the state machine is replaced with the control chart of the monitored system and the fault trees are maintained as the diagnostic models.

The main limitation of these monitors is that they are based on a monolithic concept in which all monitoring of a plant is delegated to a single object or device. This does not align well with the distributed nature of most modern systems. Systems are typically implemented as a set of sub-systems which exist in a complex cooperative structure and coordinate to accomplish system functions. Systems are also typically large and complex and show dynamic behaviour that includes complex mode and state transitions.

As a result, such systems need a distributed mechanism for safety monitoring. First, it is essential to minimise the time of online failure detection, diagnosis and hazard control; second, a distributed monitoring scheme can help focus and rationalise the monitoring process and cope with complexity.

In Ref. [57], a number of agents are deployed on two levels: lower level and higher level. Each agent is provided with a corresponding portion of the monitoring model; agents of the lower level are provided with functional models, and the higher-level agent has a Markov model. Agents are able to exchange messages to integrate their models and observations and deliver safety monitoring tasks. In a similar concept [58,59], agents are provided with monitoring models (functional models) and deployed to monitor the deliverable functionality of systems. Agents are also able to collaborate with each other to integrate their models and observations and deliver consistent monitoring tasks.

Multi-agent systems have also been exploited in a different monitoring concept. In Ref. [60], for example, a number of agents are deployed to monitor the whole functionality of the monitored system and each agent is provided with a different reasoning algorithm and monitoring model, such as self-organisation maps, principal component analysis, neural network or non-parametric approaches. Agents are also able to collaborate with each other to decide consistently on whether the monitored conditions are normal or abnormal. In Ref. [61], a number of agents are also deployed to monitor the entire functionality of the monitored system, but every agent monitors the functionality of the system from different sensory data sources and the same monitoring model and reasoning algorithm which couples Bayesian network and the method of majority voting.

Despite the monitoring success of multi-agent systems, two limitations have also been highlighted: (a) the typical lack of collaboration protocols that can support effective integration among the deployed agents [62]; (b) the logical omniscience problem in which some monitored conditions may fall beyond the knowledge of the agents [63,64].

13.1.5 Motivation

Despite the above-discussed efforts and wide variety of monitoring concepts, still there have been numerous instances of accidents that could have been averted with better monitors. The explosion and fire at the Texaco Milford Haven refinery in 1994, for instance, was attributed to late fault detection, poor alarm presentation and inadequate operator training for dealing with a stressful and sustained plant upset [65]. The Kegworth Air disaster occurred in 1989 because of (a) delay in alerting the crew pilot of the occurrence of the fault and its underlying causes; (b) ineffective alarm annunciation; (c) the lack of automated fault controlling [66]. Recently, monitoring problems contributed to a fatal accident to Air France flight AF447, in which an Airbus A330 crashed in the Atlantic on the 1 June 2009 and all 228 people on board were killed. The technical investigation partly attributed the accident to late fault detection, misleading alarm annunciation and the absence of clear guidance on emergency conditions, which fell beyond the skills and training of the pilot and co-pilot [67].

Motivated by addressing the monitoring problems of such accidents, this chapter develops a distributed safety monitor by synthesising the benefits of two strands. The first is the exploitation of knowledge obtained from the application of a model-based safety assessment technique architecture analysis and design language (AADL). The second is the distributed reasoning of multi-agent systems. Specifically, this work looks at:

- The development of an effective formalisation and distribution approach to bring the offline safety assessment model of AADL forward to serve in online safety as a distributed monitoring model.
- Addressing limitations that have faced the development of multi-agent monitors. Issues of interest are selecting a suitable reasoning paradigm for the multi-agent system and the development of an effective deployment approach, collaboration protocols and monitoring algorithms.

The ultimate aim is the achievement of a spectrum of monitoring merits ranging from the delivery of effective safety monitoring tasks to the development of a scalable and cost-effective monitor.

The rest of this chapter is organised as follows: Section 13.2 describes the nature of the monitored system, that is, modern critical systems. Section 13.3 presents the position, role and constituents of the monitor. Section 13.4 tests the monitor through the application to an aircraft brake system (ABS). Section 13.5 draws a conclusion and proposes further work.

13.2 The Monitored System

Large-scale and dynamic behaviour are two common aspects of modern critical systems, for example, phased-mission systems. While the former aspect calls into question the ability of the monitor to deliver consistent monitoring tasks over a huge number of components, the latter calls into question the ability of the monitor to distinguish between normal and abnormal conditions. A typical example of such systems is an aircraft, which delivers a trip mission over achieving a number of phases: pre-flight, taxiing, take-off, climbing, cruising, approaching and landing. Thorough knowledge about the architectural components and the dynamic behaviour is essential to monitor such systems and deliver correct monitoring tasks.

To model the mutual relations among the components of a system, a hierarchical organisation is commonly used to arrange them in a number of levels. Across the levels, components appear as parents, children and siblings. To facilitate an architectural view of systems, we introduce a components classification as shown in Figure 13.1. Hierarchical levels are classified into three types: the lowest level (level0) is classified as the basic components (BC) level. Levels extending from level1 to level$n-1$ are classified as sub-system (Ss) levels. The top level (level$n$) is classified as the system (S) level.

To model the behaviour of the monitored system, it might be necessary to understand the way in which behavioural transitions are initiated. Typically, transitions are outcomes of, firstly, normal conditions in which the system engages its components in different structures, so it delivers different functionalities. Signals upon which that structure is altered are always initiated by the basic components. For example, during the cruising of an aircraft, navigation sensors may convey signals to the navigator sub-system (NS), which in turn calculates those signals and notifies the flight control computer (FCC). Assuming that it is time to launch the approaching phase, FCC accordingly instructs the power plant

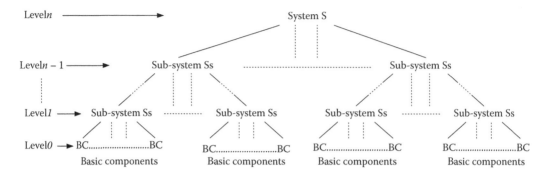

FIGURE 13.1
Architectural view of the monitored system.

sub-systems (PPS) to achieve the required thrust and the surface hydraulic controller (SHC) to achieve the required body motions. The case in which the system uses a certain structure to deliver certain functionality is called a *mode*.

Secondly, dynamic behaviour could be an outcome of the fault or fault tolerance of the basic components. Fault tolerance is typically implemented by two strategies: active fault-tolerant controlling (AFTC) and passive fault-tolerant controlling (PFTC). In the former strategy, faults cannot be corrected totally, but the consequent effects can be controlled as the system adapts to faults of its components, for example, the fault of one engine of a two-engine aircraft can be compensated by the other engine. In the latter strategy, the system has the ability to tolerate faults for a while until they are controlled totally, for example, faults that are caused by software error, ionisation radiation, electromagnetic interference, or hardware failure can be corrected within a short interval by restarting the relevant component or by isolating the faulty component and activating a redundant one.

It could, therefore, be said that during a mode, a system may appear in different health states which can be classified into two types. The first is the *Error-Free State (EFS)* in which the system or a sub-system functions healthily. The second type is the *Error State (ES)*, which in turn is classified into three different states: (a) a *Temporary Degraded or Failure State (TDFS)*, in which there is one or more functional failure, but corrective measures can be taken to transit to another state; (b) a *Permanent Degraded State (PDS)*, in which an uncontrollable fault occurs, but the safe part of the functionality can be delivered; (c) the *Failure State (FS)*, in which the intended function is totally undeliverable.

Thus, it can be said that events that are initiated by the basic components trigger and make the normal and abnormal behaviour of systems. According to the behaviour they trigger, events are classified into three types. The first type is normal events whose occurrences result in transitions from EFS to any other EFS of a different mode, for example, transition from the EFS of cruising mode to the EFS of approaching mode of an aircraft. The second type is failure events whose occurrences result in transition from EFS to any ES, due to a fault. The last type is corrective events whose occurrences result in transition from a TDES to a PDS or EFS, due to applying corrective measures.

To track the behaviour of the monitored system, such events should be continuously monitored. To achieve that, the best hierarchical level at which to monitor these events should be identified. Three monitoring factors can practically identify that level: early fault detection, computational cost and behavioural understanding of the monitored system. Achieving a trade-off among these factors could help in identifying the targeted level. Figure 13.2 illustrates the relationships among the architectural levels and those factors.

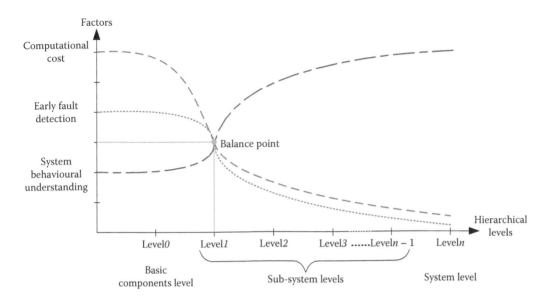

FIGURE 13.2
Balance point between three monitoring factors and architectural levels.

At level1, the occurrence of events could be identified as either normal or abnormal, for example, the decrease of aircraft velocity and altitude seem normal when the flight control computer has already launched the approaching phase of the aircraft. Excluding knowledge about the modes and focusing only on the measurements provided by the relevant sensors would certainly result in misinterpreting system behaviour, that is, decreasing velocity and altitude would appear as a malfunction and a misleading alarm would accordingly be released. Having that fact, level1 would be the best monitoring level rather than any higher level since a malfunction is detected while in its early stages. Moreover, owing to the potentially huge number of basic components, monitoring events at level0 is computationally expensive or even unworkable, whereas level1 offers the required rationality. Without loss of generality, it is assumed that primary detection of the symptoms of failure occurs at level1.

13.3 Distributed Online Safety Monitor

The monitor takes a position between the monitored system and the operators' interface. During normal conditions, the monitor provides simple feedback to confirm faultless operation. Its actual role is during abnormal conditions, which are triggered by and follow the occurrence of faults. The monitor delivers three safety tasks: prompt fault detection and diagnosis, effective alarm annunciation and fault controlling.

The term prompt, associated with fault detection and diagnosis, refers to the timeliness of detecting faults while in their early stages and before they develop into real hazards, in parallel with diagnosing the underlying causes. This is supported by selecting an appropriate hierarchical level (level1) to monitor the operational parameters and also by setting and monitoring those parameters against well-defined thresholds.

Effective alarm annunciation involves setting well-defined thresholds whose violation represents actual deviations of the monitored parameters. It also involves suppressing unimportant and false alarms whose release would overwhelm and confuse the operators. This is achieved by the following:

- Tracking the behaviour of the monitored system and distinguishing among the occurrence of normal, corrective and failure events.
- Releasing alarm only on the occurrence of genuine symptoms of faults and not on other events, such as consequent, precursor or causal events.
- Incorporating alarm information that could help the operators to control the abnormal conditions. Information is presented as assessment of the operational conditions following the occurrence of the fault and guidance on the corrective actions that should be taken manually by the operators.
- Prioritising alarm presentation. This can be achieved by highlighting the important alarms through different colours, vibration or alerting sounds, and hiding the less important alarm information, for example, optional access to the diagnostics list on the operator's interface.

Fault controlling is achieved by both active fault-tolerant controlling and passive fault-tolerant controlling and also by announcing assessment and guidance to support the manual fault controlling of abnormal conditions that may fall beyond the trained skills of the operators.

Figure 13.3 shows an illustrative view of the AADL assessment model from which the *distributed monitoring model*, a constituent of the monitor, is derived. It consists of a behavioural model as a hierarchy of state machines and fault propagation models as a number of state machines. To bring the assessment model forward to serve online monitoring, the achievement of two processes is needed. The first is formalising events that trigger transitions in the behavioural model and symptoms that associate the error propagation paths of faults as *monitoring expressions*.

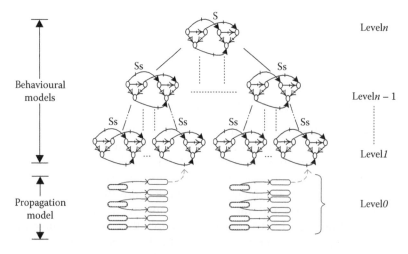

FIGURE 13.3
Illustrative view of the AADL safety assessment model.

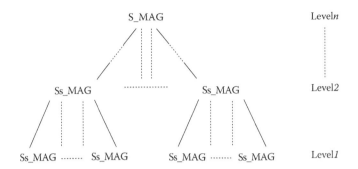

FIGURE 13.4
Illustrative view of the hierarchical deployment of agents.

In its simple form, a monitoring expression is a constraint that consists of three main parts:

- An observation which is either a state of a child or the parent or sensory measurement defined by the identifier of the relevant sensor.
- A relational operator—equality or inequality.
- A threshold whose violation results in evaluating that expression with a true truth value, that is, the relevant event or symptom occurs. Thresholds might appear as a numerical or Boolean value, such that the occurrence of the events and symptoms can be verified computationally by instantiating and evaluating monitoring expressions with real-time conditions.

Events verification supports tracking the behaviour of the monitored system, and similarly, symptoms verification supports tracking the error propagation path from the detected faults at level1 towards the underlying causes at level0. The second process is distributing the formalised model into a number of models without violating the consistency of the encoded knowledge.

Figure 13.4 illustrates the hierarchical deployment of the multi-agent system, the second constituent of the monitor, over the monitored system. According to their deployment, monitoring agents appear as follows: a number of agents deployed to monitor sub-systems, which appear as Ss_MAG and an agent that monitors the system, which appears as S_MAG.

13.3.1 Distributed Monitoring Model

In the light of the three intended monitoring tasks, agents should:

- Track the behaviour of the monitored components over different states, that is, error-free states (EFSs) and error states (ESs).
- Distinguish between normal and abnormal conditions.
- Provide the operators with information that confirms whether the conditions are normal or not; in abnormal conditions, agents should provide alarm, assessment, guidance and diagnostics.
- Be able to apply corresponding corrective measures to control faults.

Tracking the behaviour of the monitored system and its components requires informing the reasoning of the agent with behavioural knowledge. This knowledge can be derived from the hierarchy of the behavioural state machines of the AADL model (Figure 13.3). In the state machine of the sub-systems of level1, trigger events are originated by (a) the BCs of level0, which might be failure, corrective or normal events; (b) parent states (EFSs and ESs). In the state-machine of a sub-system of the levels extending from level2 to level$n-1$, trigger events appear as EFSs and ESs of the parent and children. Finally, in the state machine of the system (level$n$), events appear as EFSs and ESs of the children. Such a communication among the hierarchical state machines can be illustrated by the following example: the failure state (FS) of an engine of a two-engine aircraft triggers a transition to the permanent degraded state (PDS) in the state-machine of the power plant sub-system. The PDS, in turn, triggers a transition to another error-free state EFS of the operative engine in which the lost functionality of the faulty engine is compensated.

To distinguish among normal, fault and corrective events, the applied principle is that an alarm should be released on the occurrence of failure events only. Thus, corresponding alarm clauses should be associated with the failure events that can be verified at level1; the level at which events are monitored. Computationally, if an occurred event is associated with a 'none', then it is either a normal or a corrective event; on the contrary, the otherwise clause means that it is a failure event and the associated clause should be quoted and released as an alarm. Assessment is a description of the given conditions and guidance is about the best actions to be applied in those conditions by the operators; their clauses should thus be associated with the states.

To find the appropriate place for incorporating corrective measures, further consideration of the nature of those measures is needed. Typically, there are two different types of corrective measures. The first should be taken after diagnosing the underlying causes. This is appropriate when the verified failure event and its underlying causes are in a one-to-many relationship. In practice, measures to correct any of those causes may vary from one cause to another, so they should be incorporated in the diagnostic model (e.g., fault propagation state-machines), precisely in association with the potential causes.

The second type of corrective measures should be taken at level1, when level1's sub-systems supported by higher level components (sub-systems or system) apply measures to respond to deviations that have a clear cause. At level1, corrective measures are mostly applied with directions coming from higher levels. For example, in the flight control system (FCS) of modern aircraft, switching to the backup computer sub-system at level1 is instructed directly by the FCS at level2, whenever the primary computer sub-system (at level1) fails. Corrective measures should also be taken at level1, when level1's sub-systems supported by level0's basic components apply measures to respond to deviations that have a clear cause, that is, the detected fault and its underlying cause are in a one-to-one relationship.

State-transition tables are suggested to hold the behavioural knowledge derived from the AADL model. A state-transition table is usually defined as an alternative, executable and formal form to present a state machine and it also offers the required capacity and flexibility to incorporate knowledge about the operational conditions [68].

According to the aforementioned monitoring needs, state-transition tables of levels extending from level2 to level$n-1$ record the following:

- State transitions (current state, trigger event and new state)
- Assessment and guidance clauses

State-transition tables of level1 would record the following:

- State transitions (current state, trigger event and new state)
- Alarm, assessment and guidance clauses
- Corrective measures
- Diagnosis status, which confirms whether a diagnostic process is needed depending on the relationships between the failure events and the underlying causes

To demonstrate the derivation of the monitoring knowledge from the AADL model, the safety assessment model of the aircraft break system, the case study presented lately in this chapter (Figure 13.12), is used here as an example. Figure 13.5 shows the generic AADL error model of the brake system.

```
package Generic_Errors
public
annex Error_Model
{**
error model Basic
    features
        NM_EFS, AM_EFS, ACM_EFS, EFS : initial error state;
        NM_TDFS, AM_TDFS, NM_FS, AM_FS, ACM_FS, FS: error state;
        fault, fault1, fault2, fault3, fault4, fault5, fault6, fault7, fault8, fault9, fault_toleranceA, fault_toleranceACC: error event;
        low_pressure: in out error propagation;
end Basic;

error model implementation Basic.BasicComponent
    transitions
        EFS – [fault, in low_pressure] – > FS;
        FS – [out low_pressure] – > FS;
end Basic.BasicComponent;

error model implementation Basic.BasicSubSystem
    transitions
        NM_EFS – [fault1] – > NM_FS;
        NM_EFS – [fault2] – > NM_FS;
        NM_FS – [fault_toleranceA] – > AM_EFS;
        NM_EFS – [fault_toleranceA] – > AM_EFS;
        AM_EFS – [fault3] – > AM_FS;
        AM_EFS – [fault4] – > AM_FS;
        AM_EFS – [fault_toleranceAC] – > ACM_EFS;
        AM_FS – [fault_toleranceAC] – > ACM_EFS;
        ACM_EFS – [fault5] – > ACM_FS;
        ACM_EFS – [fault6] – > ACM_FS;
end Basic.BasicSubSystem;

error model implementation Basic.BasicSystem
    transitions
        NM_EFS – [fault7] – > NM_TDFS;
        NM_TDFS – [fault_toleranceA] – > AM_EFS;
        AM_EFS – [fault8] – > AM_TDFS;
        AM_TDFS – [fault_toleranceAC] – > ACM_EFS;
        ACM_EFS – [fault9] – > ACM_FS;
end Basic.BasicSystem;
**}
end Generic_Errors;
```

FIGURE 13.5
The generic AADL error model of ABS and its components.

The hierarchical architecture of ABS consists of three hierarchical levels:

- Level0 of the basic components, valves and sensors
- Level1 consists of two sub-systems, left-side wheel brake (LWB) and right-side wheel brake (RWB)
- Level3 or the top level, which is the system level

Figure 13.5 identifies the error-free states and the potential error states of the entire brake system and its sub-systems and basic components. The figure also shows how the mode and state transitions of the entire system are triggered by the error states of the two sub-systems and how the mode and state transitions of the sub-systems are triggered by the error states of the basic components.

Figure 13.6 shows an excerpt of the error and architectural model of the entire ABS. The 'subcomponents' part of the model shows that the lower hierarchical level of ABS consists of two sub-systems, LWB and RWB.

The 'mode' part of the model shows that there are three behavioural modes, normal, alternative and accumulative. The normal mode (NM) has two different states, error-free state (NM_EFS) and temporary degraded or failure state (NM_TDFS). Similar states can be seen with the alternative mode, error-free state (AM_EFS) and temporary degraded or failure state (AM_TDFS). The accumulative mode differently has an error-free state (ACM_EFS) and a failure state (ACM_FS). The 'annex Error_Model' part shows the way in which the states of the two sub-systems trigger state and mode transitions in the behaviour of the entire brake system. Table 13.1 shows the derived state-transition table of ABS.

In Table 13.1, it can be seen how the states of those sub-systems trigger the behavioural transitions of the brake system. Consider, for example, the first monitoring expression (event) in the table:

$$\text{LWB\_NM\_FS} == \text{true} \quad \textbf{OR} \quad \text{RWB\_NM\_FS} == \text{true}$$

The expression can be interpreted as: if the LWB sub-system or RWB sub-system is in a failure state (FS) of the normal mode (NM), then ABS should transit from the error-free state (EFS) of the normal mode (NM), that is, ABS_NM_EFS, to the temporary degraded or failure state (TDFS) of the same mode, that is, ABS_NM_TDFS. The table also incorporates assessment and guidance on the conditions at the system level.

Figure 13.7 shows an excerpt of the error and architectural model of the RWB sub-system. The 'subcomponent' part of the model shows that the RWB sub-system consists of a number of the basic components of the system.

The 'modes' part shows that the sub-system has three different behavioural modes, normal, alternative and accumulative. This part also shows that each mode consists of two different states. More specifically, the normal mode (NM), alternative mode (AM) and accumulative mode (ACM) consist identically of two states, an error-free state (EFS) and a failure state (FS). Thus, the states of those modes appear as NM_EFS, NM_FS, AM_EFS, AM_FS, ACM_EFS and ACM_FS, respectively. The 'annex Error_Model' part of the model shows how the state and mode transitions of the RWB sub-system can be triggered by the error states of the basic components.

Table 13.2 shows the derived state-transition table of the RWB sub-system. It can be seen that the state-transition table of level1 additionally incorporates alarm, controlling and diagnosis attributes.

```
system ABS
end ABS;

system implementation ABS.imp
  subcomponents
    LWB: system LWB.imp;
    RWB: system RWB.imp;

  modes
    Normal: initial mode;
    Alternative, Accumulative: mode;
    Normal: Normal – [NM_EFS] – > Normal;
    Normal1: Normal – [NM_TDFS] – > Normal;
    ToAlternative: Normal – [AM_EFS] – > Alternative;
    Alternative1: Alternative – [AM_TDFS] – > Alternative;
    ToAccumulative: Alternative – [ACM_EFS] – Accumulative;
    Accumulative1: Accumulative – [ACM_FS] – Accumulative;

  annex Error_Model {**
    model => Generic.model : : Basic.System;
    Model_hierarchy => derived;
    Derived_State_Maping =>
    Guard_Transition =>
    (LWB[NM_EFS] and RWB[NM_EFS] applies to Normal);
    Guard_Transition =>
    (LWB[NM_FS] or RWB[NM_FS] applies to Normal1);
    Guard_Transition =>
    (LWB[AM_EFS] and RWB[AM_EFS] applies to ToAlternative);
    Guard_Transition =>
    (LWB[AM_FS] or RWB[AM_FS] applies to Alternative1);
    Guard_Transition =>
    (LWB[ACM_EFS] and RWB[ACM_EFS] applies to ToAccumulative);
    Guard_Transition =>
    (LWB[ACM_FS] or RWB[ACM_FS] applies to Accumulative1);
    **}
end ABS.imp
```

FIGURE 13.6
Architectural model and error model of ABS presented by AADL.

In real time, monitoring agents evaluate only *active events*, which represent exits from the current states. Agents cyclically update expressions of those events with up-to-date sensory measurements and evaluate them, thereby achieving a *monitoring cycle*. After achieving every cycle, a new cycle is launched in which up-to-date measurements are collected and every expression is evaluated again. This state focus effectively reduces the workload of the agents and rationalises the monitoring process.

Monitoring knowledge held by the state-transition tables of level1's subsystems and the higher levels could satisfy the following:

- Tracking the behaviour of the system and its sub-systems
- Announcement of alarm and multi-level assessment and guidance
- Fault controlling at those levels

TABLE 13.1

State-Transition Table of the ABS

Current State	Conditions	Event	New State
ABS_NM_EFS	Assessment: normal line is operative and brake could be applied automatically or manually. Guidance: switching between manual and auto-brake is possible.	LWB_NM_FS == true OR RWB_NM_FS == true	ABS_NM_TDFS
ABS_NM_TDFS	Assessment: brake system is in a temporary failure. Guidance: fault controlling is in progress.	LWB_AM_EFS == true AND RWB_AM_EFS == true	ABS_AM_EFS
ABS_AM_EFS	Assessment: LWB and RWB are pressured by the alternative line. Guidance: only manual brake is applicable.	LWB_AM_FS == true OR RWB_AM_FS == true	ABS_AM_TDFS
ABS_AM_TDFS	Assessment: brake system is in a temporary failure. Guidance: fault controlling is in progress.	LWB_ACM_EFS == true AND RWB_ACM_EFS == true	ABS_ACM_EFS
ABS_ACM_EFS	Assessment: LWB and RWB are pressured by the accumulative line. Guidance: apply manual brake but anti-skid is unavailable.	LWB_ACM_FS == true OR RWB_ACM_FS == true	ABS_ACM_EFS
ABS_ACM_FS	Assessment: brake system has failed permanently. Guidance: emergency conditions.	None	None

To support diagnosing the underlying causes of the detected failure events and controlling the faults of the basic components (level*0*), a diagnostic model is needed. Such model can be derived from the fault propagation state-machines of the AADL model. Figure 13.8 shows the derived diagnostic model of the failure event 'normal brake failed'; it can be seen as the first event in Table 13.2.

For every failure event that is in a one-to-many relationship with its underlying causes, there is a diagnostic model. Links between failure events and their relevant diagnostic models are established through the appearance of the failure event at the top of the corresponding diagnostic model (see the first event in Table 13.2 and the 'FailurEvent' field in the diagnostic model).

Agents initiate the monitoring process by traversing, interpreting and uploading the state-transition tables and diagnostic models to interrelated data structures. Structure type and arrays are declared for this purpose. Arrays support direct addressing to structures that hold the knowledge, so fast access during the monitoring time is established.

13.3.2 Multi-Agent System

In addition to the common ability of intelligent agents to achieve integrated reasoning among distributed processes [69], two more reasons underpin the particular adoption of Belief-Desire-Intention (BDI) agents as monitoring agents. Firstly, as the reasoning model of these agents is based on human reasoning, effective automation of the crucial responsibilities of system operators can be facilitated. Secondly, the informative communication as well as the semi-independent reasoning of the BDI agents can support the effective collaboration and integration of two different deployment approaches. The first is spatial deployment in which agents are installed on a number of distributed

```
system RWB
  features
    NormalLine: in event port;
    AlternativeLine: in event port;
    AccumulativeLine: in event port;
end RWB;

system implementation RWB.imp
  connections
    event port self.ToAlter – > AlternativeLine in modes Normal;
    event port self.ToAcc – > AccumulativeLine in modes Alternative;

  subcomponents
    N_PS: device N_PS.imp in modes Normal;
    A_PS: device A_PS.imp in modes Alternative;
    A_CV: device A_CV.imp in modes Alternative;
    AC_PS: device AC_PS.imp in modes Accumulative;
    RN_ASV: device RN_ASV.imp in modes Normal;
    RN_MV: device RN_MV.imp in modes Normal;
    RN_CV: device RN_CV.imp in modes Normal;
    RA_ASV: device RN_ASV.imp in modes Alternative;
    RA_MV: device RN_MV.imp in modes Alternative;
    RA_CV: device RN_CV.imp in modes Alternative;
    RAC_MV: device RN_MV.imp in modes Accumulative;
    RAC_CV: device RN_CV.imp in modes Accumulative;

  modes
    Normal: initial mode;
    Alternative, Accumulative: mode;
    Normal – [self.NM_EFS] – > Normal;
    Normal – [self.NM_FS] – > Normal;
    Normal – [self.ToAlter] – > Alternative;
    Alternative – [self.AM_EFS] – > Alternative;
    Alternative – [self.AM_FS] – > Alternative;
    Alternative – [self.ToAcc] – > Accumulative;
    Accumulative – [self.ACM_EFS] – > Accumulative;
    Accumulative – [self.ACM_FS] – > Accumulative;

  annex Error_Model {**
    model => Generic.model : : Basic.SubSystem;
    Model_Hierarchy => derived;
    Derived_State_Maping =>
    NM_EFS when N_PS[EFS] and RN_CV[EFS] and RN_MV[EFS] and RN_ABV[EFS] and RN_ASV[EFS],
    NM_FS when RN_CV[FS] or RN_ABV[FS] or RN_ASV[FS],
    NM_FS when N_PS[FS],
    AM_EFS when A_PS[EFS] and RA_CV[EFS] and RA_MV[EFS] and RA_ASV[EFS],
    AM_FS when RA_CV[FS] or RA_MV[FS] or RA_ASV[FS],
    AM_FS when A_PS[FS],
    ACM_EFS when AC_PS[EFS] and RAC_CV[EFS] and RAC_MV[EFS],
    ACM_FS when RAC_CV[FS] or RAC_MV[FS],
    ACM_FS when AC_PS[FS];
    **}
end RWB.imp
```

FIGURE 13.7
Architectural model and error model of RWB sub-system presented by AADL.

computational machines. Such deployment is needed when the sub-systems of the monitored system are distributed over a geographical area, for example, a chemical plant. The second approach is semantic deployment in which monitoring agents are installed on one computational machine. Such deployment is appropriate when the sub-systems of the monitored system, although distributed, are close to each other, for example, an aircraft system.

Figure 13.9 shows the reasoning model of the BDI agent. By perceiving the operational conditions and exchanging messages with each other, each agent obtains the up-to-date belief, deliberates among its desires to commit to an intention and achieves a means-ends process to select a course of action, that is, plan. The selected plan is implemented

TABLE 13.2

State-Transition Table of the RWB Sub-System of the ABS

Current State	Condition	Event	Alarm	Controlling	Diagnosis	New State
RWB_NM_EFS	Assessment: RWB operates normally Guidance: none	RN_ASV_OP > RN_ASP_C + 50 OR RN_ASV_OP < RN_ASV_C − 50;	Normal brake failed	−RA_CV_C = 1; −RN_CV_C = 0;	Needed	RWB_NM_FS
		ABS_NM_TDFS == true	None	−RA_CV_C = 1; −RN_CV_C = 0;	Not_needed	RWB_AM_EFS
		N_PS_P − 1300 < 50	Low pressure on normal line	−RA_CV_C = 1; −RN_CV_C = 0;	Pressure at the normal line is low	RWB_NM_FS
RWB_NM_FS	Assessment: normal line of RWB has failed Guidance: fault controlling is in progress	ABS_NM_TDFS == true	None	Not_needed	Not_needed	RWB_AM_EFS
RWB_AM_EFS	Assessment: RWB is pressured by the alternative line Guidance: apply manual brake	RA_ASV_OP > RA_ASV_C + 50 OR RA_ASV_OP < RA_ASV_C − 50;	Alternative brake failed	−RAC_CV_C = 1; −RA_CV_C = 0;	Needed	RWB_AM_FS
		ABS_AM_TDFS == true	None	−RAC_CV_C = 1; −RA_CV_C = 0;	Not_Needed	RWB_ACM_EFS
		A_PS_P − 1300 < 50	Low pressure at alternative line	−RAC_CV_C = 1; −RA_CV_C = 0;	Pressure at the alternative line is low	RWB_AM_FS
RWB_AM_FS	Assessment: alternative line of RWB has failed Guidance: fault controlling is in progress	ABS_AM_TDFS == true	None	Not_needed	Not_needed	RWB_ACM_EFS
RWB_ACM_EFS	Assessment: RWB is pressured by the accumulative line Guidance: apply manual brake	RAC_MV_OP > RAC_MV_C + 50 OR RAC_MV_OP < RAC_MV_C − 50;	Accumulative brake failed	Impossible	Needed	RWB_ACM_FS
		AC_PS_P − 1300 < 50	Low pressure on accumulative line	Impossible	Pressure on the accumulative line is low	RWB_ACM_FS
RWB_ACM_FS	Assessment: RWB fails; there is no brake on the right-side landing gear Guidance: no brake is available	None	None	Not_needed	Not_needed	None

FailureEvent: RN_ASV_OP > RN_ASV_C + 50 **OR** RN_ASV_OP < RN_ASV_C - 50.
Propagator: RN_ASV **OR** RN_ABV **OR** RN_CV.

EStateName: RN_ASV_FS.
Symptom: RN_CV_PO == 1 AND RN_ABV_IP − RN_ABV_OP < 100.
Fault: anti-skid valve RN_ASV is faulty.
Controlling: none.

EStateName: RN_ABV_FS.
Symptom: RN_ABV_IP − RN_ABV_OP > 100.
Fault: auto-brake valve RN_ABV is faulty.
Controlling: none.

EStateName: RN_CV_FS.
Symptom: RN_CV_PO == 0.
Fault: control valve RN_CV is faulty.
Controlling: none.

FIGURE 13.8
The diagnostic model derived from the fault propagation model of the ADDL assessment model.

as actions towards achieving the monitoring tasks locally and as messages sent to other agents towards achieving global integration. Upon having a new belief, an agent achieves a reasoning cycle, deliberation and a means-ends process.

As agents are deployed hierarchically, each Ss_MAG of level1 updates its belief base by perceiving (a) its own portion of the monitoring model which consists of a state-transition table and a number of the diagnostic models; (b) sensory measurements that are taken to instantiate and evaluate monitoring expressions; (c) messages that are received from the parent to inform the given Ss_MAG about the new states and from the siblings, in which they either ask for or tell the given Ss_MAG about global measurements, as there is a potential need to share measurements globally. The main desires of an Ss_MAG of level1 are to monitor the local conditions of the assigned sub-system and to collaborate globally with its parent and siblings. On the achievement of the local desire, the intentions are to track the behaviour of the sub-system and to provide the operators with alarms, assessment and guidance and control faults. On the achievement of the global desire, the intentions are to exchange messages to (a) inform the parent about the new states; and (b) tell or ask the siblings about global measurements.

Each Ss_MAG of the intermediate levels (levels extending from level2 to level$n$ − 1) updates its belief by (a) perceiving its own portion of the monitoring model, which consists of a state-transition table of the assigned sub-system; (b) messages received from the parent and children to inform about their new states. The main desires of each of these

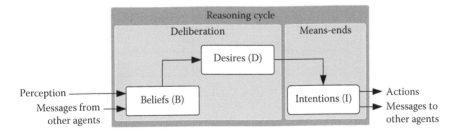

FIGURE 13.9
BDI agent reasoning model.

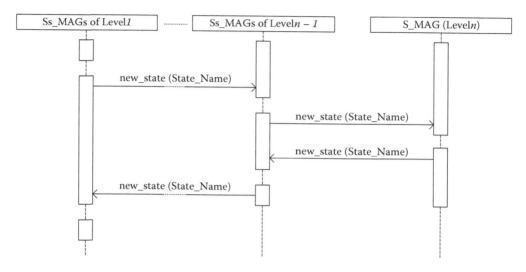

FIGURE 13.10
Collaboration protocol across the hierarchical levels.

Ss_MAGs are to monitor the local conditions of the assigned sub-system and to collaborate globally with its parent and child agents. On the local desire, the intentions are to track the behaviour of the sub-system and to provide the operators with assessment and guidance. On the global desire, the intention is to exchange messages with the parent and child agents to inform each other about the new states. The perceptions, desires and intentions of the S_MAG of level$n$ are similar to those of the Ss_MAGs of the intermediate levels. The only difference, however, is that S_MAG has no parent to exchange messages with it.

According to the Prometheus approach and notation for developing multi-agent systems [19], Figure 13.10 shows the collaboration protocols among agents to track the behaviour of the monitored system. Figure 13.11 shows the collaboration protocol among the Ss_MAGs of level1 in which they share their sensory measurements.

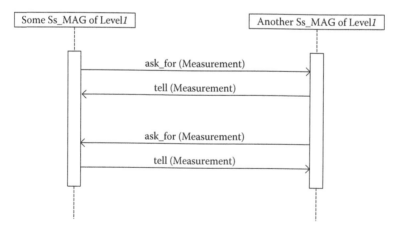

FIGURE 13.11
Collaboration protocol among Ss_MAGs of level1.

13.4 Case Study: ABS

Figure 13.12 shows a physical illustration of a hypothetical ABS. The main function of ABS is to slow down the aircraft during the taxiing and landing phases and achieve safe retardation in the case of a rejected take-off. The brake function of ABS is supported by anti-skid and an optional selection between auto-brake in which the pilot pre-arms the rate of deceleration before the landing phase and manual brake in which the pilot applies the brake manually by depressing two pedals.

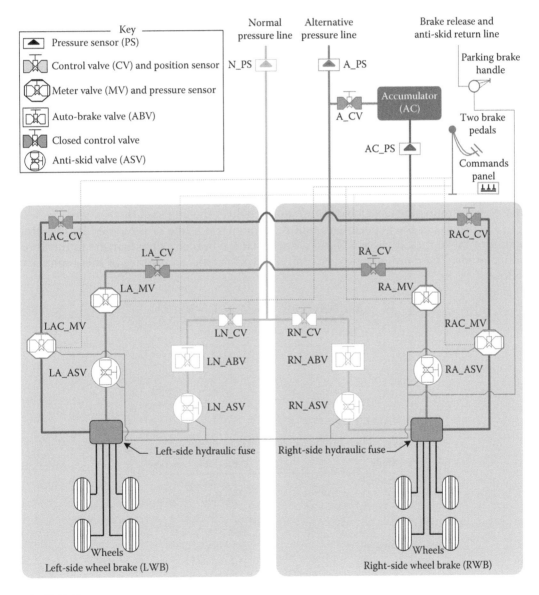

FIGURE 13.12
Physical illustration of an ABS.

The basic components of ABS include valves, sensors and three redundant pressure lines: normal, alternative and accumulative lines. The components are arranged in two subsystems: LWB and RWB sub-systems. Passive fault-tolerant controlling is implemented to control faults of ABS. Initially, the brake is pressured by the normal line; should this fail, it is isolated and the alternative line is activated. Should the alternative line fail, it too is isolated and the accumulative line is activated. A pressure line may fail due to a drop in the pressure to less than 1300 PSI or due to a fault of a basic component of that line. A brake system control unit (BSCU) is incorporated to control ABS.

During flying, the hydraulic pressure at the lines is monitored by BSCU. Once the pressure drops close to 1300 PSI, a warning light 'BRAKE SOURCE' illuminates and the pilot is advised by a message of the Engine Indication and Crew Alerting System (EICAS) to switch to another brake line.

ABS has three modes: normal, alternative and accumulative modes, according to the operative line. Over the three modes, different functions are delivered. For example, the optional selection between the auto and manual brake is only possible during the normal mode. Table 13.3 abstracts the deliverable functions over the three modes of ABS.

According to Prometheus [70], Figure 13.13 shows the multi-agent system that is deployed to monitor ABS. Two agents monitor the two sub-systems and appear as LWB_MAG and RWB_MAG and one agent monitors the entire system, appearing as ABS_MAG. The multi-agent system is implemented by a Jason interpreter, an extended version of the AgentSpeak programming language [71].

Table 13.4 shows the state-transition table of the LWB sub-system. Together, Tables 13.1, 13.2 and 13.4, in addition to Figure 13.8, contribute to achieving the monitoring experiment and demonstrating the ability of the monitor to deliver the intended three safety tasks.

In the context of the experiment, failure of the normal line is simulated by injecting a fault of the anti-skid valve RN_ASV, such that the brake system transits to the alternative mode. The monitoring expression that verifies the occurrence of this event is as follows:

$$RN\_ASV\_OP > RN\_ASV\_C + 50$$

OR

$$RN\_ASV\_OP < RN\_ASV\_C - 50;$$

TABLE 13.3

Functions Provided Over Each Mode of ABS

Mode	Deliverable Function
Normal	1-Autobrake
	2-Manual brake
	3-Anti-skid
	4-Parking brake
Alternative	1-Manual brake
	2-Anti-skid
	3-Parking brake
Accumulative	1-Manual brake
	2-Parking brake

Distributed Online Safety Monitor

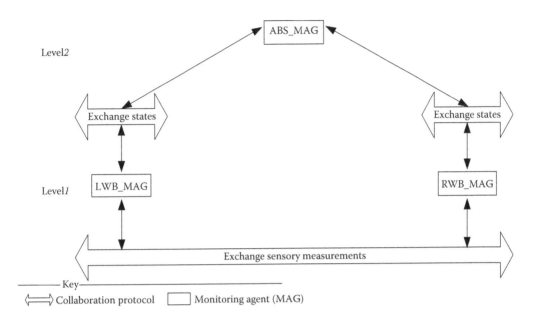

FIGURE 13.13
Multi-agent system to monitor ABS.

The expression is verified as true when the pressure measured at the output of the anti-skid valve of the normal line (RN_ASV_OP) is less or greater than the pressure commanded by BSCU to the anti-skid valve (RN_ASV_C). Plus and minus 50 PSI is added as a possible bias of the sensors. Once the occurrence of the above expression is verified, agent RWB_MAG perceives its state-transition table (Table 13.2) and achieves the following procedure:

- From the relevant ALARM attribute, agent RWB_MAG quotes 'normal brake failed' and alarms the pilot.
- From the relevant CONTROLLING attribute, agent RWB_MAG opens valve RA_CV and closes valve RN_CV to switch to the alternative line.
- From the relevant DIAGNOSIS attribute, agent RWB_MAG verifies the need for a diagnostic process. At this point and before applying the corrective measures, agent RWB_MAG updates the symptoms of the relevant diagnostic model (Figure 13.8) with the relevant measurements.
- From the relevant NEW STATE attribute, agent RWB_MAG transits to a new state, which is the failure state RWB_NM_FS. From this state, the pilot is provided with assessment, 'normal line of RWB has failed' and guidance, 'fault controlling is in progress'.
- Agent RWB_MAG communicates state RWB_NM_FS to the parent agent (ABS_MAG).

Since a diagnostic process is needed, then before launching a new monitoring cycle, agent RWB_MAG retrieves the position of the relevant diagnostic model. For the purpose of diagnosis, agent RWB_MAG exploits a diagnostic algorithm that couples between blind-depth-first and heuristic traverses. Figure 13.14 shows alarm information announced to the operator on the given conditions.

TABLE 13.4

State-Transition Table of the LWB Sub-System of ABS

Current State	Conditions	Event	Alarm	Controlling	Diagnosis	New State
LWB_NM_EFS	Assessment: LWB operates normally Guidance: none	LN_ASV_OP > LN_ASV_C + 50 OR LN_ASV_OP < LN_ASV_C − 50; ABS_NM_TDFS == true	Normal brake failed	−LA_CV_C = 1; −LA_CV_C = 0;	Needed	LWB_NM_FS
			None	−LA_CV_C = 1; −LA_CV_C = 0;	Not_needed	LWB_AM_EFS
LWB_NM_FS	Assessment: normal line of LWB has failed Guidance: fault controlling is in progress	ABS_NM_TDFS == true	None	Not_needed	Not_needed	LWB_AM_EFS
LWB_AM_EFS	Assessment: LWB is pressured by the alternative line Guidance: apply manual brake	LA_ASV_OP > LA_ASV_C + 507 OR LA_ASV_OP < LA_ASV_C − 50; ABS_AM_TDFS == true	Alternative brake failed	−LA_CV_C = 1; −LA_CV_C = 0;	Needed	LWB_AM_FS
			None	−LA_CV_C = 1; −LA_CV_C = 0;	Not_needed	LWB_ACM_EFS
LWB_AM_FS	Assessment: the alternative line of RWB has failed Guidance: fault controlling is in progress	ABS_AM_TDFS == true	None	Not_needed	Not_needed	LWB_ACM_EFS
LWB_ACM_EFS	Assessment: LWB is pressured by the accumulative line Guidance: apply manual brake	LAC_MV_OP > LAC_MV_C + 50 OR LAC_MV_OP < LAC_MV_C − 50;	Accumulative brake failed	Impossible	Needed	LWB_ACM_FS
LWB_ACM_FS	Assessment: LWB fails; there is no brake on the left-side landing gear Guidance: no brake is available	None	None	Not_needed	Not_needed	None

Distributed Online Safety Monitor

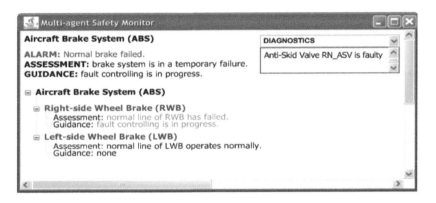

FIGURE 13.14
Operator's interface after the failure of the normal line of RWB.

When agent ABS_MAG receives a message that conveys RWB_NM_FS, it perceives its state-transition table (Table 13.1) and achieves the following procedure:

- As the current state is ABS_NM_EFS, the received state results in verifying the occurrence of 'LWB_NMFS == true OR RWB_NMFS == true'.
- From the relevant NEW STATE attribute, agent ABS_MAG transits to a new state which is the temporary degraded/failure state ABS_NM_TDFS. From this state, the pilot is provided with assessment, 'brake system is in a temporary failure' and guidance, 'fault controlling is in progress' (as shown in Figure 13.14).
- Agent ABS_MAG communicates the new state to the children (RWB_MAG and LWB_MAG).

When agents RWB_MAG and LWB_MAG receive the messages, each achieves a certain procedure as follows:

Agent RWB_MAG perceives the state-transition table (Table 13.2) and achieves the following procedure:

- While the current state is RWS_NM_FS, the received state results in verifying the occurrence of ABS_NM_TDFS == true.
- As the relevant ALARM, CONTROLLING and DIAGNOSIS attributes require no action, then from the relevant NEW STATE attribute, agent RWB_MAG transits to a new error-free state RWB_AM_EFS. From this state, the pilot is provided with assessment, 'RWB is pressured by the alternative line', and guidance, 'apply manual brake'.
- Agent RWB_MAG communicates that state to the parent agent (ABS_MAG) and launches a monitoring cycle.

Agent LWB_MAG perceives its state-transition table (Table 13.4) and achieves the following procedure:

- While the current state is LWS_NM_EFS, the received state results in verifying the occurrence of ABS_NM_TDFS == true.

- As the relevant ALARM attribute shows 'none', no alarm is released.
- From the relevant CONTROLLING attribute, agent LWB_MAG opens valve LA_CV and closes valve LN_CV, to switch to the alternative line.
- As the relevant DIAGNOSIS attribute holds 'not_needed', no action is taken.
- From the relevant NEW STATE attribute, agent LWB_MAG transits to a new state which is the error-free state LWB_AM_EFS. From this state, the pilot is provided with assessment, 'LWB is pressured by the alternative line' and guidance, 'apply manual brake'.
- Agent LWB_MAG communicates that state to the parent agent (ABS_MAG) and launches a monitoring cycle.

When agent ABS_MAG receives messages sent by agents LWB_MAG and RWB_MAG, it accordingly perceives its state-transition table (Table 13.1) and achieves the following procedure:

- While the current state is ABS_NM_TDFS, the received state results in verifying the occurrence of LWB_AM_EFS == true AND LWB_AM_EFS == true.
- From the relevant NEW STATE attribute, agent ABS_MAG transits to a new state, which is the error-free state ABS_AM_EFS. From this state, the pilot is provided with assessment, 'LWB and RWB are pressured by the alternative line' and guidance, 'only manual brake is applicable'.

After achieving the above procedures, the alternative mode of the entire brake system would be launched without asking the pilot to switch to another model. That includes the automatic isolation of the normal line by closing the control valves LN_CV and RN_CV and activation of the alternative line by opening the controlling valves LA_CV and RA_CV. The pilot is updated with the new conditions (activation of the alternative line) as shown in Figure 13.15. The interface informs the pilots about the current conditions and also advises them to apply the manual brake as the only applicable option.

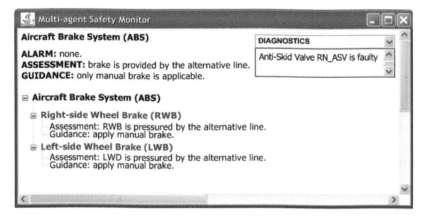

FIGURE 13.15
Operator's interface after controlling the failure of the normal line of RWB.

13.5 Discussion, Validation and Conclusion

This chapter proposed a distributed online safety monitor based on a multi-agent system and knowledge derived from design models and a safety assessment model of the monitored system. Agents exploit that knowledge to deliver a range of safety tasks. The monitor can detect symptoms of failure on process parameters as violations of simple constraints, or deviations from more complex relationships among process parameters, and then diagnose the causes of such failures. With appropriate knowledge about dynamic behaviour, the monitor can also determine the functional effects of low-level failures and provide a simplified and easier-to-comprehend functional view of failure, that is, assessment and guidance. By knowing the scope of a failure, the monitor can apply successive corrections at increasingly abstract levels in the hierarchy of a system. The achieved experiment has additionally demonstrated the ability of the monitor to deliver the following tasks:

- Prompt fault detection and diagnosis
- Effective alarm announcement that is presented as a well-organised alarm and multi-level assessment and guidance
- Automatic control of the injected faults, removing some of the demands for control that could otherwise have been unnecessarily posed on operators

Despite encouraging results, a research issue remains to be investigated. The quality of the monitoring tasks and the correctness of the inferences drawn by the monitor depend mainly on the validity of sensory measurements. The validation of the sensory measurements is, therefore, an area for further research.

References

1. C. Billings, *Human-Centred Aircraft Automation: A Concept and Guidelines*, Field CA, United States: NASA Technical Memorandum TM-103885, NASA Ames Research Centre, Moffett, 1991. Available: http://www.archive.org/details/nasa_techdoc_19910022821 (Accessed 15 July 2010).
2. I. Kim, Computer-based diagnostic monitoring to enhance the human–machine interface of complex processes, In *Proceedings of Power Plant Dynamics, Control and Testing Symposium*. Knoxville, TN, United States, 27–29 May 1992.
3. V. Venkatasubramanian, R. Rengaswamy, K. Yin and S. Kavuri, A review of process fault detection and diagnosis: Part I: Quantitative model-based methods, *Computers and Chemical Engineering*, 27(3), 2003, 293–311.
4. Y. Zhang and J. Jiang, Bibliographical review on reconfigurable fault-tolerant control systems, *Annual Reviews in Control*, 32(2), 2008, 229–252.
5. D. Pumfrey, The principled design of computer system safety analyses, DPhil thesis, University of York, 1999.
6. J. Ma and J. Jiang, Applications of fault detection and diagnosis methods in nuclear power plants: A review, *Progress in Nuclear Energy*, 53(3), 2011, 255–266.
7. I. Kim and M. Modarres, Application of goal tree success tree model as the knowledge-base of operator advisory systems, *Nuclear Engineering and Design*, 104(1), 1987, 67–81.

8. M. Modarres and S. Cheon, Function-centred modelling of engineering systems using the goal tree success tree technique and functional primitives, *Reliability Engineering & System Safety*, 64(2), 1999, 181–200.
9. D. Chung, M. Modarres and R. Hunt, GOTRES: An expert system for fault detection and analysis, *Reliability Engineering & System Safety*, 24(2), 1989, 113–137.
10. L. Felkel, R. Grumbach and E. Saedtler, Treatment, analysis and presentation of information about component faults and plant disturbances, *Proceedings of the Symposium on Nuclear Power Plant Control Instrument*, IAEA-SM-266/40, 1978, 340–347.
11. Y. Papadopoulos, Model-based system monitoring and diagnosis of failures using state-charts and fault trees, *Reliability Engineering and System Safety*, 8(3), 2003, 325–341.
12. H. Peng, W. Shang, H. Shi and W. Peng, On-line monitoring and diagnosis of failures using control charts and fault tree analysis (FTA) based on digital production model. *Proceedings of 2nd International Conference on Knowledge Science, Engineering and Management (KSEM'07). Lecture Notes in Computer Science (4798/2007)*. Berlin, Heidelberg: Springer, 2007, pp. 544–549.
13. M. Maurya, R. Rengaswamy and V. Venkatasubramanian, A signed directed graph-based systematic framework for steady-state malfunction diagnosis inside control loops, *Chemical Engineering Science*, 61(6), 2006, 1790–1810.
14. G. Dong, W. Chongguang, Z. Beike and M. Xin, Signed directed graph and qualitative trend analysis based fault diagnosis in chemical industry, *Chinese Journal of Chemical Engineering*, 18(2), 2010, 265–276.
15. S. Narasimhan, P. Vachhani and R. Rengaswamy, New nonlinear residual feedback observer for fault diagnosis in nonlinear systems, *Automatica*, 44(9), 2008, 2222–2229.
16. A. Zolghadri, D. Henry and M. Monsion, Design of nonlinear observers for fault diagnosis: A case study, *Control Engineering Practice*, 4(11), 1999, 1535–1544.
17. T. Chen and R. You, A novel fault-tolerant sensor system for sensor drift compensation, *Sensors and Actuators A: Physical*, 147(2), 2008, 623–632.
18. T. El-Mezyani, D. Dustegor, S. Srivastava and D. Cartes, Parity space approach for enhanced fault detection and intelligent sensor network design in power systems, *Proceedings of IEEE'2010 Conference on Power and Energy Society General Meeting*. Minneapolis, MN, 25–29 July 2010, pp. 1–8.
19. M. Borner, H. Straky, T. Weispfenning and R. Isermann, Model based fault detection of vehicle suspension and hydraulic brake systems, *Mechatronics*, 12(8), 2002, 999–1010.
20. M. Abdelghani and M. Friswell, A parity space approach to sensor validation, *Proceedings of the International Society for Optical Engineering (SPIE'2001)*. USA, Bellingham: Society of Photo-Optical Instrumentation Engineers, ISSN: 0277-786X CODEN, 4359(1), 2001, pp. 405–411.
21. W. Nelson, REACTOR: An expert system for diagnosis and treatment of nuclear reactor accidents, *Proceedings of AAAI 82*, August, 1982, pp. 296–301.
22. T. Ramesh, S. Shum and J. Davis, A structured framework for efficient problem-solving in diagnostic expert systems, *Computers and Chemical Engineering*, 12(9–10), 1988, 891–902.
23. T. Ramesh, J. Davis and G. Schwenzer, Catcracker: An expert system for process and malfunction diagnosis in fluid catalytic cracking units, *Proceedings of Annual Meeting of the American Institute of Chemical Engineering (AIChE)*, November 1989, San Francisco, CA.
24. S. Rich, V. Venkatasubramanian, M. Nasrallah and C. Matteo, Development of a diagnostic expert system for a whipped toppings process, *Journal of Loss Prevention in the Process Industries* 2(3), 1989, 145–154.
25. M. Maurya, R. Rengaswamy and V. Venkatasubramanian, Fault diagnosis by qualitative trend analysis of the principal components, *Chemical Engineering Research and Design*, 83(9), 2005, 1122–1132.
26. M. Maurya, R. Rengaswamy and V. Venkatasubramanian, A signed directed graph and qualitative trend analysis-based framework for incipient fault diagnosis, *Chemical Engineering Research and Design*, 85(10), 2007, 1407–1422.
27. M. R. Maurya, P. K. Paritosh, R. Rengaswamy and V. Venkatasubramanian, A framework for on-line trend extraction and fault diagnosis, *Engineering Applications of Artificial Intelligence*, 23(6), 2010, 950–960.

28. L. A. Rusinov, I. V. Rudakova, O. A. Remizova and V. Kurkina, Fault diagnosis in chemical processes with application of hierarchical neural networks, *Chemometrics and Intelligent Laboratory Systems*, 97(1–15), 2009, 98–103.
29. N. Kaistha and B. Upadhyaya, Incipient fault detection and isolation of field devices in nuclear power systems using principal component analysis, *Nuclear Technology*, 136, 2001, 221–230.
30. J. Miller, Statistical signatures used with principal component analysis for fault detection and isolation in a continuous reactor, *Journal of Chemometrics*, 20(1–2), 2006, 34–42.
31. S. Wold, A. Ruhe, H. Wold and W. Dunn, The collinearity problem in linear regression, the partial least squares (PLS) approach to generalized inverses, *SIAM Journal of Science Statistical Computer*, 5, 1984, 735–743.
32. S. Hwang, J. Lin, G. Liang, Y. Yau, T. Yenn and C. Hsu, Application control chart concepts of designing a pre-alarm system in the nuclear power plant control room, *Nuclear Engineering and Design*, 238(12), 2008, 3522–3527.
33. G. Jang, D. Seong, J. Keum, H. Park and Y. Kim, The design characteristics of an advanced alarm system for SMART, *Annals of Nuclear Energy*, 35(6), 2008, 1006–1015.
34. W. Brown, J. O'Hara and J. Higgins, *Advanced Alarm Systems: Revision of Guidance and its Technical Basis*, US Nuclear Regulatory Commission, Washington, DC (NUREG-6684), 2000. Available: http://www.bnl.gov/humanfactors/files/pdf/NUREG_CR-6684.pdf (Accessed 22 February 2011).
35. B. Oulton, Structured programming based on IEC SC 65 A, using alternative programming methodologies and languages with programmable controllers, *Proceedings of IEEE Conference on Electrical Engineering Problems in the Rubber and Plastic Industries, IEEE no: 92CH3111-2, IEEE service centre*, 31–14 April, Akron, OH, USA, 1992, pp. 18–20.
36. A. Ghariani, A. Toguyeni and E. Craye, A functional graph approach for alarm filtering and fault recovery for automated production systems, *Proceedings of 6th International Workshop on Discrete Event Systems (WODES'02)*. 02–04 October 2002, Zaragoza, Spain, pp. 289–294.
37. J. Lee, J. Kim, J. Park, I. Hwang and S. Lyu, Computer-based alarm processing and presentation methods in nuclear power plants, *Proceedings of World Academy of Science, Engineering and Technology. Library of Congress, Electronic Journals Library*, 65, 2010, 594–598.
38. C. Yu and B. Su, Eliminating false alarms caused by fault propagation in signal validation by sub-grouping, *Progress in Nuclear Energy*, 48(4), 2006, 371–379.
39. P. Baraldi, A. Cammi, F. Mangili and E. Zio, An ensemble approach to sensor fault detection and signal reconstruction for nuclear system control, *Annals of Nuclear Energy*, 37(6), 2010, 778–790.
40. I. Kim, M. Modarres and R. Hunt, A model-based approach to on-line disturbance management: The models, *Reliability Engineering and System Safety*, 28(3), 1990, 265–305.
41. I. Kim, M. Modarres and R. Hunt, *A Model-Based Approach to On-line Process Disturbance Management: The Models*, USA: System Research Centre: University of Maryland, SRC TR 88-111, 1988. Available from: http://drum.lib.umd.edu/bitstream/1903/4837/1/TR_88-111.pdf (Accessed 4 August 2011).
42. R. Clark, Instrument fault detection, *IEEE Transactions on Aerospace and Electronic Systems*, 14(3), 1978, 456–465.
43. J. O'Hara, W. Brown, P. Lewis and J. Persensky, *Human-System Interface Design Review Guidelines*, Energy Sciences & Technology Department and Brookhaven National Laboratory, Upton, NY, 2002. Available: http://www.nrc.gov/reading-rm/doc-collections/nuregs/staff/sr0700/nureg700.pdf (Accessed 22 February 2011).
44. J. O'Hara, W. Brown, B. Halbert, G. Skraaning, J. Wachtel and J. Persensky, The use of simulation in the development of human factors guidelines for alarm systems, *Proceedings of 1997 IEEE 6th Conference on Human Factors and Power Plants, Global Perspectives of Human Factors in Power Generation*. 08–13 June 1997, Orlando, FL, USA, pp. 1807–1813.
45. E. Roth and J. O'Hara, *Integrating Digital and Conventional Human-System Interfaces: Lessons Learned from a Control Room Modernization Program*, Washington, DC: Division of Systems Analysis and Regulatory Effectiveness Office of Nuclear Regulatory Research U.S. Nuclear

Regulatory Commission. NRC Job Code W6546, 2002. Available: http://www.nrc.gov/reading-rm/doc-collections/nuregs/contract/cr6749/6749-021104.pdf (Accessed 23 February 2011).

46. J. Anderson, *Alarm Handler User's Guide*, The University of Chicago, as Operator of Argonne National Laboratory, Deutches Elektronen-Synchrotron in Der Helmholtz-Gemelnschaft (DESY) and Berliner Speicherring-Gesellschaft fuer Synchrotron-Strahlung mbH (BESSY), 2007. Available: http://www.slac.stanford.edu/comp/unix/package/epics/extensions/alh/alhUserGuide.pdf (Accessed 23 February 2011).

47. J. Jiang, Fault-tolerant control systems an introductory overview, *Automatica SINCA*, 31(1), 2005, 161–174.

48. R. Patton, Fault-tolerant control: The 1997 situation, *Proceedings 3rd IFAC Symp on Fault Detection, Supervision and Safety for Technical Processes (SAFEPROCESS'97)*. August 1997, Hull, United Kingdom, pp. 1033–1055.

49. C. Seo and B. Kim, Robust and reliable H∞ control for linear systems with parameter uncertainty and actuator failure, *Automatica*, 32(3), 1996, 465–467.

50. R. Srichander and B. Walker, Stochastic stability analysis for continuous-time fault tolerant control systems, *International Journal of Control*, 57(2), 1993, 433–452.

51. I. Lopez and N. Sarigul-Klijn, A review of uncertainty in flight vehicle structural damage monitoring, diagnosis and control: Challenges and opportunities, *Progress in Aerospace Sciences*, 46(7), 2010, 247–273.

52. Delta Virtual Airlines, *Boeing 767-200/300ER/400ER Operating Manual*, 3rd ed., Delta Virtual Airlines, 2003. Available: http://www.deltava.org/library/B767%20Manual.pdf (Accessed 5 March 2011).

53. Q. Zhao and J. Jiang, Reliable state feedback control system design against actuator failures, *Automatica*, 34(10), 1998, 1267–1272.

54. W. Heimerdinger and C. Weinstock, *A Conceptual Framework for System Fault Tolerance*, Technical Report CMU/SEI-92-TR-033 ESC-TR-92-033, Software Engineering Institute, Carnegie Mellon University, Pittsburgh, Pennsylvania, 1992.

55. M. Shooman, *Reliability of Computer Systems and Networks: Fault Tolerance, Analysis and Design*, USA, New York: John Wiley & Sons, Inc, 2002.

56. Y. Papadopoulos, J. McDermid, R. Sasse and G. Heiner, Analysis and synthesis of the behaviour of complex programmable electronic systems in conditions of failure, *Reliability Engineering & System Safety*, 71(3), 2001, 229–247.

57. X. Ren, H. Thompson and P. Fleming, An agent-based system for distributed fault diagnosis, *International Journal of Knowledge-Based and Intelligent Engineering Systems*, 10, 2006, 319–335.

58. S. Eo, T. Chang, B. Lee, D. Shin and E. Yoon, Function-behaviour modelling and multi-agent approach for fault diagnosis of chemical processes, *Computer Aided Chemical Engineering*, 9 2001, 2001, 645–650.

59. S. Eo, T. Chang, B. Lee, D. Shin and E. Yoon, Cooperative problem solving in diagnostic agents for chemical processes, *Computers & Chemical Engineering*, 24(2–7), 2000, 729–734.

60. Y. Ng and R. Srinivasan, Multi-agent based collaborative fault detection and identification in chemical processes, *Engineering Applications of Artificial Intelligence*, 23(6), 2010, 934–949.

61. G. Niu, T. Han, B. Yang and A. Tan, Multi-agent decision fusion for motor fault diagnosis, *Mechanical Systems and Signal Processing*, 21(3), 2007, 1285–1299.

62. C. Wallace, G. Jajn and S. McArthur, Multi-agent system for nuclear condition monitoring, *Proceedings of the 2nd International Workshop on Agent Technologies for Energy System (ATES'11), a workshop of the 10th International Conference of Agent and Multi-agent System (AAMAS'11)*, 2nd of May 2011, in Taipei, Taiwan.

63. A. Sayda, Multi-agent systems for industrial applications: Design, development and challenges, In: Alkhateeb F., Al Maghayreh E., Abu Doush L., eds. *Multi-Agent Systems–Modeling, Control, Programming, Simulations and Applications*. Rijeka, Croatia: InTech, 2011, pp. 469–494.

64. E. Mangina, Intelligent agent-based monitoring platform for applications in engineering, *International Journal of Computer Science & Applications*, 2(1), 2005, 38–48.

65. HSE, *Better Alarm Handling, HSE Information Sheet*, UK: Health and Safety Executive, chemical sheet No 6, 2000. Available from: http://www.hse.gov.uk/pubns/chis6.pdf, (Accessed 13 February 2011).
66. E. J. Trimble, Report on the Accident to Boeing 737-400 G-OBME near Kegworth, Leicestershire on 8 January 1989, Department of Transport Air Accidents Investigation Branch, Royal Aerospace Establishment. London: HMSO, 1990. Available from: http://www.aaib.gov.uk/cms_resources.cfm?file=/4-1990%20G-OBME.pdf (Accessed 5 October 2010).
67. BEA, *Safety Investigation into the Accident on 1 June 2009 to the Airbus A330-203, flight AF447*, France: Bureau of Investigations and Analysis for the safety of civil aviation (BEA), 2011. Available: http://www.bea.aero/fr/enquetes/vol.af.447/note29juillet2011.en.pdf (Accessed 2 August 2011).
68. M. Breen, Experience of using a lightweight formal specification method for a commercial embedded system product line, *Requirements Engineering Journal*, 10(2), 2005, 161–172.
69. S. McArthur, E. Davidson, J. Hossack and J. McDonald, Automating power system fault diagnosis through multi-agent system technology, *Proceedings of 37th Annual Hawaii International Conference on System Sciences (HICSS'04)*, 5–8 January 2004, Big Island, Hawaii, pp. 1–4.
70. L. Padgham and M. Winikoff, *Developing Intelligent Agent Systems: A Practical Guide*. Chichester: John Wiley & Sons, 2004.
71. R. Bordini, J. Hubner and M. Woorldridge, *Programming Multi-Agent Systems in AgentSpeak Using Jason*. UK, Chichester: Wiley, 2007.

14

State of the Art of Service-Level Agreements in Cloud Computing

Mohammed Alhamad

CONTENTS

14.1 Introduction 347
14.2 Definition 348
14.3 Taxonomy of Cloud Computing 350
14.4 Service-Level Agreements 350
    14.4.1 SLAs for Web Services 351
    14.4.2 SLAs for Grid Computing 351
    14.4.3 SLAs for Cloud Computing 352
        14.4.3.1 Shortcomings of the Proposed Works for SLAs in the Context of Distributed Services 353
14.5 Performance Measurement Models 353
    14.5.1 SOA Performance Models...353
    14.5.2 Distributed Systems Performance Models..................................354
    14.5.3 Cloud Computing Performance Models 355
        14.5.3.1 Shortcomings of the Proposed Works for the Above Performance Models 355
14.6 Conclusion 355
References..356

14.1 Introduction

Cloud computing has been the focus of active and extensive research since late 2007. Before the term 'cloud' was coined, there was grid technology. Now, the hot topic of research is cloud, and more proposed frameworks and models of various solutions for the new technology have started to be applied to cloud architecture. In this section, we survey the literature to determine the most appropriate definition of 'cloud computing'. Also, we review the different architectural frameworks and the common challenges that may present major problems for providers and customers who are interested in understanding this type of distributed computing.

The Google trends report shows that cloud computing had surpassed grid computing by late 2007 (Figure 14.1).

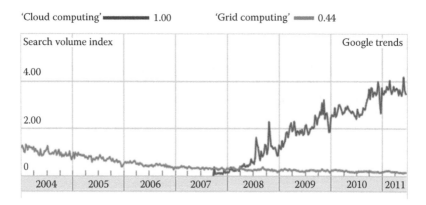

FIGURE 14.1
Cloud computing trend, source and Google search engine.

14.2 Definition

Experts and developers who investigate issues and standards related to cloud computing do not necessarily have the same technology background. In research projects, professionals from grid technology, service-oriented architecture (SOA), business and other domains of technology and management have proposed several concepts to define cloud computing. These definitions of cloud computing still need to be presented in a common standard to cover most technology and other aspects of cloud computing architecture.

In the context of networking and communication, the term 'cloud' is a metaphor for the common Internet concept [1]. The cloud symbol is also used to present the meaning of network connection and the way cloud technology is provided by the Internet infrastructure. 'Computing' in the context of the cloud domain refers to the technology and applications that are implemented in the cloud data centres [2].

In Ref. [3], Vaquero et al. comment on the lack of a common definition of cloud computing. They state that developers and business decision-makers confuse the understanding of the technology with the features of cloud data centres. Hence, large budgets may be allocated to implement private or even public cloud data centres. However, these data centres face several problems when users or public customers want to connect the interfaces of their legacy systems with the new technology of cloud architecture. Vaquero et al. link the challenge of maximising the revenue of building cloud technology to professionals who are involved in distributed services. Since they come from a traditional computing domain, they are confused about the other concepts of distributed services such as grid and web services. The definition used in Ref. [3] is as follows:

> Clouds are a large pool of easily usable and accessible virtualized resources (such as hardware, development platforms and/or services). These resources can be dynamically re-configured to adjust to a variable load (scale), allowing also for an optimum resource utilization. This pool of resources is typically exploited by a pay-per-use model in which guarantees are offered by the Infrastructure Provider by means of customized SLAs.

Although this definition presents the main features of cloud computing, it does not encompass other important components of cloud architecture, which include the method of establishing and managing network, applications, and supporting services.

Wang et al. [4] define cloud computing as follows:

> A computing Cloud is a set of network enabled services, providing scalable, QoS guaranteed, normally personalized, inexpensive computing infrastructures on demand, which could be accessed in a simple and pervasive way.

Wang's definition of cloud focuses on the technical aspects of services. Business and functional characteristics are absent from the proposed definition. On the other hand, Knorr and Gruman [5] explain the main technical concepts of a cloud services model and define cloud computing from the developers' perspective. The authors show how the cloud computing architecture takes advantage of the way different distributed services (mainly web services and SOA) are implemented. Two types of cloud services are presented along with this definition: they define software as a service (SaaS) and platform as a service (PaaS). Despite the importance of infrastructure as a service (IaaS) as a main component of cloud architecture, they do not adequately discuss this type of cloud delivery model.

In this chapter, we adopted and considered the definition provided by the U.S. National Institute of Standards and Technology (NIST) [6], according to which 'Cloud computing is a model for enabling convenient, on demand network access to a share pool of configurable computing resources (e.g., networks, servers, storage, applications, and services) that can be rapidly provisioned and released with minimal management afford or service provider interaction' [6].

The shortcomings of the proposed definitions of cloud computing given above are as follows:

1. None of the definitions consider cloud computing from the technical and business perspectives. This would cause confusion to decision makers in large organisations, especially when they want to define the parameters of a costing model of cloud services.
2. Existing cloud definitions do not specify the onus of responsibility in cases of poor quality of service (QoS) delivery.
3. Most of the proposed definitions consider specific types of cloud services, whereas a comprehensive definition of cloud should clearly define all classes of cloud services.
4. The proposed definitions do not consider a definition of cloud users.

Table 14.1 includes the scope of definitions and lists the main shortcomings of the definitions discussed above.

TABLE 14.1

Conclusion of Cloud Definitions

Reference	Scope of Definition	Missing
Vaquero et al. [3]	Define architecture and service model	Management, supporting and trust concepts
Wang et al. [4]	Technical concepts	Business and functional characteristics
Knorr and Gruman [5]	Comparison between cloud computing, web services and SOA	Definition of IaaS and DaaS
Mell and Grance [6]	Technical features, management and security concepts	Costing and billing model

14.3 Taxonomy of Cloud Computing

Buyya et al. [7] present more than 15 characteristics to distinguish cloud computing from other distributed systems. Buyya uses scalability, automatic adaption, virtualisation and a dynamic model of billing as the main concepts to construct the architecture of cloud computing. Moreover, he explains how cloud services can be delivered to different types of users. For instance, users who want to develop small-sized applications may connect to one of the PaaS such as Microsoft Azure [8] without the need to instal any of the development tools. Hoefer et al. [9] identify a clear taxonomy framework for existing categories of cloud services. The class of cloud services is described in a tree-structured taxonomy and the unique characteristics of each model of service are used to identify each node of the proposed tree structure. Hoefer's classification provides a clear comparison of cloud services at a high level on the tree structure. However, at the base of the structure, the taxonomy of cloud services is not enough to distinguish services in more detail. The taxonomy presented by Laird in Ref. [10] defines cloud technology from the perspective of service providers. The proposed taxonomy presents the common vendors of cloud services. Laird presents two classifications of services. The first classification defines the infrastructure of cloud services and the second classification defines the services based on cloud features such as security, billing and applications that are built into the system. Rimal et al. [11] present a comprehensive framework for the architecture of cloud computing. They describe the taxonomy of cloud services with more focus on the management domain of cloud contents. The concepts of management, business, billing and support of cloud services are defined in great depth to present cloud architecture as a new business model. The main advantage of the proposed work by Rimal is that relationships between security features and cloud components are provided as a part of the comparison of service models in cloud computing. The taxonomy proposed by Oliveira et al. [12] classifies the concepts of cloud computing according to the dimensions of cloud architecture, business model, technology infrastructure, pricing, privacy and standards. The proposed taxonomy is provided in a hierarchical tree with parent and child relationships. Oliveira uses SaaS, PaaS, IaaS and database as a service (DaaS) as sub-taxonomy for the business model. This classification is used in the literature of cloud computing to distinguish the service delivery for end users of cloud services. These sub-taxonomy terms may cause confusion in understanding the way various business models are constructed for cloud services. The taxonomy proposed by Oliveira describes the concepts of cloud architecture from the perspective of e-science. Therefore, many of the technical aspects of cloud computing are missing from the proposed taxonomy.

14.4 Service-Level Agreements

A service-level agreement (SLA) is a document that includes a description of the agreed service, service-level parameters, guarantees and actions for all cases of violation. SLA is very important as a contract between the consumer and the provider. The main idea of SLAs is to give a clear definition of the formal agreements about service terms such as performance, availability and billing. It is important that SLA includes the obligations and the actions that will be taken in the event of any violation, with clearly expressed and shared semantics between each party involved in the online contract.

This section discusses works related to SLAs in three domains of distributed services. First, we discuss the proposed SLA structure for web services. Second, the frameworks of SLAs designed for grid computing are reviewed. Third, we discuss the main works that specifically focus on cloud computing. Finally, in this section, we include the main shortcomings of these SLA frameworks.

14.4.1 SLAs for Web Services

Several specifications for defining SLAs have been proposed for web services. Web service-level agreement (WSLA) language [13] introduces a mechanism to help users of web services to configure and control their resources to meet the service level. Also, the service users can monitor SLA parameters at run time and report any violation of the service. WSLA was developed to describe services under three categories: (1) Parties: in this section, information about service consumers, service providers and agents is described. (2) SLA parameters: in this section, the main parameters that are measurable are presented in two types of metrics. The first is resource metrics, a type of metrics used to describe a service provider's resources as row information. The second one is composite metrics. This metric is used to calculate the combination of information about a service provider's resources. (3) The final section of the WSLA specification is service-level objective (SLO). This section is used to specify the obligations and all actions when service consumers or service providers do not comply with the guarantees of services. WSLA provides an adequate level of online monitoring and contracting but does not clearly specify when and how a level of service can be considered a violation. Web service offering language (WSOL) [14] is a service-level specification designed mainly to specify different objectives of web services. The defining concepts of service management, cost and other objectives of services can be presented in WSOL. However, WSOL cannot adequately meet the objectives of the new paradigm of cloud computing.

WS-Agreement [15] is created by an open-grid forum (OGF) to create an official contract between service consumers and service providers. This contract should specify the guarantees, the obligations and penalties in case of violations. Also, the functional requirements and other specifications of services can be included in SLA. The WS-Agreement has three main sections: name, context and terms. A unique ID and optional names of services are included in the name section. The information about the service consumer and service provider, domain of service and other specifications of service are presented in the context section. The terms of services and guarantees are described in greater detail in the terms section. These types of online agreements were developed for use with general services. For cloud computing, service consumers need more specific solutions for SLAs to reflect the main parameters of the visualisation environment; at the same time, these SLA solutions should be dynamically integrated with the business rules of cloud consumers.

The primary shortcomings of these approaches are that they do not provide for dynamic negotiation, and various types of cloud consumers need a different structure for the implementation of SLAs to integrate their own business rules with the guarantees that are presented in the targeted SLA.

14.4.2 SLAs for Grid Computing

In the context of grid computing, there are a number of proposed specifications that have been developed especially to improve security and trust for grid services. In Ref. [16], an SLA-based knowledge domain has been proposed by Sahai to represent the measurable metrics for business relationships between all parties involved in the transaction of grid services. Also, the author proposed a framework to evaluate the management proprieties of grid services in the life cycle. In this work, business metrics and a management evaluation framework are combined to produce an estimated cost model for grid services. In our research, we extend this approach to build a general costing model based on the technical and business metrics of the cloud domain. The framework proposed in this work lacks a dynamic monitoring technique to help service customers know who takes responsibility when a service level is not provided as specified in SLA documents. Leff et al. [17] conducted a study of the main requirements to define and implement SLAs for the grid community. The author provides an ontology and a detailed definition of grid computing. Then a scientific discussion is presented about the requirements that can help developers and decision-makers to deploy trusted SLAs in a grid community. A basic prototype was implemented to validate the use of SLAs as a reliable technique when the grid service provider and customer need to build a trusting relationship. The implementation of the framework in this study does not consider important aspects of security and trust management in grid computing. Keung et al. [18] proposed an SLA-based performance prediction tool to analyse the performance of grid services. Keung uses two sources of information as the main inputs for the proposed model. The source code information and hardware modelling are used to predict the value of performance metrics for grid services. The model proposed by Keung can be used in other types of distributed computing. But in the cloud environment, this model cannot be integrated with a dynamic price model of cloud services. It needs to be improved by using different metrics for cost parameters to reflect the actual price of cloud services. The system proposed by Padget et al. in Ref. [19] considers the response time of applications in the grid systems. The main advantage of the proposed system is that it can predict the central processing unit (CPU) time for any node in the grid network before conducting the execution. When Padget tested the adaptation of the SLA model using a real experiment on the grid, the prediction system produced values for response time close to the values obtained when users executed the same application on the grid. Noticing the delay recorded for the large size of executed files, the author claims that the reason for this delay is the external infrastructure such as Internet connections. The author also discusses the impact of the time delay caused by external parties on the reputation of service providers when using SLA management systems. Although the author provides a good method for calculating the response time for grid resources, other metrics such as security and management metrics are absent in this work.

14.4.3 SLAs for Cloud Computing

The context of this research is the management of SLAs in cloud communities. In the sections above, we presented the frameworks and models in the current literature that are designed mainly for managing SLAs in traditional distributed systems. In this section, SLAs and approaches to agreement negotiations in the cloud community are presented.

Stantchev et al. [20] describe the QoS related to cloud services and different approaches applied to map SLA to the QoS. Services ontology for cloud computing is presented to define service capabilities and the cost of service for building a general SLA framework. The proposed framework does not consider all types of cloud services; it is general and was tested only on the Amazon EC2. It also needs to consider other types of cloud providers such as PaaS, DaaS and SaaS. Our framework in this research considers this issue in the validation phase of the research. The framework developed by Wen et al. [21] focuses on the SaaS model of delivery in cloud computing. More details are provided on how the services can be integrated to support the concept of stability of the cloud community, especially for SaaS.

14.4.3.1 Shortcomings of the Proposed Works for SLAs in the Context of Distributed Services

The frameworks and structures that were discussed in the previous sections have the following problems:

1. The existing frameworks focus more on the technical attributes than on the security and management aspects of services.
2. The proposed structures of SLAs in the above domains do not include a clear definition of the relationship between levels of violation and the cost of services.
3. Most of the above studies do not integrate a framework of trust management of the service provider with the collected data from monitoring systems of SLAs.
4. The concepts and definitions of service objectives and service descriptions included in SLAs are not easy to understand, especially for business decision-makers.
5. The proposed works for cloud environments focus more on the evaluation of virtualisation machines on local servers than on existing cloud service providers.
6. Most of the proposed structures of SLAs are defined by technical experts.

14.5 Performance Measurement Models

Cloud providers have been increased to deliver different models of services. These services are provided at different levels of QoS. Cloud customers need to have a reliable mechanism to measure the trust level of a given service provider. Trust models can be implemented with various measurement models of services. As a part of this research, we investigate the use of a measurement approach to develop a general trust model for a cloud community. In this section, the measurement model of SOA, distributed and grid services will be reviewed.

14.5.1 SOA Performance Models

Kounev and Buchmann in Ref. [22] propose an analytical approach to modelling performance problems in SOA-based applications. The authors discuss the different realistic J2EE applications for large systems of SOA architecture. A validated approach has been tested for capacity planning of the organisations that use distributed services as an outsourcing infrastructure. The advantage of the proposed method is its ability to predict the number of application servers based on the collected information of SLA metrics. Chandrasekaran and Senthilanand [23] implemented a simulation tool to analyse the performance of composite services. The authors used an online bookstore as a case study to simulate experimental scenarios. They focus on measuring communication latency and transaction completion time. Real data sets were compared with the simulation results. The authors state that the simulation tool presents results that approximate those of the real data. This type of simulation can be extended and applied to other distributed services. For cloud computing, more efforts are required to make this technique compatible with existing interfaces of cloud providers. Rud et al. in Ref. [24] use the web services business process execution language (WS-BPEL) composition approach to evaluate the performance of utilisation and throughput of SOA-based systems in large organisations. They developed the proposed methodology using a mathematical model to improve the processes of SLAs in the SOA environment. The main focus of Rud's method is on the management aspects of services. However, this approach does not consider the performance issues of response time, data storage and other metrics of technical infrastructure. For the optimisation of total execution time and minimisation of business processes cost, Menasce et al. in Ref. [25] provide an optimised methodology based on the comparison of performance metrics of SOA-based services. In this study, Menasce developed the proposed method to estimate the cost level of all services that are registered in the SOA directory under medium-sized organisations. Then the cost metric is compared to the real performance of services. The parameters of the performance metrics can be selected by service customers. Hence, the proposed model can be used for different types of services. Although the proposed method produces a high level of reliability and usability, issues such as risk management and trust mechanisms of the relationship between service providers and service customers are not discussed in more detail.

14.5.2 Distributed Systems Performance Models

Kalepu et al. [26] propose a QoS-based attribute model to define the nonfunctional metrics of distributed services. Availability, reliability, throughput and cost attributes are used in their work to define the performance of resources of a given service provider. Two approaches of resources are used to calculate the final value of reputation. The first resource is the local rating record. Ratings of services that are invoked by local customers are stored in this record. In the second resource, global ratings of all services that are executed on resources of a given service provider are stored. Although Kalepu et al. discuss the need to use SLA parameters to calculate the value of performance metrics, they do not explain how these parameters can be linked to the local global resources of a rating system. In Ref. [27], Yeom et al. provide a monitoring methodology of the performance parameters of service. The proposed methodology uses the broker monitoring systems to evaluate the performance of resources of a service provider. The collected data of performance metrics are not maintained on the service consumer database. This method incurs low cost in terms of implementing measurement architecture but presents more risk in terms of privacy, availability of data and security. Such risks are not easy to control, especially in the case of multi-tenant distributed systems. Kim et al. in Ref. [28] analyse the quality factors of the performance level of services and propose a methodology to assign priorities to message processing of distributed web services based on the quality factors of services. This assigning aspect of their framework is a dynamic process in different service domains. They claim that their framework satisfies the agreement regarding service level in web services. The validation methodology of the proposed work lacks a clear definition of the evaluation criteria and a description of the way in which the experiment was conducted to produce the claimed results. The work proposed by Guster et al. in Ref. [29] provides an evaluation methodology for distributed parallel processing. In the proposed method, the authors use a parallel virtual machine (PVM) and real hosting servers to compare the results of their experiments. The efficiency of the evaluation method performed better in PVM for the processing time. In the real server environment, the experiments presented better performance in terms of communication time. The evaluation of this work does not include the implementation processes and the experiment results are not clearly explained.

14.5.3 Cloud Computing Performance Models

Several studies already exist on the scalability of virtual machines. Most of these studies considered the measurement of performance metrics on the local machines. The background loads of tested machines are controlled to compare the results of performance with a different scale of loads. Evangelinos and Hill [30] evaluated the performance of Amazon EC2 to host high-performance computing (HPC). They use a 32-bit architecture for only two types of Amazon instances. In our study, we run various experiments on most types of Amazon EC2 instances. These instances are: small, large, extra large, high CPU, medium and high CPU extra large. Jureta et al. [31] propose a model called quality-value-dependency priority (QVDP) that has three functions: specifying the quality level, determining the dependency value, and ranking the quality priority. These functions consider the QoS from the customer's perspective. However, the performance issues related to cloud resources are not discussed and details are missing regarding the correlation of the quality model with the costing model of services. Cherkasova and Gardner [32] use a performance benchmark to analyse the scalability of disk storage and CPU capacity with Xen virtual machine monitors. They measure the performance parameters of the visualisation infrastructure that are already deployed in most data centres. But they do not measure the scalability of cloud providers using the visualisation resources. However, our proposed work profiles the performance of virtualisation resources that are already running on the infrastructure of existing cloud providers.

14.5.3.1 Shortcomings of the Proposed Works for the Above Performance Models

1. The above proposed models for evaluating the virtualisation services focus on how to measure the performance of virtual machines using local experiments. However, the techniques used for measuring the actual resources of cloud providers need further refinement to ensure some level of trust between the service providers and the customers.
2. Most of the proposed works on performance evaluation do not allow service customers to specify the parameters of performance metrics. In cloud computing, service customers need a more flexible and dynamic approach to modify the parameters of performance metrics to solve the problem of dynamic changes of service requirements and business models of customers.
3. The experiments using the above proposed models do not specify the benchmarks for the performance evaluation.
4. In cloud computing architecture, the relationship between performance monitoring and cost metric is very important. The proposed models do not link the results of performance monitoring with the actual cost metric of services. Hence, service customers are not able to build a trust relationship with service providers without having a real cost model of services.

14.6 Conclusion

In this chapter, we have reviewed the proposed architectures for cloud computing and discussed the differences between cloud computing and other distributed services.

The discussion of cloud definitions, taxonomy and shortcomings of the existing definitions of cloud computing are presented to provide a brief overview of the main concepts of the new paradigm of cloud computing. Then we have analysed the existing proposed framework of SLAs in the context of web services, grid computing and cloud computing. We have provided the main shortcomings of the existing frameworks of SLAs. Finally, we discussed the existing models of performance measurements for SOA, distributed services and cloud computing followed by the list of the main shortcomings of these models.

References

1. H. Katzan Jr., On an ontological view of cloud computing, *Journal of Service Science (JSS)*, 3, 1–6, 2011.
2. D. C. Wyld, Moving to the cloud: An introduction to cloud computing in government, IBM Center for the Business of Government, 2009.
3. L. M. Vaquero et al., A break in the clouds: Towards a cloud definition, *ACM SIGCOMM Computer Communication Review*, 39, 50–55, 2008.
4. L. Wang et al., Cloud computing: A perspective study, *New Generation Computing*, 28, 137–146, 2010.
5. E. Knorr and G. Gruman, What cloud computing really means, *InfoWorld*, 37, 2008.
6. P. Mell and T. Grance, Draft NIST working definition of cloud computing, Referenced on June 3, 2009.
7. R. Buyya et al., Cloud computing and emerging IT platforms: Vision, hype, and reality for delivering computing as the 5th utility, *Future Generation Computer Systems*, 25, 599–616, 2009.
8. R. Jennings, *Cloud Computing with the Windows Azure Platform*, Wrox, Indianapolis, IN, 2010.
9. C. Hoefer and G. Karagiannis, Taxonomy of cloud computing services, Globecom workshops CCG workshops, pp. 1345–1350, Miami, USA, 2010.
10. P. Laird, Different strokes for different folks: A taxonomy of cloud offerings, *Enterprise Cloud Submit, INTEROP*, 2009.
11. B. P. Rimal et al., A taxonomy, survey, and issues of cloud computing ecosystems, *Cloud Computing*, 21–46, 2010.
12. D. Oliveira et al., Towards a taxonomy for cloud computing from an e-science perspective, In: N. Antonopoulos, and L. Gillam, (eds.) *Cloud Computing: Principles, Systems and Applications*, Computer Communications and Networks, Springer-Verlag, London, 2010.
13. H. Ludwig et al., Web service level agreement (WSLA) language specification, IBM Corporation, 2003.
14. V. Tosic, *WSOL Version 1.2*, Carleton University, Department of Systems and Computer Engineering, 2004.
15. A. Andrieux et al., Web services agreement specification (WS-Agreement), Global Grid Forum (vol.2), pp. 1–81, 2004.
16. A. Sahai et al., Specifying and monitoring guarantees in commercial grids through SLA, Cluster computing and the grid (CCGrid) (vol. 3), pp. 292–299, 2003.
17. A. Leff et al., Service-level agreements and commercial grids, *IEEE Internet Computing*, 7, 44–50, 2003.
18. H. N. L. C. Keung et al., Self-adaptive and self-optimising resource monitoring for dynamic grid environments, *15th International workshop on Database and Expert Systems Applications*, Zaragoza, Spain, pp. 689–693, 2004.
19. J. Padgett et al., Predictive adaptation for service level agreements on the grid, *International Journal of Simulation: Systems, Science and Technology*, 7, 29–42, 2006.

20. V. Stantchev and C. Schröpfer, Negotiating and enforcing QoS and SLAs in grid and cloud computing, *Advances in Grid and Pervasive Computing*, 5529, 25–35, 2009.
21. C. H. Wen et al., A SLA-based dynamically integrating services SaaS framework, *IET International Conference on Frontier Computing, Theory, Technologies and Applications*, January 2010, pp. 306–311.
22. S. Kounev and A. Buchmann, Performance modeling and evaluation of large-scale J2EE applications, *CMG-CONFERENCE*, Dallas, TX, vol. 1, Computer Measurement Group, 273–284, 2003.
23. S. Chandrasekaran et al., Composition performance analysis and simulation of web services, Technical report, Georgia, pp. 002–005, 2002.
24. D. Rud et al., Performance modeling of WS-BPEL-based web service compositions, *Services Computing Workshops, SCW'06*, Chicago, Illinois, 140–147, 2006.
25. D. A. Menascé et al., A heuristic approach to optimal service selection in service oriented architectures, *Proceedings of the 7th International Workshop on Software and Performance*, ACM, New York, NY, 13–24, 2008.
26. S. Kalepu et al., Verity: A QoS metric for selecting web services and providers, *Fourth IEEE International Conference on Web Information Systems Engineering*, Rome, Italy, 131–139, 2003.
27. G. Yeom and D. Min, Design and implementation of web services QoS broker, *IEEE International Conference on Next Generation Web Services Practices, NWeSP 2005*, Seol, Korea, pp. 2–3, 2005.
28. D. Kim et al., Improving web services performance using priority allocation method, *IEEE International Conference on Next Generation Web Services Practices, NWeSP 2005*, Seol, Korea, pp. 6–11, 2005.
29. D. Guster et al., Computing and network performance of a distributed parallel processing environment using MPI and PVM communication methods, *Journal of Computing Sciences in Colleges*, 18, 246–253, 2003.
30. C. Evangelinos and C. Hill, Cloud computing for parallel scientific HPC applications: Feasibility of running coupled atmosphere–ocean climate models on Amazon's EC2, *Ratio*, 2, 2–34, 2008.
31. I. Jureta et al., A comprehensive quality model for service-oriented systems, *Software Quality Journal*, 17, 65–98, 2009.
32. L. Cherkasova and R. Gardner, Measuring CPU overhead for I/O processing in the Xen virtual machine monitor, *Proceedings of the USENIX Annual Technical Conference*, Anaheim, CA, 24, 2005.

15

Used Products Return Service Based on Ambient Recommender Systems to Promote Sustainable Choices

Wen-Jing Gao, Bo Xing and Tshilidzi Marwala

CONTENTS

15.1 Introduction 360
15.2 Background 361
    15.2.1 Product Recovery Management 361
    15.2.2 Ambient Intelligence 361
    15.2.3 Recommender Systems 362
    15.2.4 Summary 362
15.3 Emerging Technologies and Platforms 363
    15.3.1 Near-Field Communication 363
    15.3.2 Product-Embedded Information Device 363
        15.3.2.1 Radio-Frequency Identification 363
    15.3.3 Internet of Things 364
15.4 Agent-Based Used Products Return Service 365
    15.4.1 Scenario Description 365
    15.4.2 Agent-Based Modelling 365
        15.4.2.1 User Agent 365
        15.4.2.2 Collector Agent 365
        15.4.2.3 Directory Agent 366
        15.4.2.4 Coordinator Agent 366
15.5 Ambient Recommender Systems for Used Product Return Service System Design 366
    15.5.1 Design Environment 366
    15.5.2 System Working Principles 366
        15.5.2.1 Registration Module 367
        15.5.2.2 Communication Module 367
        15.5.2.3 Migration Module 368
        15.5.2.4 Enquiry Module 369
    15.5.3 System Implementation 370
        15.5.3.1 Step 1: Registration with Directory Agent 370
        15.5.3.2 Step 2: List of Candidate Collectors Acquiring 373
        15.5.3.3 Step 3: Candidate Collectors' System Load Enquiry 374
        15.5.3.4 Step 4: Disposal Request Sending 374
15.6 Conclusions 375
15.7 Future Work 375
References 376

15.1 Introduction

Studies by the U.S. Academy of National Engineering show that in the United States, 93% of exploited resources are never transformed into final products, 80% of all products are one-way products, and 99% of the material contents of goods become waste within 6 weeks [1]. In addition to poor resource productivity, end-of-life (EoL) products, for example, waste electrical and electronic equipment (WEEE), end-of-life vehicle (ELV) and end-of-life tyres (ELT), have become a major environmental problem in the world. For example, in Europe alone, the annual volume of WEEE generated is estimated to be around 6–7 million tons per year [2] and is expected to rise to 95 million tons in 2010 all over the world if current development trends persist [3]. Meanwhile, every year, approximately 800 million scrap tyres are disposed around the globe. This amount is expected to increase by approximately 2% every year [4]. Their hazardous contents put human health at risk and pollute the earth. This fact triggers the awareness of environmentally conscious societies to manage an effective product recovery process.

Product recovery requires that EoL products are acquired from the end users so that the value-added operations can be processed and the products (or parts from them) can be resold in the market. But the authors of Ref. [5] argued that the company never accepts individual returns as the channel returns from the end user have too high a cost to be effective. Nowadays, one of the most important trends in this area is to outsource EoL products collection to a specialised collector or provider. However, end users as one of the society parties in the product recovery are obliged to deal with issues concerning the environment. For example, in the case of return of products under warranty, end users are usually responsible for the transportation of products or are required to contact the responsible original equipment manufacturer (OEM) to dispose of their products. Unfortunately, in spite of this way of product return, people are often unaware of how to discard or return a product properly [6].

Recently there have been great advances in some technologies such as wearable computers and devices, wireless networks, sensor devices and control appliances. The improvements regarding their connectivity, battery life, weight and size make it feasible to integrate computer systems in our daily lives, making ubiquitous environments a reality. However, since the advent of the web, more than 10,000 terabytes of information have gone online. In addition, more than 513 million people around the world have access to this global information resource [7]. So, how do all these people find the information they are most interested in from among all those terabytes? Also, how do they find the other people they would most like to communicate with, work with and play with?

Increasingly, people are turning to recommender systems to help them find the information that is most valuable to them. Recommender systems, such as Netflix [8], using collaborative filtering, are a popular technique to dominate their respective market segments online. The systems hold the promise of delivering high-quality recommendations on palmtop computers, even when disconnected from the Internet. Furthermore, they can protect the user's privacy by storing personal information locally, or by sharing it in encrypted form. These systems support a broad range of applications now, including recommending movies, books, relevant search results and even pets. Unfortunately, there is little research assessing the effectiveness of these systems in influencing used products return service. In addition, articulating the end user's preferences will not only make the EoL product return service more customer focused but also make the business more successful. We essentially aim at improving the bridge between end users and EoL product disposal flows to form a seamless, synchronous network that functions to support the product recovery process from the end user's point of view. Further, the proposed model will behave as an added personalised recommender to the end users in their return choices. In the process, the end users will have the impression that the system is entirely for them, thereby increase their self-confidence in the product return system.

The remainder of this chapter is organised as follows. In Section 15.2, we will give a brief review of product recovery management (PRM), ambient intelligence (AmI) and recommender systems. This is followed by a description of the emerging technologies and platforms that our research is going to build on in Section 15.3. In Section 15.4, we will describe our focal scenario and define the roles for each of the agents. Next, in Section 15.5, our proposed framework is explained. Finally, the conclusions and future work are highlighted in Sections 15.6 and 15.7, respectively.

15.2 Background

15.2.1 Product Recovery Management

PRM consists of two key steps, namely, collecting EoL products from end users (i.e., reverse logistics [RL]) and reprocessing the returns (e.g., remanufacturing) to capture their remaining value. The objective of PRM is to make effective use of returns so as to maximise the value of this resource. One of the first publications that shaped the area of RL [9] defined the five product recovery systems: remanufacturing, repairing, refurbishing, cannibalisation and recycling. Their study included a comprehensive discussion of the product design approach for recovery, the importance of reducing disposal waste, the preparation of customers for green products and environmental legislation issues for recovery systems. Refs. [10] and [11] categorised recovery simply into material recovery [12] (recycling) and added value recovery [9] (remanufacturing). Comprehensive reviews on the issues in environmentally conscious manufacturing and product recovery were provided in Ref. [13].

One of the key concerns of the companies involved in product recovery is the EoL product collection as mentioned in Ref. [5], especially the acquisition of the data required to make accurate analyses on PRM issues [14–17], such as information on the composition of products and information on the magnitude and uncertainty of return flows [9]. In fact, it is the first activity of product recovery and triggers the other activities of the recovery system. Logistics literature is remarkably rich in papers that deal with the collection operations in the context of product recovery and recycling. For example, Refs. [18–25] are some of the studies that consider the collection facility of the location–allocation problem. Another issue about this activity is the need to foster and manage EoL product acquisition via incentive mechanisms (see Refs. [5,26,27] for instance). Meanwhile, in an impact assessment on WEEE, the collection rate seems to be an agreed indicator worthy of further study. In particular, how to make a good estimation of collection rates will become a new issue [28].

15.2.2 Ambient Intelligence

AmI, proposed by the European Commission Information Society Technologies Advisory Group, refers to a seamless and invisible computing environment, which is able to provide users with proactive and adaptive services [29]. In light of this statement, the key features of AmI systems are embedded, context-aware, personalised, adaptive and anticipatory [30]. Practically, AmI has potential applications in many areas of life, including in the home, office, transport, industry, entertainment, tourism, recommender systems, safety systems, e-health and supported living of many different variations. For example, the authors of Refs. [31–33] described an architecture called SALSA (a middleware designed to support the development of AmI environments based on autonomous agents), for health care, which uses agents as abstractions, to act on behalf of users, to represent services and to provide wrapping of complex functionality to be hidden from the user.

Nowadays, the development of AmI is a very complex task because this technology must often adapt to contextual information as well as unpredictable behaviours and environmental features. Some authors (e.g., Ref. [29]) have stressed the need to combine distributed intelligence paradigms within the architectural levels identifiable in AmI systems, namely, ubiquitous computing devices (in particular, pervasive computing devices, such as production data acquisition [PDA]), ubiquitous wireless communication, intelligent multi-modal interfaces, artificial intelligence and multi-agent systems. Furthermore, beyond e-business and e-commerce, a number of trends have emerged in technology and in the ways in which technology is deployed in the field of AmI-related applications. These trends represent remarkable new opportunities for the development of new functions and services, especially in the area of marketing and advertising.

15.2.3 Recommender Systems

Recommender systems are designed to make recommendations (typical products and services) to users of the Internet based on prior user actions and a model of user preferences. Often, this model is derived from cross-similarities among activity profiles across a collection of users, in which case it is termed collaborative filtering. A familiar example of collaborative filtering is Amazon.com's 'customers who bought' feature. In addition, the authors of Ref. [34] suggested the use of recommendation agents as a promising approach in future e-learning systems, where a recommender agent saw what a student was doing and recommended actions. Meanwhile, the recommender system can be used to provide the user with access to political documents and to political information according to the user profile [35].

The work given in Ref. [36] presented an overview of the field of recommender systems and the recommendations were classified as content based, collaborative and hybrid. Research into those systems, for example, Ref. [37], attempted to extend the functionality of conventional recommender systems. This involved designing proactive systems that were intelligent and ubiquitous advisors in everyday contexts, which reduce the information overload for the user, and were sensitive, adaptive and responsive to user needs, habits and emotions.

Also, there are some agent-based e-commerce systems as well as fuzzy-based recommended systems that are available in the literature [38–40]. In addition, the authors of Ref. [41] provided an Internet-based recommender system, which employs a particle swarm optimisation (PSO) algorithm to learn the personal preferences of users, as well as tailored suggestions. They found that, in general, the PSO-based system was faster, particularly when compared with the genetic algorithm (GA), and more accurate than its rivals in matching user profiles. In a similar vein, Ref. [42] presented a new collaborative filtering algorithm along with five peer-to-peer architectures for finding neighbours.

15.2.4 Summary

As consumer demand and government regulations for more environmentally sustainable products continue to grow, design-effective EoL products that return service must include support for end users' decision making. According to the literature review, most papers focus on EoL product collection from the practitioner's perspective; very few studies pay attention to EoL product issues from the end user's point of view. This research is trying to fill this gap. In this chapter, we adopt the multi-agent paradigm for designing and implementing AmI recommender systems. If successful, this pipeline will loop back on itself to become a sustainable production–collection cycle. Successfully balancing collection and production will produce a new ballpark figure for waste and environmental damage: zero.

15.3 Emerging Technologies and Platforms

15.3.1 Near-Field Communication

In recent years, near-field communication (NFC), a radio-frequency identification (RFID)–related standard that has emerged from the telecommunication industry, denotes a technology that enables the integration of RFID functionality into personal devices such as mobile phones, thus making them both an RFID reader and an RFID transponder device [43]. Predictions indicate that there could be as many as 1.95 billion NFC-enabled devices by 2017. Many applications take advantage of the touch-like interactions between them to facilitate mobile payment, ticketing or information retrieval. The goal of this programme is to create a user-friendly, convenient product recovery experience for the customers and to improve the economics and efficiency of the retail operation.

15.3.2 Product-Embedded Information Devices

Product-embedded information devices (PEIDs), a new generation of products, are available in the market. Among other innovative features, these products allow the monitoring of new parameters of the product and its environment along its whole life cycle. Unlike the actual approach of data creation, management and use, which focuses on product type, new emerging technologies allow focusing on product items. This is a new paradigm in that it is possible to monitor each single item of a product type. Instead of gathering information for the next version of a product, the gathered data can be analysed and transformed into information and knowledge that can then be used to optimise the whole life cycle, including the EoL stage [44]. Among the various PEIDs, RFID stands at the forefront of the technologies driving the vision, mainly due to its maturity and low cost and consequently its strong support from the business community. Furthermore, tags will communicate only with readers, not with other tags; so, instead of peer-to-peer types of communications, the nature of RFID protocols will tend to be one-to-many, with one reader communicating with many tags [45].

15.3.2.1 Radio-Frequency Identification

RFID is a semiconductor-based technology that can be used to identify objects that are composed of three elements: an RFID tag formed by a chip connected with an antenna, an RFID reader that tracks the physical movement of the tag and passes its digital identity and other relevant information to a computer system, and finally a middleware that bridges RFID hardware and enterprise applications [46]. Compared with the traditional bar code systems, it has many benefits; for example, accurate and fast data gathering without human intervention, operations in a harsh environment, and its ability to sense the surroundings [47,48].

Owing to these benefits of RFID technology, recently, concepts, standards and solutions for the integration of RFID in enterprise applications have been dramatically highlighted from industry, such as Wal-Mart, and the U.S. Department of Defense has mandated that by 2005, suppliers must deliver ePC-tagged cases and pallets that are ready for RFID tackling [49]. Intel identified RFID as a potential technology to track inventory in the return process. Hewlett-Packard (HP) also started tagging its printers, which allows identification of individual units that have been returned, repaired or recycled. In a similar way, Dutch telecom carrier KPN (in full Koninklijke KPN N.V., also Royal KPN N.V., the largest telephone company in Netherlands) is undertaking a trial whereby KPN stores apply tags to returned or unsold phones. The tag on a returned product carries an identification number as well as the reasons for its return. Then, upon arrival at the distribution center (DC), phones that are fitted for use are immediately directed for resale while the others are sent for rework [50].

In addition, technologies such as RFID enable the tracking and tracing of products that are recently launched to obtain more product information through the whole product life cycle. Thanks to these techniques, the uncertainty in the product recovery caused by a lack of information and control mechanisms regarding quantity, timing and product compositions, and the quality of returns can be largely reduced. There have been some published researches dealing with necessary information and the overall framework for EoL product recovery optimisation. For example, Ref. [51] described product information availability and requirements during product recovery decision making, and addressed the role of networked RFID systems in delivering the requisite product information to improve product recovery decisions. Another example of applying condition monitoring to power tools to enhance economically beneficial remanufacturing was presented [52]. Instead of monitoring individual products, we can also monitor just inventories of products. So, in terms of how chips in crates are used to optimise the product recovery process, interested readers please refer to Ref. [53].

Furthermore, the authors of Ref. [54] believed that information management is the key to creating an efficient PRM and that RFID can be actively used to manage product returns. Ref. [55] mentioned using RFID to improve the quality of data and reduce the amount of manual data transfer of information in the supply chain to enable a product data management system. Ref. [56] felt that future improvements in PRM will be driven by technological developments such as RFID that will allow low-cost remote monitoring of information for a wide range of products and their processes. The details of RFID applications with respect to enhancing product recovery value in closed-loop supply chains are discussed in Ref. [57].

15.3.3 Internet of Things

Internet of things (IoT) can be defined as: 'A global network infrastructure, linking physical and virtual objects through the exploitation of data capture and communication capabilities. This infrastructure includes existing and evolving Internet and network developments. It will offer specific object-identification, sensor, actuator and connection capability as the basis for the development of independent federated services and applications. These will be characterized by a high degree of autonomous data capture, event transfer, network connectivity and interoperability' [58].

From the point of view of an end user, the most obvious effects of introducing IoT will be visible in EoL product disposal activity. In this context, it improves the EoL product management in the product recovery context and allows the linkage of products identity information from a tag to other information stored on networked databases [59].

15.4 Agent-Based Used Products Return Service

15.4.1 Scenario Description

The aforementioned technologies produce new opportunities for practitioners to achieve a more environmental friendly ambient recommender system of EoL product disposal and collection. Bear this in mind; we are going to consider the following envisioned scenario as our research focus:

Imagine a person who wants to discard an EoL product. To do so, he/she will only have to follow several steps: first, he/she scans the RFID tag on the EoL product and chooses the disposal option; shortly afterwards, the EoL product collectors' information will be forwarded to him/her together with the EoL product treatment options as well; once he/she confirms one of these options displayed on his/her cell phone, a certain EoL product collector will contact him/her to sort out the EoL product collection issues.

The core nature of this scenario is that the end users get their EoL products treatment information from the Internet by simply touching an RFID tagged product with their mobile devices. We think in a perfect case of a recommender system, advertisements will become irrelevant. You would have a direct mapping between customer desire and service awareness. We chose this scenario since it provides realistic benefits for the end user and the society, appears frequently enough in everyday life, is possible to be deployed, and can be explained and understood quickly. On this basis, we are trying to model how information flow works, underlying our envisioned scenario.

15.4.2 Agent-Based Modelling

As outlined in Ref. [29], the multi-agent approach somehow finds a 'natural' application with AmI since it enables the development of systems in which the intelligent and autonomous functions are exhibited both by the individual devices and by their interaction in a network of devices. In this section, the roles for each of the agents within the e-disposal framework were defined and a detailed description of the responsibilities for each of these roles was provided. We assume that each EoL product has been tagged with an RFID-like tag, that is, what we called 'smart product' in this research, and in the meantime, each EoL product holder has an inherent willingness to return it.

15.4.2.1 User Agent

In our e-disposal scenario, we assume that each end user will be in possession of one kind of mobile devices (e.g., cell phones, personal digital assistants, laptops or tablets) that works as a user agent. The function of a user agent is twofold: first, it can help its owner to be registered with the local directory service provider; second, by reading different RFID tags embedded in smart products, a master agent (i.e., a user agent) can generate a series of slave agents in which each agent is responsible for a single RFID-related event and sends these agents to various destinations in the network to perform specific tasks such as migration and enquiry.

15.4.2.2 Collector Agent

A collector agent represents an entity (e.g., dealer, dismantler or third party) that is physically located inside the coverage area of the local directory service provider. There can be multiple collectors inside a given area that are specialised in different kinds of EoL product collection activities. Consequently, one of the collector agent's tasks is to get its representative collector registered with the local directory service provider. Apart from this, the main task of a collector agent is to provide appropriate services to every residing slave agent.

15.4.2.3 Directory Agent

The role of a directory agent is more like a yellow page service provider that primarily provides lookup services for user agents and collector agents. Therefore, we have divided our whole target environment into several clusters and each one of these clusters is composed of one directory agent, multiple user agents and multiple collector agents. In any given cluster, for instance, a user agent may submit a disposal request to the directory agent. The request consists of some requirements that specify the EoL product collection services that are needed by a user agent. When such a lookup service is finished, the directory agent will reply the user agent with a list of suitable collector agents. Then the user agent can contact the selected collector agent(s) directly. By using this architectural design concept, it is not necessary for a collector agent to establish a knowledge database about all other peer agents. This approach will significantly reduce the communication overhead and simplify the internal structure of each individual collector agent.

15.4.2.4 Coordinator Agent

It is responsible for all types of communications between agents. In common terms, it facilitates the information exchange between different players in an e-disposal scenario, for example, the communication between an end user and an EoL product collector.

15.5 Ambient Recommender Systems for Used Product Return Service System Design

15.5.1 Design Environment

To examine our hypothesis, we are actively implementing the simulation system with NetLogo. NetLogo is a cross-platform and programmable multi-agent modelling environment for natural and social phenomena simulation [17]. Each agent is created through a turtle keyword in NetLogo and can follow instructions independently and simultaneously. The NetLogo platform offers advantages for simulating a generic mobile agent framework. The interfaces provided by NetLogo are user friendly. This platform is ideal for fast prototyping, for modelling and simulating frameworks that are based on concurrently active agents. A simple and efficient scripting language allows entities to be controlled and their interaction with the environment to be described.

15.5.2 System Working Principles

To build our framework in a modular style and also to make our explanation more user friendly, we have devised the following modules that are interlinked with each other during our current implementation: registration module, communication module, migration module and enquiry module.

15.5.2.1 Registration Module

The goal of a registration module is to let master agents (i.e., user agent and collector agent in our framework) register with the directory agent via the coordinator agent. Therefore, the registrations can be done as follows:

1. User agent registration: The concept behind the user agent registration is that whenever a user agent enters a new cluster (although our discussions are mainly carried out within the same cluster, this is definitely the case for a wider geographical distribution scenario), it will submit its registration details to the responsible directory agent. Thereafter, the directory agent will store such registration details in its corresponding database. The whole process or communication is processed through the coordinator agent.

2. Collector agent registration: In general, since a collector agent is not as mobile as a user agent, this type of registration is more like a once-off registration (although the collector's profile updating might occasionally cause re-registration, this is temporarily out of the scope of this chapter). So, the process of collector agent registration is that a collector agent submits its registration details to the directory agent and gets back registered information before it can join the iteration of our e-disposal framework. Collector registration is also routed through the coordinator agent.

15.5.2.2 Communication Module

Communication is the most fundamental type of interaction between agents. To ease our discussion, the agent who is sending the message is often referred to as the sender and the receiver means the agent who is receiving such a message, while the coordinator agent stays in between the sender and the receiver to make such communication possible (see Figure 15.1).

The role of the sender and the receiver is often swapped, depending on the specific situation. For example, when a new user agent enters a cluster, it has to send a registration request message to the local directory agent via the coordinator agent. Then the message will be stored at the requesting queue, waiting for further treatment. By now, the sender is the user agent and the directory agent acts as the receiver. Once the registration is done, the directory agent will send back a confirmation message through the coordinator agent with a text such as 'Welcome new user, your registration has been completed!' So, for now, the directory turns to the sender and the user agent becomes the receiver.

FIGURE 15.1
Communication through the coordination agent.

15.5.2.3 Migration Module

The key players involved in migration module are a set of slave agents that are created by their corresponding master agents. The migration module is activated only when a slave agent is going to move to its destination agent. Taking a user agent, for instance, it can create a wide range of slave agents and dispatch them via the user's mobile device, and each one of these slave agents is responsible for a particular disposal task. For the rest of this section, we are going to depict how the migration module works by using this example.

First, in the NetLogo environment, we have assumed that each slave agent is represented in the form of a randomly predefined number called 'running time'. The sample code at the user side is shown in Figure 15.2.

Then, when dispatching a slave agent, we include the user agent's details, unique slave agent's ID, size of the slave agent and the running time in the message. We have created two lists: *SlaveAgentInfoList* and *JobPendingList*. The user agent has to save the unique slave agent's ID, receiver's ID and the size of the message in the *SlaveAgentInfoList* as soon as a slave agent is created so that it could be retrieved in future. In general, each user agent holds one such list. The *JobPendingList* is to execute the job request, and each collector holds one such list.

Next, on the receiver side, a collector agent recognises a user slave agent and sends an acknowledgement (ACK) message back to the user agent about the arrival of a specific slave agent. Thereafter, unless its master agent (i.e., user agent) recalls it from the collector agent regardless of the completion of the task, the user slave agent will stay at the collector agent until its allocated running time is completed. This process is illustrated through the following sample code shown in Figure 15.3.

```
let running_time random 50
let updated_time running_time
set Object_ID random 100
set Object_ID Object_ID + [ID] of Sender

set Receiver one-of collectors
 .
 .
 .
ask Coordinator [forward_msg Sender Receiver send_msg random-msg-id
size_of_object]
```

FIGURE 15.2
Sample code for a slave agent migrating from a user agent.

```
if (item 0 msg?item = 2)
[ let ack_msgs item 6 msg-item
    ifelse (member? "ACK" ack_msgs = false)
    [ set input-size 1
      set inpout-msg-id input-msg-id +76
      let rem but?first msg-item
       .
       .
       .
    ]
    ask Coordinator [forward_msg receiver-item sender-item reply-msg
input-mg-id input-size]
```

FIGURE 15.3
Sample code for a slave agent migrating to a collector agent.

At the end of each round of execution, the collector agent will invoke a NetLogo procedure that is named *to execute* to deal with the remaining tasks in their *JobPendingList*. We have followed the round-robin approach to process these remaining tasks, that is, at every tick, each job in the queue of *JobPendingList* will be revisited. This implies that each remaining job shares the same priority of being assigned to be processed by the central processing unit (CPU). Therefore, we divided the capability of the CPU into equal timeslots by using Equation 15.1:

$$\text{Timeslot} = \frac{\text{CPU\_Capability}}{\text{number of jobs in the RemainingJobList}} \quad (15.1)$$

And the following two cases will be treated separately:

1. Running time ≤ available timeslot: In this case, the waiting task will be assigned to the CPU and the completed task will be moved to another list called *JobDoneList*. If a portion of the timeslot of this finished task is left available, it will be added to the next job's timeslot under the same tick.
2. Running time > available timeslot: Under this situation, the running time is split into two parts: The first part of running time that equals a standard timeslot will be pulled out and assigned to the CPU, whereas the second part of running time, which is equivalent to the remainder of (running time–available timeslot), will be assigned to the CPU for further treatment in a round-robin fashion.

In both cases, the collector agent needs to manage its system load by itself. Here, we use Equation 15.2 to measure the load of the system, that is, how much load the system is currently handling.

$$\text{System load} = \frac{\text{CPU\_Used}}{\text{CPU\_Capability} \times \text{ticks}} \quad (15.2)$$

where the CPU_Used is the sum of the execution time of the jobs that have been assigned to the CPU up to now. Therefore, for the first case, the CPU_Used equals (CPU_Used + running time), whereas in the second case, the CPU_Used equals (CPU_Used + time slot).

After sending a slave agent to a collector agent, our framework will run for several ticks and then move onto the following module to make sure that such slave agent has reached its destination.

15.5.2.4 Enquiry Module

In the migration module, we have mentioned about *SlaveAgentInfoList*. This list stores the unique slave agent ID, receiver's details and the size of the message packet. The functionality of the enquiry module is performed through the following two steps:

- The user agent pulls the first item from the *SlaveAgentInfoList* and extracts the slave agent ID and receiver's details out of it. The sender enquires about the status of such slave agent by sending a request containing 'slave agent ID' to the receiver. This process repeats until all the items in the *SlaveAgentInfoList* are pulled out and enquired about the completion of their tasks under the same tick.

- At the receiver side, the receiver checks the slave agent ID in its *JobDoneList*. If the searching object exists in the *JobDoneList*, it will reply with 'Task is Finished or ACK' status; otherwise, it will send a message with 'Task is incomplete or non-ACK' status, which means that the slave agent ID is still in the *JobPendingList*.

Our implementation mainly deals with communication among the three major agent types: user agent, collector agent and directory agent of our proposed framework in Section 15.7. One user agent can use his/her mobile device to generate multiple slave agents and dispatches them to various collectors where different facilities can be provided for slave agents to execute. The proposed architecture is based on the assumption that all products have been tagged with the item-level RFID. They can be identified in each step of the business process, such as during disposal/collection.

15.5.3 System Implementation

In this section, some implementation details are explained with several sample code sections (see Figure 15.4) and a screenshot in NetLogo (see Figure 15.5).

- First, in the procedure of *breed*, different types of agents and their corresponding variables are defined
- Second, in the procedure of *to setup*, the specific properties of each type of agent will be further created
- Third, in the procedure of *to go*, the NetLogo will execute the code within this procedure repeatedly once the go button has been activated

Next, some plotting functions such as network utilisation and system load will also be invoked right after the *to go* procedure.

Our implementation mainly deals with the communication among the three major agent types: user agent, collector agent and directory agent of our proposed framework in Section 15.4.2. One user agent can use his/her mobile device to generate multiple slave agents and dispatch them to various collectors where different facilities can be provided for slave agents to execute. The proposed architecture is based on the assumption that all products have been tagged with item-level RFID. They can be identified in each step of the business process, such as during disposal/collection.

It can be observed that, in Figure 15.5, several turtle shapes in NetLogo are used to represent different agent types: 'computer server' stands for directory agent, mobile agent is modelled as 'person' and 'house' shape has been chosen to represent collector agent. Meanwhile, the white background behind each 'computer serve' means the coverage area of the directory agent. In our model, the directory and collector agents are static entities, whereas, the mobile agents are all movable around the environment.

15.5.3.1 Step 1: Registration with the Directory Agent

The e-disposal process cannot proceed unless the user agent(s) and the collector agent(s) are correctly registered with the directory agent. Therefore, in our proposed framework, the function of the registration module needs to be performed first on top of anything else. The implementation of a user agent's or a collector agent's registration in NetLogo environment can be done by following the stages below. Owing to the fact that both registrations

```
breed [ users user ]
breed [ collectors collector ]
breed [ coordinators coordinator ]
breed [ directories directory]
collectors-own [ collector-color ]
turtles-own [ ID ]
globals [ ServiceCoverage ... ]

to setup
  clear-all
  .
  .
  .
  set-default-shape coordinators "computer server"
  set-default-shape users "person"
  set-default-shape collectors "house"
  create-collectors number_of_collectors
  [ let tempDirectory one-of directories
    setxy xcor ([xcor] of tempDirectory) + random
Directory_radio_coverage)
    setxy ycor ([ycor] of tempDirectory) + random
Directory_radio_coverage)
    ask collectors
    [ set ID who + (random 9007199254740992) ]
    ask patches in-radius collector_service_coverage
    [ set pcolor serviceCoverage ]
  ]
  create-users number_of_users
  [ setxy random-xcor random-ycor
    ask users [ set ID who + (random 9007199254740992) ]
  ]
  .
  .
  .
  create-directories 1
  [ setxy 0 0 ]
  .
  .
end

to go
  .
  .
  .
  ask users
  [ let rand (random 4)
    if rand = 0 [ fd 2.1 ]
    if rand = 1 [ bk 2.1 ]
  ]
  .
  .
end

to plot-system-load
  set-current-plot "System-Load"
  .
  .
end
```

FIGURE 15.4
Sample code of prototype implementation in NetLogo.

share the similar characteristics, hereafter, we only describe a collector agent's registration for demonstration purpose.

1. Registration message sending: As shown in Figure 15.1, since the coordinator agent is a mediator for communications actually taking place between the sender and the receiver, to complete the registration, first, a collector agent has to send a registration request message to the coordinator agent, and the latter will forward

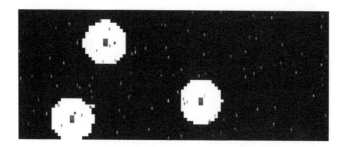

FIGURE 15.5
Screenshot of prototype implementation in NetLogo.

such message to the directory agent for the purpose of registration. In NetLogo, we can translate such process into a NetLogo procedure as shown in Figure 15.6.

While 'ask' is a NetLogo command that can be used to give an order to an agent (i.e., coordinator agent in this case), *forward_msg* is also the name of a NetLogo procedure and it can accept five types of variables (i.e., sender ID, receiver ID, message content, unique message ID and the size of the message) as its input.

2. Registration message pretreatment: After receiving a registration message containing five parameters from a collector agent, the coordinator agent will save this message in a waiting list. Meanwhile, according to the size of the message, the following two cases will be treated:

   a. Request message size ≤ network capacity: Network capacity is the maximum size of the message that a coordinator can handle. In this condition, the coordinator agent will pull out the details of the message from the waiting list and forward it to the directory agent by using a NetLogo procedure called *send_msg*. The following piece of code (see Figure 15.7) shows you how it works.

   b. Request message size > network capacity: In this case, the coordinator agent cannot handle the size of the message. Therefore, it will split the message into two parts where the first part is of size equivalent to the network capacity value and the second part holds the remaining size of the message. Then, the first part of the message will be distributed over the network as usual, whereas the second part will be held temporarily. Next, the coordinator agent will verify

```
ask collector
[ask coordinator
  [forward_msg hosts-reg-sender host-reg-receiver host-msg random-msg-id1 random_size]
```

FIGURE 15.6
Communication through the coordinator agent.

```
ask coordinator
[ask current-receiver
  [send_msg current-sender curren-receiver curren-msg current-id curren-id current-size]
```

FIGURE 15.7
Sample code of *send_msg*.

```
ask receiver-item
[ask coordinator
    [forward_msg receiver-item sender-item directory-reply-msg input-
msg-id input-size]
```

FIGURE 15.8
Sample code of registration message processing.

```
ask coordinator
[ask current-receiver
    [send_msg current-sender current-receiver current-msg current-id
crrent-size]
```

FIGURE 15.9
Sample code of registration message feedback.

> the size of the second part again by referring to the criteria of situations (a) and (b) until the message can be eventually forwarded.

3. Registration message processing: After receiving the registration request message from a collector agent, the directory agent will search the incoming message ID in its registration queue. If the directory agent can match the incoming message ID with any of the existing message IDs in the registration queue, it will simply discard the new message since such message is being processed. Otherwise, if the new message ID cannot be located in the registration queue, the directory agent will add such message into the queue waiting for further treatment. Once the registration request has been processed, a piece of the ACK information will be sent back to the corresponding collector agent via the coordinator agent. The following code script (see Figure 15.8) illustrates this process.

   The functionality and input parameters of the method *forward_msg* here is similar to the *forward_msg* method in step 1. The forward_msg method accepts five parameters as an input: sender ID, receiver ID, message content, unique message ID and size of the message.

4. Registration message feedback: The coordinator agent stores the details of the ACK message from the directory agent and then forwards it to the corresponding collector agent. The sample code for this process is shown in Figure 15.9.

5. Registration message checking: In the NetLogo procedure of *send_msg*, the collector agent checks whether the feedback message ID matches the registration request message ID stored in its own database when such request was initially submitted. If the answer is yes, the collector agent will remove such message ID from its database. By now, a typical process of a collector agent registering with the director agent is done. In a similar way, the same process is also applied to other registration cases such as a user agent registering with the director agent.

15.5.3.2 Step 2: List of Candidate Collectors Acquiring

After the registration step, on each execution of our proposed e-disposal strategy, any user agent can send a randomly chosen EoL product disposal request to the directory agent. On receiving such requirement, the directory agent will verify the details of registered EoL product collectors stored in its registration database. As soon as a collector, with the required EoL product processing capabilities that are needed by an enquirer, is found, it will be promptly put into a list. When no more collectors can be found, the directory agent will return the user agent with a final accumulated list of candidate collectors. In case the same user agent might request different lists of collectors based on various EoL product disposal requirements, we follow a random picking rule to give equal weight to all the requirements belonging to an individual user agent.

15.5.3.3 Step 3: Candidate Collectors' System Load Enquiry

After making sure that the list of candidate collectors has been received, the user agent will compare the network utilisation with a load-balancing threshold value (LBTV). Here, the network utilisation means the ratio of current network traffic to the maximum traffic that the port can handle. It indicates usage of the bandwidth in the network. In our framework, we calculate the value of network utilisation by using Equation 15.3

$$\text{Network utilisation} = \frac{\text{Packet\_Sum}}{\text{ticks} \times \text{Network\_Capacity}} \quad (15.3)$$

where Packet_Sum is the summation of sizes of packets assigned to the network till now, Network_Capacity is the maximum size of the message packets that the network can handle, and tick is a single time unit in the NetLogo environment.

While LBTV is a value varying from 0 to 1, it is regarded as a threshold value for the coordinator agent to balance the message packets. The value of LBTV can be adjusted by using a slider button. After adjusting LBTV, we will compare the network utilisation with the adjusted LBTV by considering the following situations:

1. Network utilisation ≤ LBTV: which means that the network is not overloaded with messages or requests; hence, the sender (user) will directly communicate with the collectors in the candidate list by sending a message to them. At the receiver side, the receiver checks the request and replies back with a relevant message such as ACK to the sender.
2. Network utilisation > LBTV: that is, the network is overwhelmed with messages or requests; in that case, the user agent needs to send the message or request to the collector with the underutilisation and having the shortest response time among the candidate collectors. To locate the collector with the above qualities among the candidate collectors' list, the user agent broadcasts a message to each collector in the candidate collectors' list to acquire a list of system loads of each collector.

After receiving all the requests for the system loads from the same user, the last collector in the candidate collectors' list takes the responsibility to gather the system loads of each collector to compile a list. The compiled list will be sent back to the user. The communication module plays a key role to facilitate this interaction.

15.5.3.4 Step 4: Disposal Request Sending

The functionality of the migration module will be performed in this step. By using the system loads list obtained in the previous step, the user agent builds its plan considering the lowest system load and sends out a slave agent to that particular collector agent.

Meanwhile, two lists will be created: one is the *SlaveAgentInfoList* on the user agent side; the other is the *JobPendingList* on the collector agent side. Then the landing slave agent will continue to work at the collector agent until it finishes its task, unless the user agent wants to retrieve it. By now, a simulated networking mechanism is illustrated. With the user's movement, information should always be available to him/her at any new location. Therefore, the framework presented here is to be flexible, high performing and easy to implement.

15.6 Conclusions

From a business perspective, the product recovery process begins when the end user returns the product and ends when the company has recovered the maximum value. Thanks to the rapid development of information technology, the efficient management of product recovery-dependent EoL product information generated in reverse logistic networks across end users, collectors, redistributors and re-processors (remanufacturers, recyclers, etc.) has surfaced as an important factor.

The conceptual framework presented in this chapter is a contribution to the request in the literature [26] to formulate concepts and theories for analysing and understanding the implementation process of sustainability in used products return services. In detail, we are currently developing a new version of used products return services intended to support the broader aspects of end users' positive environmental consciousness. Particularly, we are interested in exploring the possibilities enabled by treating e-disposal/collection as a two-way channel between the collectors and end users. The proposed framework is a recommender system for personalisation based on multi-agent architectures, especially the agents' design and the agents' communication. As web content and users' interests are both highly evolving, the proposed system filters a complete listing of programmes to only those matching a user's personal profile. By utilising this information, we believe that a more intelligent and responsive interface between users and collectors and the computing infrastructure can be implemented in a seamless environment.

The main contribution relies in addressing used products return services in the context of pervasive environment, by providing the end users with comprehensive support for enabling disposal-aware AmI systems to obtain a globally sustainable development. Furthermore, companies that use e-disposal strategy as an opportunity to enhance business will prosper by maintaining customer support, the ultimate issue for profitability.

15.7 Future Work

Additional empirical research is necessary to test this framework. The relationship between the convenience characteristics for the end user and used products return service with the implementation strategy for sustainability is an essential element in this study. A multiple case study approach [60] would be a desirable method to test this framework. Furthermore, while this chapter has discussed agent-based filtering primarily in the context of the disposal/collection task, we are confident that this technique can help to form the basis for agents that provide an information channel to people engaged in tasks in the physical world. Future developments will focus on providing further capabilities. Meanwhile, making the information of all the EoL products easily achievable without the risk of downstream information overflow has also proved to be especially challenging. We believe that the results will add a new 'innovation' dimension and stimulate the further development of theory building and conceptual development within the interdisciplinary field of sustainability and product recovery research.

References

1. Mont, O. 2008. Innovative approaches to optimising design and use of durable consumer goods. *International Journal of Product Development 6*, 227–250.
2. Zuidwijk, R. and Krikke, H. 2008. Strategic response to EEE returns: Product eco-design or new recovery processes? *European Journal of Operational Research 191*, 1206–1222.
3. Walther, G., Steinborn, J., Spengler, T.S., Luger, T. and Herrmann, C. 2010. Implementation of the WEEE-directive– economic effects and improvement potentials for reuse and recycling in Germany. *International Journal of Advanced Manufacturing Technology 47*, 461–474.
4. Sasikumar, P., Kannan, G. and Haq, A.N. 2010. A multi-echelon reverse logistics network design for product recovery—A case of truck tire remanufacturing. *International Journal of Advanced Manufacturing Technology 49*, 1223–1234.
5. Guide, V.D.R., Teunter, R.H. and Wassenhove, L.N.V. 2003. Matching demand and supply to maximize profits from remanufacturing. *Manufacturing & Service Operations Management 5*, 303–316.
6. Hanafi, J. 2008. Modeling of collection strategies for end-of-life products using coloured Petri net. In School of Mechanical and Manufacturing Engineering, Doctoral thesis. (Sydney, Australia: The University of New South Wales).
7. Lyman, P. and Varian, H. 2000. How much information? In http://www.sims.berkeley.edu/how-much-info.
8. Singh, N., Bartikowski, B.P., Dwivedi, Y.K. and Williams, M.D. 2009. Global megatrends and the web: Convergence of globalization, networks and innovation. *The Database for Advances in Information Systems 40*, 14–27.
9. Thierry, M., Salomon, M., Nunen, J.V. and Wassenhove, L.V. 1995. Strategic issues in product recovery management. *California Management Review 37*, 114–135.
10. Fleischmann, M., Bloemhof-Ruwaard, J.M., Dekker, R., Laan, E.V.D., Nunen, J.A.E.E.V. and Wassenhove, L.N.V. 1997. Quantitative models for reverse logistics: A review. *European Journal of Operational Research 103*, 1–17.
11. Güngör, A. and Gupta, S.M. 1999. Issues in environmentally conscious manufacturing and product recovery: A survey. *Computers & Industrial Engineering 36*, 811–853.
12. Porada, T. 1994. Materials recovery: Asset alchemy. In *Proceedings of the IEEE International Symposium on Electronics and the Environment*, 2–4 May, pp. 171–173. (San Francisco, CA).
13. Ilgin, M.A. and Gupta, S.M. 2010. Environmentally conscious manufacturing and product recovery (ECMPRO): A review of the state of the art. *Journal of Environmental Management 91*, 563–591.
14. Jun, H.-B., Cusin, M., Kiritsis, D. and Xirouchakis, P. 2007. A multi-objective evolutionary algorithm for EOL product recovery optimization: Turbocharger case study. *International Journal of Production Research 45*, 4573–4594.
15. Rogers, D.S. and Tibben-Lembke, R.S. 1999. *Going Backwards: Reverse Logistics Trends and Practices*. (Pittsburgh, PA: Reverse Logistics Executive Council).

16. Daugherty, P.J., Richey, R.G., Genchev, S.E. and Chen, H. 2005. Reverse logistics: Superior performance through focused resource commitments to information technology. *Transportation Research Part E 41*, 77–92.
17. Kumar, S. and Putnam, V. 2008. Cradle to cradle: Reverse logistics strategies and opportunities across three industry sectors. *International Journal of Production Economics 115*, 305–315.
18. Jayaraman, V., Guide, V.D.R. and Srivastava, R. 1999. A closed-loop logistics model for remanufacturing. *Journal of the Operational Research Society 50*, 497–508.
19. Jayaraman, V., Patterson, R.A. and Rolland, E. 2003. The design of reverse distribution networks: Models and solution procedures. *European Journal of Operational Research 150*, 128–149.
20. Fleischmann, M., Beullens, P., Bloemhof-Ruwaard, J.M. and Wassenhove, L.N.V. 2001. The impact of product recovery on logistics network design. *Production and Operations Management 10*, 156–173.
21. Salema, M.I.G., Barbosa-Povoa, A.P. and Novais, A.Q. 2007. An optimization model for the design of a capacitated multi-product reverse logistics network with uncertainty. *European Journal of Operational Research 179*, 1063–1077.
22. Louwers, D., Kip, B.J., Peters, E., Souren, F. and Flapper, S.D.P. 1999. A facility location allocation model for reusing carpet materials. *Computers & Industrial Engineering 36*, 855–869.
23. Min, H., Ko, H.J. and Ko, C.S. 2006. A genetic algorithm approach to developing the multi-echelon reverse logistics network for product returns. *Omega 34*, 56–69.
24. Barros, A.I., Dekker, R. and Scholten, V. 1998. A two-level network for recycling sand: A case study. *European Journal of Operational Research 110*, 199–214.
25. Lu, Z. and Bostel, N. 2007. A facility location model for logistics systems including reverse flows: The case of remanufacturing activities. *Computers & Operations Research 34*, 299–323.
26. Guide, V.D.R. and Wassenhove, L.N.V. 2001. Managing product returns for remanufacturing. *Production and Operations Management 10*, 142–155.
27. Ray, S., Boyaci, T. and Aras, N. 2005. Optimal prices and trade-in rebates for durable, remanufacturable products. *Manufacturing & Service Operations Management 7*, 208–228.
28. Wen, L., Lin, C.-H. and Lee, S.-C. 2009. Review of recycling performance indicators: A study on collection rate in Taiwan. *Waste Management 29*, 2248–2256.
29. Sadri, F. 2011. Ambient intelligence: A survey. *ACM Computing Surveys 43*(36), 31–36, 66.
30. Aarts, E. 2004. Ambient intelligence: A multimedia perspective. *IEEE MultiMedia 11*, 12–19.
31. Favela, J., Rodriguez, M., Preciado, A. and Gonzalez, V.M. 2004. Integrating context-aware public displays into a mobile hospital information system. *IEEE Transactions on InformationTechnology in Biomedicine 8*, 279–286.
32. Rodriguez, M.D., Favela, J., Martinez, E.A. and Munoz, M.A. 2004. Location-aware access to hospital information and services. *IEEE Transactions on Information Technology in Biomedicine 8*, 448–455.
33. Rodriguez, M., Favela, J., Preciado, A. and Vizcaino, A. 2005. Agent-based ambient intelligence for healthcare. *AI Communication 18*, 201–216.
34. Romero, C. and Ventura, S. 2007. Educational data mining: A survey from 1995 to 2005. *Expert Systems with Applications 33*, 135–146.
35. Lappas, G. 2008. An overview of web mining in societal benefit areas. *Online Information Review 32*, 179–195.
36. Adomavicius, G. and Tuzhilin, A. 2005. Towards the next generation of recommender systems: A survey of the state-of-the-art and possible extensions. *IEEE Transactions on Knowledge and Data Engineering 17*, 734–749.
37. González, G., López, B. and Rosa, J.L.L.D.L. April 2004. Managing emotions in smart user models for recommender systems. In *Proceedings of the 6th International Conference on Enterprise Information Systems (ICEIS)*, Universidade Portucalense, 14–17, pp. 187–194. (Porto, Portugal).
38. Schafer, J.B., Konstan, J.A. and Riedl, J. 2001. E-commerce recommendation applications. *Data Mining and Knowledge Discovery 5*, 115–153.

39. Freiheit, T.I. 2003. Reliability and productivity of reconfigurable manufacturing systems. In Department of Mechanical Engineering, Doctor of Philosophy. (The University of Michigan).
40. Yager, R.R. 2003. Fuzzy logic methods in recommender systems. *Fuzzy Sets and Systems 136*, 133–149.
41. Ujjin, S. and Bentley, P. 2003. Particle swarm optimization recommender system. In *Proceedings of the IEEE Swarm Intelligence Symposium (SIS)*, (Indianapolis, Indiana), pp. 124–131.
42. Miller, B.N., Konstan, J.A. and Riedl, J. 2004. PocketLens: Toward a personal recommender system. *ACM Transactions on Information Systems 22*, 437–476.
43. Falke, O., Rukzio, E., Dietz, U., Holleis, P. and Schmidt, A. 2007. *Mobile Services for Near Field Communication.* (Munich, Germany: Media Informatics Group, Department of Computer Science, University of Munich), p. 15.
44. Kiritsis, D. 2011. Closed-loop PLM for intelligent products in the era of the Internet of things. *Computer-Aided Design 43*, 479–501.
45. Bolić, M., Simplot-Ryl, D. and Stojmenović, I. 2010. *RFID Systems: Research Trends and Challenges*. (West Sussex, United Kingdom: John Wiley & Sons Ltd.).
46. McFarlane, D. and Sheffi, Y. 2003. The impact of automatic identification on supply chain operations. *International Journal of Logistics Management 14*, 1–17.
47. Jun, H.-B., Shin, J.-H., Kim, Y.-S., Kiritsis, D. and Xirouchakis, P. 2009. A framework for RFID applications in product lifecycle management. *International Journal of Computer Integrated Manufacturing 22*, 595–615.
48. Ferrer, G., Dew, N. and Apte, U. 2010. When is RFID right for your service? *International Journal of Production Economics 124*, 414–425.
49. Tajima, M. 2007. Strategic value of RFID in supply chain management. *Journal of Purchasing & Supply Management 13*, 261–273.
50. Karaer, Ö. and Lee, H.L. 2007. Managing the reverse channel with RFID-enabled negative demand information. *Production and Operations Management 16*, 625–645.
51. Kulkarni, A.G., Parlikad, A.K.N., McFarlane, D.C. and Harrison, M. 2005. *Networked RFID Systems in Product Recovery Management.* (Cambridge, UK: Auto-ID Lab, University of Cambridge), p. 12.
52. Klausner, M., Grimm, W.M. and Hendrickson, C. 1998. Reuse of electric motors in consumer products: Design and analysis of an electronic data log. *Journal of Industrial Ecology 2*, 89–102.
53. Flapper, S.D.P., Nunen, J.A.E.E.V. and Wassenhove, L.N.V. eds. 2005. *Managing Closed-Loop Supply Chains.* (Berlin, Heidelberg: Springer-Verlag).
54. Fleischmann, M., Nunen, J.A.E.E.V. and Gräve, B. 2003. Integrating closed-loop supply chains and spare-parts management at IBM. *Interfaces 33*, 44–56.
55. Krikke, H., Blanc, I.L. and Velde, S.V.D. 2004. Product modularity and the design of closed-loop supply chains. *California Management Review 46*, 23–38.
56. Nunen, J.A.E.E.V. and Zuidwijk, R. 2004. E-enabled closed-loop supply chains. *California Management Review 46*, 40–54.
57. Visich, J.K., Li, S. and Khumawala, B.M. 2007. Enhancing product recovery value in closed-loop supply chains with RFID. *Journal of Managerial Issues 19*, 436–452.
58. Moisescu, M.A., Sacală, I.Ş. and Stănescu, A.M. 2010. Towards the development of Internet of things oriented intelligent systems. *University of Polytech Bucharest Science Bulletin, Series C 72*, 115–124.
59. Amaral, L.A., Hessel, F.P., Bezerra, E.A., Corrêa, J.C., Longhi, O.B. and Dias, T.F.O. 2011. eCloud RFID—A mobile software framework architecture for pervasive RFID-based applications. *Journal of Network and Computer Applications 34*, 972–979.
60. Yin, R.K. 1994. *Case Study Research: Design and Methods.* (2nd Thousand Oaks, CA: Sage Publications).

Index

A

AADL, *see* Architecture analysis and design language (AADL)
ABS, *see* Aircraft brake system (ABS)
ABSINTH, *see* Abstract instruction extraction helper (ABSINTH)
ABSOLUT
  mapping and co-simulation, 96, 97
  performance model, 94
  platform modeling, 96
  workload models, 95–96, 99–100
  Y-chart approach, 94
ABSOLUT OS_Service modelling
  ABSOLUT MAC, 108
  data link and transport layer services, 106
  Generic_Serv_IF, 106
  New OS_Services deriving, 103, 104
  OS Services accessment, 104
  OS_Service base class, 103, 104, 105
  registering OS_Services to OS models, 104, 106
Abstract instruction extraction helper (ABSINTH), 96, 113–114
Access control list (ACL), 303
Accumulative mode (ACM), 328
  state-transition table, 330, 332, 338
ACK, *see* Acknowledgement (ACK)
Acknowledgement (ACK), 177, 368
ACL, *see* Access control list (ACL)
ACM, *see* Accumulative mode (ACM)
Active fault-tolerant controlling (AFTC), 319, 322
Active learning method
  random score function, 160
  RankBoost_Active, 159
  for ranking functions, 152, 159
Activity context, 245, 248
Activity ontology, 253
Additive white Gaussian noise (AWGN), 108
ADSL, *see* Asynchronous digital subscriber loop (ADSL)
AFTC, *see* Active fault-tolerant controlling (AFTC)
Agent-based used products return service
  agent-based modelling, 365–366
  scenario description, 365
Aggregation operations, 41, 48, 50, 51

Aircraft brake system (ABS), 321, 335
  AADL error model, 327
  ABS_MAG, 339
  architectural and error model, 329
  brake function, 335
  hierarchical architecture, 328
  modes, 336
  multi-agent system, 337
  RWB_MAG, 337, 339
  state-transition table, 330
Alarm annunciation, 318–319
Alternative mode (AM), 328
ALTO, *see* Application layer traffic optimization (ALTO)
AM, *see* Alternative mode (AM)
Ambient intelligence (AmI), 361
  multi-agent approach, 365
  proactive and adaptive services, 361
  ubiquitous computing devices, 362
Ambient recommender systems
  design environment, 366
  system working principles, 366–370
  system implementation, 370
  candidate collectors, 373–374
  disposal request sending, 374–375
  NetLogo, 370, 371
  prototype implementation in NetLogo, 372
  registration with directory agent, 370–373
  system load enquiry, 374
AmI, *see* Ambient intelligence (AmI)
Anaesthesia workstation
  innovative clinical applications, 309, 310
  medical devices interfacing, 308
  SomnuCare software, 307
  vital signs and alarms, 306
  workplace, 306
API, *see* Application programming interface (API)
Application execution phase, 113
Application layer traffic optimization (ALTO), 71
Application programming interface (API), 104, 220–221
  middleware technologies, 98
  POSIX, 103
  SMEPP model, 222, 223
  SomnuCare, 307, 308

Application-oriented networking (AON), 68
    application layer, 80
    query message structure, 80
    router design, 79, 80
    routing, 74
    structure, 79
Architecture analysis and design language (AADL), 321
    architectural model, 329
    behavioural state machines, 326
    error model, 327
    RWB sub-system, 331
    safety assessment model, 324
Argumentation generation, 283
AS numbers, *see* Autonomous system numbers (AS numbers)
Asynchronous digital subscriber loop (ADSL), 194
Auto MPG, 144
Autonomous system numbers (AS numbers), 70, 73
Average estimator, 33
AWGN, *see* Additive white Gaussian noise (AWGN)

B

B2B, *see* Business-to-business (B2B)
Back propagation (BP), 195
Backus–Naur form description (BNF description), 51
    AREASPAN and TIMESPAN clauses, 53
    FROM clauses, 52
    SELECT syntax, 52
    WHERE and WHILE phrases, 53
Basic component level (BC level), 321, 322
Batch algorithm, 126
    neighbourhood function, 126
    Newton's method, 125
BC level, *see* Basic component level (BC level)
BDI agent, *see* Belief-Desire-Intention agent (BDI agent)
Belief-Desire-Intention agent (BDI agent), 330
    agent reasoning model, 333
    reasoning model, 331
BER, *see* Bit error rate (BER)
Berkeley Software Distributions (BSD), 100, 104
Best matching unit (bmu), 120, 123
    Kohonen's SOM, 127
    neighbourhood function, 127, 132
    TKM, 130
    topological preservation, 149

Bipartite ranking function (BRF), 152
Bit error rate (BER), 109
    calculation accuracy analysis, 108
BitTorrent (BT), 192, 274
    file-sharing network, 193
    flow, 192
    protocol, 192, 193
    traffic detection, 193–194
bmu, *see* Best matching unit (bmu)
BNF description, *see* Backus–Naur form description (BNF description)
BP, *see* Back propagation (BP)
Brake system control unit (BSCU), 336, 337
BRF, *see* Bipartite ranking function (BRF)
BSCU, *see* Brake system control unit (BSCU)
BSD, *see* Berkeley Software Distributions (BSD)
BT, *see* BitTorrent (BT)
Business-to-business (B2B), 279

C

CAD, *see* Channel-aware detection (CAD)
Candidate collectors
    acquiring list, 373–374
    system load enquiry, 374
CARISMA, *see* Context-aware reflective middleware system for mobile applications (CARISMA)
Caterpillar's global remanufacturing business, 274
Central authority, 26
Central processing unit (CPU), 92, 369
    models configuration data, 113
    for node in grid network, 352
    performance counters, 111
    scalability of disk storage, 355
    using equations, 369
Centralised architecture
    average forecasting errors, 27
    coefficient of determination values, 27
    factors and parameters, 27
    forecast precision, 26
Channel-aware detection (CAD), 172
Classical routing, 74
Classification accuracy
    accuracy measurements and metrics, 205–206
    accuracy results, 206, 207
    classification time, 207
    training data and test cases, 204–205
Cloud computing, 347, 349
    cloud services model, 349

features, 348
frameworks and structures, 353
Google trends report, 347, 348
performance models, 355
service capabilities, 352
taxonomy, 350
Cloud domain, 348, 352
Cloud providers, 352, 353
Clustering, 75
Coalition payment division, 215–216
Coalitions and incentives (COINS), 212
    service code, 236
    service contract, 235
CoBrA system, 248
Collaborative leasing-based injection mould remanufacturing process analysis
    cooperation, 283
    d-AMR system, 285
    multi-agent supply chain, 284
    RFID technology, 284
Collector agent, 365–366
    coordinator agent communication, 372
    e-disposal process, 370
    registration, 371, 372, 373
    send_msg code, 372
Communication module, 367
Component planes, 128, 129
Computing Cloud, 349
Configuration and workload generation via code instrumentation and performance counters (CORINNA), 111
    output files, 112
    phases, 111–112
    pre-compilation and post-compilation steps, 112
    salient features, 113–114
    workload modelling via, 113
Content discovery service, 230, 232
Content distribution system, 213, 214; *see also* Peer-to-peer middleware (P2P middleware)
    checking mechanism, 215
    incentive mechanism, 213, 215–216
    protocol overview, 214
    reciprocity-based approaches, 219
    simulation results, 217–219
    task assignment, 214
Context acquisition, 255–256
Context information, 246
Context models
    on activity, 252
    context-aware system, 247

    context-related entities and relationship, 250
    high-level components, 251
    markup scheme models, 248
    ontology, 251
    semantic web, 250
Context sensing, *see* Context information
Context-aware reflective middleware system for mobile applications (CARISMA), 228
Context-aware systems, 243, 245
Continuous map generation
    continuous path projection, 137
    discrete and continuous maps, 137
    GRNN, 136
    punctured sphere data and normalised model, 136
Cooperative linear regression-based learning algorithm, 14–15
Coordinator agent, 366
CORINNA, *see* Configuration and workload generation via code instrumentation and performance counters (CORINNA)
CORR method, 50
CPU, *see* Central processing unit (CPU)
C/S-type integrated sensor networks, 42

D

d-AMR, *see* Distributed agent-based mould remanufacturing framework (d-AMR)
Data aggregation systems, 41
Data flow diagrams (DFD), 317–318
Data mining
    DDPM, 8–9
    forecasting models, travelling time, 9–10
    MAS architectures, 5–8
    problem formulation, 10–11
Data processing
    DDPM, 8–9
    forecasting models, travelling time, 9–10
    MAS architectures, 5–8
    problem formulation, 10–11
Data safety, 302
Database as a service (DaaS), 350, 352; *see also* Infrastructure as a service (IaaS)
DCF, *see* Distributed coordination function (DCF)
DDPM, *see* Distributed data processing and mining (DDPM)
DDS, *see* Distributed decision support (DDS)

Decentralised coordinated architecture
　　average forecasting errors, 31, 32
　　communication events number, 30
　　cross-validation technique, 31
　　R^2 values, 31
　　simulation parameters, 29
　　single agent system dynamics, 30
　　system parameter estimation, 30
Decentralised uncoordinated architecture
　　average estimation, 32
　　communication events, 30
　　forecasting errors, 27, 29
　　quality of agent models, 31
　　R^2 values, 28, 31
　　system parameter estimation, 28, 30
Deep flow inspection (DFI), 192, 198
　　arrangement, 197
　　classifier accuracy, 204
　　combination, 196, 197
　　DFI-based detection, 195–196
　　methods, 199, 201
　　online classifier, 204
　　techniques, 194
Deep flow inspection-based detection
　　MTU, 196
　　P2P traffic detection, 195
Deep packet inspection (DPI), 192, 198
　　combination, 196, 197
　　comparison, 202
　　DPI-based detection, 194–195
　　methods, 201
Deep packet inspection-based detection, 194–195
Degree of confidence, 262
Denial of service (DoS), 187; *see also* Infrastructure as a service (IaaS)
Derived services, 103, 104
Device interconnect protocol (DIP), 98, 201; *see also* Internet protocol (IP); Simple object access protocol (SOAP)
Devices profile for web services (DPWS), 295, 298
DFD, *see* Data flow diagrams (DFD)
DFI, *see* Deep flow inspection (DFI)
DICOM, *see* Digital imaging and communications in medicine (DICOM)
Digital imaging and communications in medicine (DICOM), 302
Digital signal processor chips (DSP chips), 93
DIP, *see* Device interconnect protocol (DIP)
Directory agent, 366
Disposal request sending, 374–375

Distributed agent-based mould remanufacturing framework (d-AMR), 278
　　complex structure, 279
　　deployment view, 285
　　objective, 278
　　RFID technology, 284
Distributed cooperative forecasting
　　cooperative kernel-based learning algorithm, 20–21
　　kernel density estimation, 17–18
　　kernel-based local regression model, 19–20
　　numerical example, 21–24
Distributed cooperative recursive forecasting
　　cooperative linear regression-based learning algorithm, 14–15
　　forecasting with multivariate linear regression, 11–13
　　local recursive parameter estimation, 13–14
　　numerical example, 15–17
Distributed coordination function (DCF), 106, 107
Distributed data processing and mining (DDPM), 4
　　algorithmic solutions, 8
　　data collection and transmission, 8
　　implementation, 9
　　in MAS, 5
　　modules, 6
　　structure, 4, 6
Distributed decision support (DDS), 6, 7
Distributed embedded systems, 92
Distributed monitoring model
　　AADL error model of ABS, 327
　　corrective measures, 326
　　fault propagation model, 333
　　monitoring cycle, 329
　　monitoring tasks, 325
　　state-transition tables, 326, 327
Distributed networks, 3
　　data processing and mining, 5–11
　　DDPM technology, 4
　　distributed cooperative forecasting, 17–24
　　distributed cooperative recursive forecasting, 11–17
　　MAS architectures, 3
Distributed online safety monitor
　　AADL safety assessment model, 324
　　distributed monitoring model, 325–330
　　fault controlling, 324
　　faultless operation, 323
　　hierarchical deployment of agents, 325
　　monitoring expression, 325
　　multi-agent system, 330–334

Index 383

Distributed service discovery (DSD), 68
   overlay and the underlay topology mismatching, 69
   P2P-based routing architecture, 72–80
   underlay routing *vs.* overlay routing, 68
Distributed systems performance models, 354
Document retrieval (DR), 151
DoS, *see* Denial of service (DoS)
DPI, *see* Deep packet inspection (DPI)
DPWS, *see* Devices profile for web services (DPWS)
DR, *see* Document retrieval (DR)
DSD, *see* Distributed service discovery (DSD)
DSP chips, *see* Digital signal processor chips (DSP chips)

E

e-commerce, *see* Electronic commerce (e-commerce)
Eco-leasing, 270
EFS, *see* Error-Free State (EFS)
EICAS, *see* Engine Indication and Crew Alerting System (EICAS)
EIED, *see* Exponential increase and linear decrease (EIED)
Electronic commerce (e-commerce), 279
ELT, *see* End-of-life tyres (ELT)
ELV, *see* End-of-life vehicle (ELV)
Embedded peer-to-peer systems (EP2P systems), 213
   attacks for, 227
   design process, 222
   security, 227
Embedded system, 92
   ABSOLUT, 94–97
   modelling ABSOLUT OS_services, 103–108
   modelling multithreaded support, 100–103
   physical layer models, 108–111
   SLPE, 97–100
End-of-life (EoL), 268
   directory agent, 366
   IoT, 364
   PRM, 361
   products, 270, 360
End-of-life tyres (ELT), 360
End-of-life vehicle (ELV), 272, 360
Engine Indication and Crew Alerting System (EICAS), 336
Enquiry module
   disposal/collection process, 370
   SlaveAgentInfoList, 368, 369

Entity ontology, 252–253
Environmental Monitoring 2.0, 43
EoL, *see* End-of-life (EoL)
EP2P systems, *see* Embedded peer-to-peer systems (EP2P systems)
EPR, *see* Extended producer responsibility (EPR)
Error State (ES), 322, 325
Error-Free State (EFS), 322, 325, 328
   operative engine, 326
   state-transition table, 330
ES, *see* Error State (ES)
Ethernet, 297
Exclusive-Or (XOR), 70
Exponential increase and linear decrease (EIED), 106
Extended producer responsibility (EPR), 268, 272
Extensible markup language (XML), 295, 297
   based WSDL, 298
   encryption and signature, 303
   MA, 59
   to OWL mapping, 249
   relationships, 249
   SMEPP service, 222
   in SOAP messages, 244–245

F

Failure mode and effect analysis (FMEA), 318
Failure state (FS), 322, 326, 328
Fault controlling, 319
Fault detection techniques
   model-based and data-based techniques, 317
   monitoring algorithm, 318
Fault-tolerant control system (FTCS), 319
FCC, *see* Flight control computer (FCC)
FCS, *see* Flight control system (FCS)
FFA, *see* Functional failure analysis (FFA)
FFBD, *see* Functional flow block diagrams (FFBD)
File transport protocol (FTP), 194, 298; *see also* Transmission control protocol (TCP)
FIPA, *see* Foundation for intelligent, physical agents (FIPA)
Flen plant, 274–275
Flight control computer (FCC), 321, 322
Flight control system (FCS), 326
FMEA, *see* Failure mode and effect analysis (FMEA)
FOAF, *see* Friend of a friend (FOAF)

Forecasting models
　algorithm, 11
　average estimator, 33
　average forecasting errors, 33
　factors, 25
　intelligent transport systems, 9–10
　TIC, 10
Foundation for intelligent, physical agents (FIPA), 284
Frame error rate calculation accuracy analysis, 109
Free-riding, 212
Friend of a friend (FOAF), 251
FS, *see* Failure state (FS)
FTCS, *see* Fault-tolerant control system (FTCS)
FTP, *see* File transport protocol (FTP)
Functional failure analysis (FFA), 318
Functional flow block diagrams (FFBD), 317–318

G

GA, *see* Genetic algorithm (GA)
Game theory
　attack model, 175
　intruder marking and decision making, 177–179
　mathematical model, 175, 176, 177
　notations and meanings, 175
　statistical model, 174
Game-theoretic model analysis
　highest dropping probability, 180
　maximum utility, 181
　packet dropping probability, 179, 180, 181, 182, 183, 184
Gateways, 45
GCC, *see* GNU compiler collection (GCC)
GDP, *see* Gross domestic product (GDP)
General regression neural network (GRNN), 136, 149; *see also* Local area network (LAN)
Generic embedded system platform (GENESYS), 98
GENESYS, *see* Generic embedded system platform (GENESYS)
Genetic algorithm (GA), 362
Global industry classification standards, 74
Global positioning system (GPS), 244, 253
Global sensor network (GSN), 43, 45
GNG, *see* Growing neural gas (GNG)
GNU compiler collection (GCC), 96
Goal Tree Success Tree (GTST), 318

GPS, *see* Global positioning system (GPS)
Graphical user interface (GUI), 295
　application, 237
　SomnuCare software, 307
　surgical workstation, 300
Grid computing, service-level agreement
　grid services performance, 352
　knowledge domain, 351
GRNN, *see* General regression neural network (GRNN)
Gross domestic product (GDP), 145, 147
Ground truth generation, 203
Growing neural gas (GNG), 126
GSN, *see* Global sensor network (GSN)
GTST, *see* Goal Tree Success Tree (GTST)
GUI, *see* Graphical user interface (GUI)

H

Hardware (HW), 103, 104; *see also* Software (SW)
HAZard and OPerability study (HAZOP), 318
HAZOP, *see* HAZard and OPerability study (HAZOP)
Hewlett-Packard (HP), 364
Hierarchically Performed Hazard Origin and Propagation Studies (HiP-HOPS), 319
High-performance computing (HPC), 355
HiP-HOPS, *see* Hierarchically Performed Hazard Origin and Propagation Studies (HiP-HOPS)
Hospital IT/Post-Operative Care interaction, 302
Host database, 200
HP, *see* Hewlett-Packard (HP)
HPC, *see* High-performance computing (HPC)
HTML, *see* Hypertext markup language (HTML)
HW, *see* Hardware (HW)
Hybrid neighbourhood function
　distance ranking in rectangular lattice, 132
　neighbourhood functions, 132
　new topographical ranking, 131
Hybrid SOM, 130
　continuous maps generation, 136–137
　hybrid neighbourhood function, 131–132
　local linear modeling, 137–140
　supervised algorithm, 141
　topology preservation constant, 134–136
　updating rule, 133–134

Index

Hybrid SOM–NG algorithm, 131
    using distance and distance ranking, 132
    distance ranking in rectangular lattice, 132
    Hybrid Neighbourhood Function, 131, 132
    quantisation and topographic error for, 136
    for rectangular and hexagonal lattices, 135
    topology preservation constant, 134–135, 136
    updating rule, 133, 134
Hypernym match, 259–260
Hypertext markup language (HTML), 161

I

IaaS, *see* Infrastructure as a service (IaaS)
IDPS, *see* Intrusion detection and prevention system (IDPS)
IDS, *see* Intrusion detection system (IDS)
IEEE, *see* Institute of Electrical and Electronics Engineers (IEEE)
IEEE 802.11 DCF, 107
IETF, *see* Internet engineering task force (IETF)
IMR, *see* Intelligent message routing (IMR)
Incentive mechanism, 213
    coalition payment division, 215–216
    responsiveness bonus computation, 216
Inference mechanism, 254
Information retrieval (IR), 151, 254–255
Infrastructure as a service (IaaS), 349; *see also* Database as a service (DaaS)
Injection moulding, 276
Institute of Electrical and Electronics Engineers (IEEE), 186
    802.11 DCF, 106, 107
    intrusion detection, 171
    WLAN standards, 186
Integrated sensor networks, 40, 41
    C/S-type, 42
    data aggregation systems, 41
    gateways, 45
    P2P-type, 42–43
    portal sites and centre servers, 45
    query description, 45–56
    sensor data aggregation systems, 56–62
    sensor network sites, 43, 44
    spatiotemporal query descriptions, 43
    system architecture, 43, 44
Integrated squared error (ISE), 18
Intelligent combination, 197
    BT packet flows, 198
    BT packets detection, 202
    DPI and DFI methods, 197
    offline training module, 199–200
    online classification module, 200–201
    performance evaluation, 202–207
    strengths, 201
Intelligent message routing (IMR), 68
Intelligent routing support
    DDPM module, 8
    structure, 7
Inter-process communication, 102
Internet engineering task force (IETF), 71
Internet of things (IoT), 185, 364
Internet protocol (IP), 68, 195; *see also* Transmission control protocol (TCP); File transport protocol (FTP)
Internet service provider (ISP), 68, 192
Intruder marking
    monitoring threshold, 184
    tampered packets, 183
    upstream and downstream nodes, 182
Intruders, 168
    marking, 182–184
Intrusion detection and prevention system (IDPS), 186
Intrusion detection system (IDS), 169
Intrusion handling, *see* Intrusion tackling
Intrusion prevention system (IPS), 169
Intrusion tackling
    game theory, 174–179
    network characteristics, 172–173
    PaS model, 173, 174
    security model, 172–173
    with tricky trap, 172
IoT, *see* Internet of things (IoT)
IP, *see* Internet protocol (IP)
IPS, *see* Intrusion prevention system (IPS)
IR, *see* Information retrieval (IR)
Iris data set classification
    map sizes, 144
    quantisation features, 143
ISE, *see* Integrated squared error (ISE)
ISP, *see* Internet service provider (ISP)

J

JADE, *see* Java agent development framework (JADE)
Java agent development framework (JADE), 284, 285
Java virtual machine (JVM), 59
JavaScript Object Notation (JSON), 56, 57
JobPendingList, 368, 369
JSON, *see* JavaScript Object Notation (JSON)
Just Kohonen's map, *see* Self-organising map (SOM)
JVM, *see* Java virtual machine (JVM)

K

K-means algorithm, 120, 121, 195
    clustering algorithm, 120
    implementation, 121
    iterations, 122
    Voronoi tessellation, 121
K-nearest neighbours (KNN), 152, 153, 155
Kahn process network (KPN), 98, 364
Kernel density estimation
    bandwidth selection methods, 18
    data-smoothing technique, 17
Kernel-based learning algorithm
    streaming data, 20
    threshold, 21
Kernel-based local regression model
    confidence interval, 20
    kernel functions, 19
Keyboard–video–mouse switches (KVM switches), 298
KNN, *see* K-nearest neighbours (KNN)
Knowledge extraction, supervised algorithm for
    cut hill data and gradient norm map, 141
    gradient component maps, 142
    Pearson's correlation, 143
    topological order, 141
KPN, *see* Kahn process network (KPN)
KVM switches, *see* Keyboard–video–mouse switches (KVM switches)

L

Laparoscopic intervention, 311–313
Lattice, 127
    rectangular and hexagonal lattices, 127
LBTV, *see* Load-balancing threshold value (LBTV)
Learning rule
    batch SOM, 129
    neighbourhood function, 127
    sequential SOM, 128
Learning to rank (LETOR), 151, 160
    experiments, 160–162
    number of queries, 152
    RankBoost algorithm, 153
    ranking of alternatives, 153
    semi-supervised ranking, 153–154
Learning to Rank for Information Retrieval (LR4IR), 160
Leasing, 269
Leasing-based injection mould remanufacturing
    injection moulding, 276
    remachining operations, 278
    reverse logistics, 277
Least-squares cross-validation function (LSCV), 18
Least-squares estimator (LSE), 12, 28
Left-side wheel brake (LWB), 328
    agents, 340
    state-transition table, 338
    sub-system, 336
LETOR, *see* Learning to rank (LETOR)
Levenberg–Marquard algorithm, 145
LILD, *see* Linear increase and linear decrease (LILD)
Linear increase and linear decrease (LILD), 106, 107
Linear regression model, 12
LiveE! integrated sensor networks, 42
Lloyd algorithm, 121
Load-balancing threshold value (LBTV), 374
Local area network (LAN), 92
Local linear modeling, 137
    gradient vectors, 139
    neighbourhood function, 125
    Newton's method, 140
    supervised batch SOM–NG, 140
    time-series prediction, 124
    updating rules, 138
Local recursive parameter estimation, 13–14
Local relevancies searching
    component planes, 146
    global models, 145
    gradient component planes, 148
    nuclear electricity sources, 149
    zones labelled, 147
Longest prefix match (LPM), 70
Loop operations, 51
Low-dimensional map, 127
LPM, *see* Longest prefix match (LPM)
LR4IR, *see* Learning to Rank for Information Retrieval (LR4IR)
LSCV, *see* Least-squares cross-validation function (LSCV)
LSE, *see* Least-squares estimator (LSE)
LWB, *see* Left-side wheel brake (LWB)

M

M^2IO switch
    KVM switches, 298
    smartOR, 299
MA systems, *see* Mobile agent systems (MA systems)
MAC, *see* Medium access control (MAC)

Index

MAC protocol data unit (MPDU), 109
MAC service data unit (MSDU), 109
MANET, see Mobile ad hoc network (MANET)
MAP, see Mean average precision (MAP)
Mapping, 96, 97
    direct match, 257, 258
    Hypernym match, 259–260
    No match, 260
    Synonym match, 258–259
    XML data extraction, 256
    XML to OWL mapping methodology, 257, 258
MAS, see Multi-agent system (MAS)
Master–worker programming model, 101
Matrix switch, see Keyboard–video–mouse switches (KVM switches)
MAX () method, 48, 50
Maximum transmission unit packets (MTU packets), 196
MB, see Mercedes-Benz (MB)
MBSA, see Mercedes-Benz South Africa (MBSA)
MC-CDMA technique, see Multicarrier code division multiple access technique (MC-CDMA technique)
Mean average precision (MAP), 161, 162
Medical devices interfacing, 308
Medium access control (MAC), 93, 188
Memory mapped file (MMF), 307, 309
Mercedes-Benz (MB), 275
Mercedes-Benz South Africa (MBSA), 275, 276
Message level intelligence
    CISCO, 72
    internet business and entertainment applications, 71
Middleware
    SMEPP, 220
    using SMEPP, 222–227
    SMEPP abstract model, 221–222
Migration module
    collector agent migration, 368
    CPU, 369
    slave agents, 368
    user agent migration, 368
MLP, see Multi-layer perceptron (MLP)
MMF, see Memory mapped file (MMF)
Mobile ad hoc network (MANET), 185; see also General regression neural network (GRNN)
Mobile agent systems (MA systems), 41
    aggregation operations, 41

generation, 59
migration, 60
PIAX, 43
sensor data aggregation architecture, 57
MOC, see Model of computation (MOC)
Modelling multithreaded support
    inter-process communication and system synchronisation model, 102–103
    POSIX threaded applications, 100
    system synchronization, 101–102
Model of computation (MOC), 97
Model quality, 12
Modern critical systems
    architectural view, 322
    aspects, 321
    fault tolerance, 322
    large-scale and dynamic behavior, 321
    monitoring factors and architectural levels balance point, 323
Monolithic monitor, 319
Moulding, 276–278
MPDU, see MAC protocol data unit (MPDU)
MQ2008-semi dataset, 161
MSDU, see MAC service data unit (MSDU)
MTU packets, see Maximum transmission unit packets (MTU packets)
Multi-agent framework
    business unit description, 280–281
    cooperation and interaction mechanism, 282–283
    e-commerce, 279
    multi-agent system, 278–279
    participants description, 280
    PPS description, 281
    proposed multi-agent framework, 279–282
    remanufacturing-centred process, 279
    supplier unit description, 282
    warehousing unit description, 281
Multi-agent system (MAS), 278, 279, 320
    architectures, 3
    collaboration protocols, 334
    complex modern information systems, 5
    DDPM and DDS modules structure, 6
    deployment, 331
    distributed systems, 6
    reasoning model, 330
    road traffic system, 7
    Ss_MAG, 333, 334
Multi-hop acknowledgement algorithm, 177, 178
Multi-layer perceptron (MLP), 130

Multicarrier code division multiple access technique (MC-CDMA technique), 108, 109
Multithreaded application modeling, 98
Multivariate linear regression, forecasting with by an individual agent, 11
   estimates efficiency, 13
   linear regression model, 12
   regression analysis, 11
Mutual reciprocity, 219

N

NACK, *see* No acknowledgement (NACK)
National Institute of Standards and Technology (NIST), 349
Navigator sub-system (NS), 321
NDCG, *see* Normalized discounted cumulative gain (NDCG)
Near-field communication (NFC), 363
Nearest-neighbours algorithm (NN algorithm), 154
Neighbourhood function, 125
NetLog, 366
Network interface card (NIC), 203
Network on terminal architecture (NoTA), 98
Network utilization, 374
Neural gas (NG), 120
   batch algorithm, 125–126
   cost function, 123
   local modeling, 124–125
   sequential algorithm, 123–124
   variations, 126
   vector quantization, 122
NFC, *see* Near-field communication (NFC)
NFP, *see* Non-functional property (NFP)
NG, *see* Neural gas (NG)
NIC, *see* Network interface card (NIC)
NIST, *see* National Institute of Standards and Technology (NIST)
NM, *see* Normal mode (NM)
NN algorithm, *see* Nearest-neighbours algorithm (NN algorithm)
No acknowledgement (NACK), 179
Nomenclatures, 301
Non-functional property (NFP), 97
Normal mode (NM), 328
Normalized discounted cumulative gain (NDCG), 161
NoTA, *see* Network on terminal architecture (NoTA)
NS, *see* Navigator sub-system (NS)

O

OA, *see* Organising agent (OA)
OASIS, *see* Organization for the Advancement of Structured Information Standards (OASIS)
OC, *see* Organisation calendar (OC)
OEM, *see* Original equipment manufacturer (OEM)
Offline training module
   flow characteristics, 200
   K-means algorithm, 199
   mapping process, 199
   offline training flow chart, 203
OGF, *see* Open-grid forum (OGF)
OGSA, *see* Open grid service architecture (OGSA)
Oil consumption estimation
   for cars, 144
   estimation error real values and histogram, 144
   MLP, 145
   output variable, 144
Online classification module
   databases, 200
   description, 200–201
Online deep flow inspection classifier, 204
Ontological model, 248
Open grid service architecture (OGSA), 229
Open-grid forum (OGF), 351
Open surgical communication bus (OSCB), 295
   event manager, 297
   IP-based network, 296
   SOA-based integration framework, 296
Open SystemC Initiative (OSCI), 97
Open systems interconnection model (OSI model), 93
OpenLIDS, 170–171
Operating room (OR), 292, 293
Operating system (OS), 95
OR, *see* Operating room (OR)
Oracle-based estimator, 34
Organisation calendar (OC), 262
Organising agent (OA), 42; *see also* Sensing agent (SA)
Organization for the Advancement of Structured Information Standards (OASIS), 303
ORIGA chip, 228
Original equipment manufacturer (OEM), 360
OS, *see* Operating system (OS)

Index

OSCB, *see* Open surgical communication bus (OSCB)
OSCI, *see* Open SystemC Initiative (OSCI)
OSI model, *see* Open systems interconnection model (OSI model)
Overlay routing, 68, 81

P

P2P, *see* Peer-to-peer (P2P)
P4P Pastry, 71
PaaS, *see* Platform as a service (PaaS)
Packet error rate accuracy analysis, 109–111
Packet flow database, 200
Packet marking, 179
PANA, *see* Protocol for carrying authentication for network access (PANA)
Parallel virtual machine (PVM), 354
Particle swarm optimisation algorithm (PSO algorithm), 362
PaS model, *see* Pay-and-stay model (PaS model)
Passive fault-tolerant controlling (PFTC), 322
Pay-and-stay model (PaS model), 167
    intrusion database, 173
    intrusion tackling model, 174
PC, *see* Personal calendar (PC)
PDA, *see* Personal digital assistant (PDA)
PDS, *see* Permanent Degraded State (PDS)
Pearson's correlation, 143
Peer-to-peer-based DSD routing architecture
    AON router design, 79–80
    AON-based query message structure, 80
    ISP, 75, 76
    layered design current, 72, 73
    layered design proposed, 73, 74
    underlay layer query routing algorithm design, 75–77
    underlay query routing process, 77–79
Peer code, 229
    behaviours and functionality, 230
    sequence diagram, 231
Peer code, 233
Peer-to-peer middleware (P2P middleware), 212
    applications, 68–70, 71
    CARISMA, 228
    security in SMEPP, 227–228
    sensor data, 42
    Sensormap, 43
    systems, 192
    type, 42
    SMEPP abstract model, 221–222
    using SMEPP, 222–227
PEID, *see* Product-embedded information device (PEID)
Performance measurement models
    cloud computing performance models, 355
    cloud providers, 353
    distributed systems performance models, 354
    SOA performance models, 353–354
Permanent Degraded State (PDS), 322, 326
Personal calendar (PC), 262
Personal digital assistant (PDA), 92, 362
    middleware, 213
    SMEPP, 220
PFTC, *see* Passive fault-tolerant controlling (PFTC)
Physical layer models, 98
    bit error rate calculation, 108
    frame error rate calculation, 109, 110
    packet error rate, 109–111
PIAX, 43, 59, 60
PIPPON, 70, 71
PKI, *see* Public key infrastructure (PKI)
Place ontology, 253
Platform as a service (PaaS), 349; *see also* Database as a service (DaaS)
Platform modeling, 96, 97
Plethora, 71
Plug-in methods, 18
Port-based detection, 194
Post-execution phase, 112
Power plant sub-systems (PPS), 321, 322
PPS, *see* Power plant sub-systems (PPS); Production planning and scheduling (PPS)
PRM, *see* Product recovery management (PRM)
Procurement agent, 282
Product recovery management (PRM), 361
Product-embedded information device (PEID), 363
Production data acquisition, *see* Personal digital assistant (PDA)
Production planning and scheduling (PPS), 279
Progressive filling algorithm, 214
Protocol, 214
    BT, 194–195
    collaboration, 334
    connectionless, 77
    converter, 300
    core P2P, 72, 73
    higher-level, 187
    IP and IP-port, 195
    MAC, 93
    OSCB, 300
    P2P file-sharing, 192

Protocol, (*Continued*)
  RFID, 363
  SOAP, 256
  STAR secure routing, 227
  TCP and UDP, 93
  transmission, 10
Protocol for carrying authentication for network access (PANA), 171
PSO algorithm, *see* Particle swarm optimisation algorithm (PSO algorithm)
Pthread_lib service process, 102, 103
Public key infrastructure (PKI), 304
PVM, *see* Parallel virtual machine (PVM)

Q

QoS-based attribute model, 354
Quality value-dependency priority (QVDP), 355
Quantisation error, 135
Query description, 55, 261
  character length comparison, 53
  comparison, 55
  conventional descriptions, 55
  design, 45–51
  implementation, 51
  issues, 45
Query description design
  aggregated sensor data, 45
  loop operations, 51
  sensor data extraction from spatiotemporal ranges, 51
  spatial units designation, 48–49
  spatiotemporal ranges data selection, 50–51
  spatiotemporal ranges designation, 46–47
  temporal units designation, 47–48
QVDP, *see* Quality value-dependency priority (QVDP)

R

RADAR, 170
Radio frequency (RF), 93
Radio frequency identification (RFID), 284, 363
  product returns management, 364
  semiconductor-based technology, 363
Random score function, 160
RankBoost_Active, 159
RankBoost algorithm, 153
  NN algorithm, 154
  RankBoostSSA, 155–157

RankBoostSSA, 155–157
  performance MAP Measures, 161, 162
  performance NDCG@n Measures, 162
Ranking, 151–152
Ranking features selection algorithm, 157–159
Rb, *see* Responsiveness bonus (Rb)
RDF, *see* Resource description framework (RDF)
Real-world data, application to
  Iris data set classification, 143–144
  local relevancies searching, 145–147
  oil consumption estimation, 144–145
Reciprocity-based approaches, 219
Recommender systems, 362
Registration module, 367
Regression analysis, 4, 11–12
Remanufacturing
  economic viability, 272–273
  embedded in leasing products, 276
  industrial process, 270
  leasing-based injection mould, 276–278
  reasons for adoption, 272
  recycling form, 271
  uncertainty of returned products management, 273
RemTec, 275
Reputation management, 170
Reputation-based mechanisms, *see* Reciprocity-based approaches
Resampling methods, 18
Resource description framework (RDF), 244
Responsiveness bonus (Rb), 213
  computation, 216
  management service, 231–232
Responsiveness bonus management service
  COINS, 231
  service operations/messages, 232
RetreadCo, 275
Reverse logistics (RL), 361
RF, *see* Radio frequency (RF)
RFID, *see* Radio frequency identification (RFID)
Right-side wheel brake (RWB), 328
  architectural model and error model, 331
  operator's interface after controlling failure, 340
  operator's interface after failure, 339
  state-transition table, 332
Risk management, 304
RL, *see* Reverse logistics (RL)
Road traffic system, 7
Round trip time (RTT), 70, 71, 80
RTT, *see* Round trip time (RTT)
RWB, *see* Right-side wheel brake (RWB)

S

SA, *see* Sensing agent (SA)
SaaS, *see* Software as a service (SaaS)
Secure middleware for embedded peer-to-peer systems (SMEPP system), 212–213
  abstract model, 221–223
  design, 229–232
  implementation, 233–234
  Java API, 222, 223
  peer code or service code, 223
  peers and groups management, 224
  security in, 227–228
  services management, 224–227
Secure shell (SSH), 203, 206
Self-organising feature map, *see* Self-organising map (SOM)
Self-organising map (SOM), 120, 127
  application to real-world data, 143–147
  component planes and unified distance matrix, 128–130
  k-means, 120–122
  lattice, 127
  learning rule, 127, 128
  NG, 122–126
  uses, 130
  vector quantisation, 127
Semi-supervised learning, 151, 152
  algorithm to traffic classification, 195
  classification accuracy results, 206
  DFI module, 204
  flow characteristics in, 200
  of ranking functions, 154, 162
Semi-supervised learning BitTorrent traffic detection, 191
  BT protocol, 192–193
  BT traffic detection, 193–194
  comparison with other methods, 201–202
  detection methods, 196–197
  DFI-based detection, 195–196
  DPI and DFI combination, 196
  DPI-based detection, 194–195
  intelligent combination, 197–199, 201
  offline training module, 199–200
  online classification module, 200–201
  performance evaluation, 202–207
  port-based detection, 194
Semi-supervised method
  RankBoost algorithm adaptation, 154–157
  selection of ranking features algorithm, 157–159
Semi-supervised ranking, 153–154

Sensing agent (SA), 42; *see also* Organising agent (OA)
Sensor data aggregation systems
  design, 56–58
  implementation, 58–61
  using MA, 56
  system evaluation, 61–62
Sensormap, 43
Sensor network information, 44, 58–59
Sensor network sites
  organisations, 43
  sensor information, 44
Sensors
  context information, 246
  continuous sensing, 246–247
  network sites, 43, 44
Sequential algorithm
  prototype vectors, 123
  sequential neural gas, 124
Service class encoding, 74
Service code, 229
  behaviours and functionality, 230
  sequence diagram, 231
Service discovery, 222
Service location problem (SLP), 72
Service oriented access protocol (SOAP), 295; *see also* Transmission control protocol (TCP)
Service oriented architecture (SOA), 72
Service provider, 296
Service set identifier (SSID), 186
Service-level agreement (SLA), 350
  for Cloud Computing, 352–353
  formal agreements, 350
  for Grid Computing, 351–352
  for Web Services, 351
Service-level objective (SLO), 351
Service-oriented architecture (SOA), 72, 294, 295, 348
  based integration, 295, 296
  patient protection, 302
  performance metrics comparison, 354
  performance models, 353–354
  simulation tool, 353
Service-oriented computing (SOC), 68
Service-specific functionality, 103, 104
Signal-to-noise ratio (SNR), 110
Simple object access protocol (SOAP), 244, 297, 298
  context acquisition, 255, 256
  messages to OWL ontologies, 249
  request message, 256
  WS-Security, 303
  XML payload in, 263

Single-use camera remanufacturing, 274
SLA, *see* Service-level agreement (SLA)
SlaveAgentInfoList, 368, 369
SLO, *see* Service-level objective (SLO)
SLP, *see* Service location problem (SLP)
SLPE
    distributed systems, 106
    enhancements, 99–100
    performance evaluation, 97–98
    performance models for, 115
SLUP, 71
smartOR Project
    distributed network intelligence, 294
    manufacturer-independent framework, 294
    medical devices smooth handling, 293
    nomenclatures, 301
    OR table parameters, 301
    OSCB, 300
SMEPP system, *see* Secure middleware for embedded peer-to-peer systems (SMEPP system)
SNR, *see* Signal-to-noise ratio (SNR)
SOA, *see* Service oriented architecture (SOA); Service-oriented architecture (SOA)
SOAP, *see* Service oriented access protocol (SOAP); Simple object access protocol (SOAP)
SOC, *see* Service-oriented computing (SOC)
Software (SW), 103; *see also* Hardware (HW)
Software as a service (SaaS), 349
SOM, *see* Self-organising map (SOM)
SomnuCare Software, 307
SPARQL, *see* SPARQL protocol and RDF query language (SPARQL)
SPARQL protocol and RDF query language (SPARQL), 254–255
    ontology, 255
    query language, 261
Spatial units designation
    AREASPAN, 48
    interpolation methods, 49
    in spatial range, 49
Spatiotemporal query descriptions, 43
Spatiotemporal ranges data selection
    aggregation methods, 50
    SELECT phrases, 50
    UDEF method, 51
Spatiotemporal ranges designation, 46
    FROM clauses, 47
    spatiotemporal units, 47
    temporal ranges, 46

SSE, *see* Sum of squares of errors (SSE)
SSH, *see* Secure shell (SSH)
SSID, *see* Service set identifier (SSID)
Ss levels, *see* Sub-system levels (Ss levels)
SSR, *see* Sum of squares regression (SSR)
SST, *see* Total sum of squares (SST)
Sub-system levels (Ss levels), 321
SUM method, 50
Sum of squares of errors (SSE), 12
Sum of squares regression (SSR), 12
Supervised algorithm
    for estimation tasks, 141
    for knowledge extraction, 141–143
Supervised algorithm, 141
Support vector machine algorithm (SVM algorithm), 195
Surgical workstation
    device operations, 305
    OR-table control panel, 304, 305
SVM algorithm, *see* Support vector machine algorithm (SVM algorithm)
SW, *see* Software (SW)
Synonym match, 258, 259
System synchronization, 101–102

T

TAC, *see* Trading agent competition (TAC)
TCP, *see* Transmission control protocol (TCP)
TDFS, *see* Temporary Degraded or Failure State (TDFS)
Telesupervision scenario, 310
Temporal Kohonen map (TKM), 130
Temporal units designation, 47–48
Temporary Degraded or Failure State (TDFS), 322, 328
TIC, *see* Traffic Information Centre (TIC)
Time ontology, 253
TKM, *see* Temporal Kohonen map (TKM)
TLM, *see* Transaction-level modelling (TLM)
TM, *see* Transductive method (TM)
TOPK method, 50
TOPLUS, 70–71
Topographical error, 135
Topology preservation constant
    iris data set, 136
    lattice effect, 134
    neighbourhoods, 135
    quantisation error, 135
Total sum of squares (SST), 12
Trading agent competition (TAC), 279
Traffic Information Centre (TIC), 10, 11, 24

Index

Traffic network simulation, 24
    factors and bandwidth values, 26
    road network, 25
    travel frequency, 26
Transaction-level modelling (TLM), 99
Transductive method (TM), 152
Transmission control protocol (TCP), 93, 194, 297; *see also* Simple object access protocol (SOAP)
    distributed applications, 113–114
    IP-based, 313
    in OSCB, 301
    port-based detection, 194
    SomnuCare interface API, 308
    transport-level services, 103
Transport layer, 297
    models, 98
    services, 106

U

U-matrix, 129, 130
UDP, *see* User datagram protocol (UDP)
UML, *see* Unified modelling language (UML)
Underlay layer query routing
    algorithm design, 76
    AON routers, 75
    IP layer, 77
    learning state and steady state, 77
    process, 82
    query routing in steady state, 78
    routing analysis, 79
    scenarios in query forwarding, 77
Underlay routing, 68
    inter-ISP traffic in, 86
    links utilisation in, 83
    scenario, 82
Underlay-aware distributed service discovery architecture
    analysis scenario, 81, 82
    AON routing, 86
    AON-based query processing, 84
    complexity analysis, 83
    inter-ISP traffic in overlay routing, 85–86
    inter-ISP traffic in underlay routing, 86
    links estimation in overlay, 83, 84
    links estimation in underlay, 83–84
    redundant traffic causes, 85
    routers estimation in overlay, 82
    routers estimation in underlay, 82–83
    time taken for query routing, 86–87
Unified modelling language (UML), 248, 317
Uniform resource identifier (URI), 297

User agent, 365
User datagram protocol (UDP), 93, 194, 297; *see also* Simple object access protocol (SOAP)
User interaction workstations
    anaesthesia workstation, 306–310
    surgical workstation, 304–305

V

Vehicular ad hoc network (VANET), 185

W

W3C, *see* World Wide Web Consortium (W3C)
Waste electrical and electronic equipment (WEEE), 268, 360
WC, *see* Work calendar (WC)
Web browser interface, 60, 61
Web service description language (WSDL), 295
Web service offering language (WSOL), 351
Web service-level agreement (WSLA), 351
Web services, 297
    agreement, 351
    discovery, 298
    security, 303
    service-level agreement, 351
Web services business process execution language (WS-BPEL), 353
WEEE, *see* Waste electrical and electronic equipment (WEEE)
WEP, *see* Wired equivalent privacy (WEP)
Wired equivalent privacy (WEP), 188
Wireless intrusion detection and prevention system
    intruder tackling, 186
    limitations, 188
    security capabilities, 187–188
    WLAN, 187
Wireless local area network (WLAN), 186; *see also* Local area network (LAN)
Wireless mesh network (WMN), 167
    achievements and motivation current status, 170–172
    architecture, 168
    IDSs and IPS status, 169–170
    intrusion tackling schemes, 168
    tackling intrusion, 168, 169
Wireless sensor network (WSN), 93, 169
WLAN, *see* Wireless local area network (WLAN)
WMN, *see* Wireless mesh network (WMN)
Work calendar (WC), 262

Workload extraction
  via CORINNA, 111
  phases, 111
  pre-compilation and post-compilation steps, 112
Workload models, 95–96
World Wide Web Consortium (W3C), 297
WS-BPEL, *see* Web services business process execution language (WS-BPEL)
WSDL, *see* Web service description language (WSDL)
WSLA, *see* Web service-level agreement (WSLA)
WSN, *see* Wireless sensor network (WSN)
WSOL, *see* Web service offering language (WSOL)

X

X-Sensor 1.0, 42
Xerox's remanufacturing system, 274
XML, *see* Extensible markup language (XML)
XOR, *see* Exclusive-Or (XOR)

For Product Safety Concerns and Information please contact our
EU representative GPSR@taylorandfrancis.com Taylor & Francis
Verlag GmbH, Kaufingerstraße 24, 80331 München, Germany